Beyond Imported Magic

Inside Technology Series

edited by Wiebe E. Bijker, W. Bernard Carlson, and Trevor Pinch

Eden Medina, Ivan da Costa Marques, and Christina Holmes, editors, *Beyond Imported Magic: Essays on Science, Technology, and Society in Latin America*

Anique Hommels, Jessica Mesman, and Wiebe E. Bijker, editors, *Vulnerability in Technological Cultures: New Directions in Research and Governance*

Amit Prasad, *Imperial Technoscience: Transnational Histories of MRI in the United States, Britain, and India*

Charis Thompson, *Good Science: The Ethical Choreography of Stem Cell Research*

Tarleton Gillespie, Pablo J. Boczkowski, and Kirsten A. Foot, editors, *Media Technologies: Essays on Communication, Materiality, and Society*

Catelijne Coopmans, Janet Vertesi, Michael Lynch, and Steve Woolgar, editors, *Representation in Scientific Practice Revisited*

Rebecca Slayton, *Arguments that Count: Physics, Computing, and Missile Defense, 1949–2012*

Stathis Arapostathis and Graeme Gooday, *Patently Contestable: Electrical Technologies and Inventor Identities on Trial in Britain*

Jens Lachmund, *Greening Berlin: The Co-Production of Science, Politics, and Urban Nature*

Chikako Takeshita, *The Global Biopolitics of the IUD: How Science Constructs Contraceptive Users and Women's Bodies*

Cyrus C. M. Mody, *Instrumental Community: Probe Microscopy and the Path to Nanotechnology*

Morana Alač, *Handling Digital Brains: A Laboratory Study of Multimodal Semiotic Interaction in the Age of Computers*

Gabrielle Hecht, editor, *Entangled Geographies: Empire and Technopolitics in the Global Cold War*

Michael E. Gorman, editor, *Trading Zones and Interactional Expertise: Creating New Kinds of Collaboration*

Matthias Gross, *Ignorance and Surprise: Science, Society, and Ecological Design*

Andrew Feenberg, *Between Reason and Experience: Essays in Technology and Modernity*

Wiebe E. Bijker, Roland Bal, and Ruud Hendricks, *The Paradox of Scientific Authority: The Role of Scientific Advice in Democracies*

Park Doing, *Velvet Revolution at the Synchrotron: Biology, Physics, and Change in Science*

Gabrielle Hecht, *The Radiance of France: Nuclear Power and National Identity after World War II*

Richard Rottenburg, *Far-Fetched Facts: A Parable of Development Aid*

Michel Callon, Pierre Lascoumes, and Yannick Barthe, *Acting in an Uncertain World: An Essay on Technical Democracy*

Ruth Oldenziel and Karin Zachmann, editors, *Cold War Kitchen: Americanization, Technology, and European Users*

Deborah G. Johnson and Jameson W. Wetmore, editors, *Technology and Society: Building Our Sociotechnical Future*

Trevor Pinch and Richard Swedberg, editors, *Living in a Material World: Economic Sociology Meets Science and Technology Studies*

Christopher R. Henke, *Cultivating Science, Harvesting Power: Science and Industrial Agriculture in California*

Helga Nowotny, *Insatiable Curiosity: Innovation in a Fragile Future*

Karin Bijsterveld, *Mechanical Sound: Technology, Culture, and Public Problems of Noise in the Twentieth Century*

Peter D. Norton, *Fighting Traffic: The Dawn of the Motor Age in the American City*

Joshua M. Greenberg, *From Betamax to Blockbuster: Video Stores tand the Invention of Movies on Video*

Mikael Hård and Thomas J. Misa, editors, *Urban Machinery: Inside Modern European Cities*

Christine Hine, *Systematics as Cyberscience: Computers, Change, and Continuity in Science*

Wesley Shrum, Joel Genuth, and Ivan Chompalov, *Structures of Scientific Collaboration*

Shobita Parthasarathy, *Building Genetic Medicine: Breast Cancer, Technology, and the Comparative Politics of Health Care*

Kristen Haring, *Ham Radio's Technical Culture*

Atsushi Akera, *Calculating a Natural World: Scientists, Engineers and Computers during the Rise of U.S. Cold War Research*

Donald MacKenzie, *An Engine, Not a Camera: How Financial Models Shape Markets*

Geoffrey C. Bowker, *Memory Practices in the Sciences*

Christophe Lécuyer, *Making Silicon Valley: Innovation and the Growth of High Tech, 1930–1970*

Anique Hommels, *Unbuilding Cities: Obduracy in Urban Sociotechnical Change*

David Kaiser, editor, *Pedagogy and the Practice of Science: Historical and Contemporary Perspectives*

Charis Thompson, *Making Parents: The Ontological Choreography of Reproductive Technology*

Pablo J. Boczkowski, *Digitizing the News: Innovation in Online Newspapers*

Dominique Vinck, editor, *Everyday Engineering: An Ethnography of Design and Innovation*

Nelly Oudshoorn and Trevor Pinch, editors, *How Users Matter: The Co-Construction of Users and Technology*

Peter Keating and Alberto Cambrosio, *Biomedical Platforms: Realigning the Normal and the Pathological in Late-Twentieth-Century Medicine*

Paul Rosen, *Framing Production: Technology, Culture, and Change in the British Bicycle Industry*

Maggie Mort, *Building the Trident Network: A Study of the Enrollment of People, Knowledge, and Machines*

Donald MacKenzie, *Mechanizing Proof: Computing, Risk, and Trust*

Geoffrey C. Bowker and Susan Leigh Star, *Sorting Things Out: Classification and Its Consequences*

Charles Bazerman, *The Languages of Edison's Light*

Janet Abbate, *Inventing the Internet*

Herbert Gottweis, *Governing Molecules: The Discursive Politics of Genetic Engineering in Europe and the United States*

Kathryn Henderson, *On Line and On Paper: Visual Representation, Visual Culture, and Computer Graphics in Design Engineering*

Susanne K. Schmidt and Raymund Werle, *Coordinating Technology: Studies in the International Standardization of Telecommunications*

Marc Berg, *Rationalizing Medical Work: Decision Support Techniques and Medical Practices*

Eda Kranakis, *Constructing a Bridge: An Exploration of Engineering Culture, Design, and Research in Nineteenth-Century France and America*

Paul N. Edwards, *The Closed World: Computers and the Politics of Discourse in Cold War America*

Donald MacKenzie, *Knowing Machines: Essays on Technical Change*

Wiebe E. Bijker, *Of Bicycles, Bakelites, and Bulbs: Toward a Theory of Sociotechnical Change*

Louis L. Bucciarelli, *Designing Engineers*

Geoffrey C. Bowker, *Science on the Run: Information Management and Industrial Geophysics at Schlumberger, 1920-1940*

Wiebe E. Bijker and John Law, editors, *Shaping Technology / Building Society: Studies in Sociotechnical Change*

Stuart Blume, *Insight and Industry: On the Dynamics of Technological Change in Medicine*

Donald MacKenzie, *Inventing Accuracy: A Historical Sociology of Nuclear Missile Guidance*

Pamela E. Mack, *Viewing the Earth: The Social Construction of the Landsat Satellite System*

H. M. Collins, *Artificial Experts: Social Knowledge and Intelligent Machines*

See http://mitpress.mit.edu/books/series/inside-technology for a complete list of titles in this series.

Beyond Imported Magic

Essays on Science, Technology, and Society in Latin America

edited by Eden Medina, Ivan da Costa Marques, and Christina Holmes
with a foreword by Marcos Cueto

The MIT Press
Cambridge, Massachusetts
London, England

© 2014 Massachusetts Institute of Technology

All rights reserved. No part of this book may be reproduced in any form by any electronic or mechanical means (including photocopying, recording, or information storage and retrieval) without permission in writing from the publisher.

MIT Press books may be purchased at special quantity discounts for business or sales promotional use. For information, please email special_sales@mitpress.mit.edu.

This book was set in ITC Stone Sans Std and ITC Stone Serif Std 9/13 pt by Toppan Best-set Premedia Limited, Hong Kong. Printed and bound in the United States of America.

Library of Congress Cataloging-in-Publication Data

Beyond imported magic: essays on science, technology, and society in Latin America / edited by Eden Medina, Ivan da Costa Marques, and Christina Holmes; with a foreword by Marcos Cueto.
    page cm — (Inside technology)
  Includes bibliographical references and index.
  ISBN 978-0-262-02745-8 (hardcover : alk. paper) — ISBN 978-0-262-52620-3 (pbk. : alk. paper)
  1. Technology transfer—Latin America. 2. Science—Latin America. 3. Technology—Social aspects—Latin America. I. Medina, Eden, 1976– editor of compilation. II. Marques, Ivan da Costa editor of compilation. III. Holmes, Christina, 1973– editor of compilation.
  T24.A1B46 2014
  338.98'06—dc23
                                   2013046628

10 9 8 7 6 5 4 3 2 1

Contents

Foreword by Marcos Cueto   vii
Preface   xi

1   Introduction: Beyond Imported Magic   1
Eden Medina, Ivan da Costa Marques, and Christina Holmes

Part I: Latin American Perspectives on Science, Technology, and Society

2   Who Invented Brazil?   27
Henrique Cukierman

3   Innovation and Inclusive Development in the South: A Critical Perspective   47
Mariano Fressoli, Rafael Dias, and Hernán Thomas

4   Working with Care: Narratives of Invisible Women Scientists Practicing Forensic Genetics in Colombia   67
Tania Pérez-Bustos, María Fernanda Olarte Sierra, and Adriana Díaz del Castillo H.

5   Ontological Politics and Latin American Local Knowledges   85
Ivan da Costa Marques

6   Technology in an Expanded Field: A Review of History of Technology Scholarship on Latin America in Selected English-Language Journals   111
Michael Lemon and Eden Medina

Part II: Local and Global Networks of Innovation

7   South Atlantic Crossings: Fingerprints, Science, and the State in Turn-of-the-Twentieth-Century Argentina   139
Julia Rodriguez

8   Tropical Assemblage: The Soviet Large Panel in Cuba   159
Hugo Palmarola and Pedro Ignacio Alonso

9   Balancing Design: OLPC Engineers and ICT Translations at the Periphery   181
Anita Say Chan

10   Translating Magic: The Charisma of One Laptop per Child's XO Laptop in Paraguay   207
Morgan G. Ames

11   Nanoscience and Nanotechnology: How an Emerging Area on the Scientific Agenda of the Core Countries Has Been Adopted and Transformed in Latin America   225
Noela Invernizzi, Matthieu Hubert, and Dominique Vinck

12   Latin America as Laboratory: The Camera and the Yale Peruvian Expeditions   245
Amy Cox Hall

Part III: Science, Technology, and Latin American Politics

13   Bottling Atomic Energy: Technology, Politics, and the State in Peronist Argentina   267
Jonathan Hagood

14   Peaceful Atoms in Mexico   287
Gisela Mateos and Edna Suárez-Díaz

15   Neoliberalism as Political Technology: Expertise, Energy, and Democracy in Chile   305
Manuel Tironi and Javiera Barandiarán

16   Creole Interferences: A Conflict over Biodiversity and Ownership in the South of Brazil   331
Ana Delgado and Israel Rodríguez-Giralt

17   The Juridical Hospital: Patient-Citizen-Consumers Claiming the Right to Health in Brazilian Courts   349
João Biehl

Contributors   373
Index   379

Foreword

This book testifies to the progress of critical investigations on how science and technology have been constructed, designed, imagined, and practiced in Latin America. This collection is truly outstanding, not only because the contributors take into consideration the philosophical, historical, sociological, anthropological, and political science dimensions of a little-studied region of the world, but also because they address issues relevant to scholars and readers interested in the phenomenon of science and technology all over the world. And it is no small feat that the literature they use is up-to-date. One issue that certainly appears clear in this publication is that scientific knowledge constructs its "universal" legitimacy depending not only on time, place, and field of study, but also on the agency of individuals, institutions, and nations.

Has Latin America always been passive and derivative in terms of scientific creativity, or has it played a central role in the making of contemporary specialized practices? The answers coming from Latin America show, with increasing emphasis, that science and technology should be understood as an arena contested by a wide variety of individuals, institutions, and actors and through complex local processes of reception, rejection, adaptation, and hybridization. From this perspective, Western science could be understood as a process of polycentric networks, and as a global dynamic interplay in ever-shifting networks.

This book offers several specific contributions to the literature. First, it addresses the complex and difficult coexistence of the significant advances made to Western science by Latin American centers of knowledge making and the asymmetry of knowledge production between Latin American nations and their counterparts in more industrialized parts of the world. This means that, at least during the twentieth and twenty-first centuries, Latin American science has contributed to the world store of knowledge and helped construct the hegemony of metropolitan Western knowledge, but at the same time has faced adverse conditions such as low cultural esteem, few full-time laboratory positions, ephemeral academic journals, underfunded universities, and limited financial resources. As a result, high-quality Latin American science and technology have survived in the international scientific community despite the asymmetry in power and financial relations. It also means that frequently "good" science and scientists exist despite the absence of consolidated scientific communities. This is a paradox, and a fascinating theme that is explored in this publication.

Second, the authors in this book illustrate that the unique combination of modernity and underdevelopment in Latin America questions the "natural" frontiers between nature and society determined by politicians and science practitioners, making them appear clearly artificial. Third, in offering a detailed analysis of the day-to-day practices of scientists, engineers, and technologists, the book shows that these practitioners are close to the public reception of scientific knowledge, which they both produce and translate as they attempt to reconcile diverse lay and expert viewpoints. This is different from what often occurs in more industrialized nations, where there is often a larger gap between the mediation of scientific knowledge and the views of the general public.

The chapters in the book are a clear by-product of a scholarly reaction in the region that can be traced to the 1980s, when researchers of the social studies of science went beyond the prevailing internalist perspectives, familiar traditional ideals of Western science, and models of scientific diffusion, social control, or scientific imperialism. The flaw of these models was that they assumed that anything in Latin American societies (modernization, development or dependence) came from outside those societies. Contextual history was emphasized later, but after a few years it was clear that this too was not sufficient. The studies in this book take the approach a step further by intertwining contextual and internal factors and demonstrating that scientific life had a life of its own. They are a splendid mirror of the noticeable increase in the quantity, quality, and international relevance of scholarship in the social studies of science in Latin America.

These new trends have also appeared in historical studies of medicine in Latin America, which have shown how European conquest, political governance, and culture in Latin America were inspired, legitimized, portrayed, and based on scientific and technological knowledge. In a continent marked by frequent revolutions, social upheavals, migrations, changing borders, and natural disasters, scientific expertise was frequently used as a catalyst for change—or a supporter of continuity—in institutions, landscapes, gender roles, races, societies, and states. The investigation of how science was grounded in cultural, social, and political currents of the past was made possible in part by novel research on new problems, using new sources, and carried out by a new historiography that matured most notably in the pages of the journal edited in Rio de Janeiro by the Casa Oswaldo Cruz, *Historia, Ciencia, Saúde—Manguinhos* (which first appeared in 1994).

The original idea for this book grew out of a meeting that took place in Buenos Aires on the social studies of science and technology, bringing together scholars from across the Americas. The geographical and thematic distribution of the chapters reflects the excellent research going on in Brazil, Argentina, Mexico, and other countries in the region. Its publication in English—the Latin of our day—will help scholars in other countries to take stock of the field, to promote a fluid dialogue between all researchers interested in the social and cultural dimensions of science and technology studies.

I would like to end by telling an anecdote that occurred when I attended a conference in Mexico with several historians of science several years ago. We were wandering through the galleries of the splendid Museum of History and Anthropology in Mexico City. After our brilliant guide gave us a detailed explanation of the sophisticated mathematical knowledge of the Mayas before the conquest, I mumbled something like "This is amazing!" to a British historian. I will never forget his response: "What is amazing is that I knew nothing about it before!" I sincerely hope that this authoritative book will help to uncover unknown valuable information and suggestive new perspectives for the social studies of science and technology in the United States and Europe. The editors, authors, and publisher should be congratulated for producing a remarkable and coherent study that will be a landmark in the social studies of science and technology of Latin America and an intriguing and nuanced inspiration for broader debates on transnational STS research. It will certainly raise new important questions. It will also illuminate the current discussions on how science and technology can help to solve long-standing challenges in the region such as the achievement of full citizenship, social equality, and social cohesion.

Marcos Cueto
Visiting Professor
Casa Oswaldo Cruz, Fiocruz, Rio de Janeiro

Preface

The idea for this book grew out of a meeting sponsored by the Society for the Social Studies of Science (4S) that took place in June 2011 in Buenos Aires, Argentina. The meeting brought together representatives from 4S and the Sociedad Latinoamericana de Estudios Sociales de la Ciencia y la Tecnología (ESOCITE). It also commenced planning for a joint 4S/ESOCITE conference and opened discussions about writing a book on science and technology studies (STS) and Latin America. The timing seemed right. Recent years had witnessed a substantial increase in the quantity and international visibility of STS scholarship on Latin America and a deepening engagement between Latin American and non-Latin American STS scholars that study the region. For example, attendance at the biennial meeting of ESOCITE has reached several hundred participants in the last few years, and 4S had run multiple tracks on STS in Latin America at four of its five most recent meetings. In addition, scholars of Latin America were becoming increasingly interested in science and technology. For example, the journal *Hispanic American Historical Review* published a special issue on the history of science in 2011, and the *Journal of Latin American and Caribbean Anthropology* has a special issue on science and technology forthcoming. Collectively, these developments show that research on the history and social study of science and technology in Latin America is a highly fertile area of inquiry and one with a growing community of interested scholars. We decided to compile an edited volume to highlight newer scholarship in this area and increase the conversation among scholars working in the north and in the south.

This volume is the product of much collaborative work and conversation. The majority of its authors traveled to Bloomington, Indiana, for a two-day, intensive peer review workshop in August 2012. Peer review processes can be tedious, but in this case a lively and supportive community formed. We thank David Hess and Julia Rodriguez for skillfully running workshop sessions, drawing out constructive criticism and new ideas, and pushing for thematic unity in the volume. We further thank those at Indiana University who willingly gave up a weekend to participate in the workshop and strengthen the volume: Kate Bishop, Nathan Ensmenger, Laura Foster, Lessie Jo

Frazier, Stephanie Kane, Samantha Merritt, David Nemer, and Alberto Samaniego. The efforts of Shane Greene, Matt Van Hoose, Deborah Cohn, Michele Dompke, Dave Cooley, Sarah Dees, and Irasema Rivera made the workshop logistics run smoothly. WFHB radio featured contributing authors on its Spanish-language radio program *Hola Bloomington*. We thank them for taking this work beyond the ivory tower and making it accessible to the Indiana Spanish-speaking public.

The workshop for the book was made possible with generous support from the National Science Foundation (grant # SES-1230211). Additional support was provided by the Mellon Foundation, as well as Indiana University's School of Informatics and Computing, Center for Latin American and Caribbean Studies, Department of American Studies, Department of Anthropology, Institute for Advanced Study, College Arts and Humanities Institute, Office of the Vice Provost for Faculty and Academic Affairs, and the Office of the Vice President for International Affairs. We are grateful for the financial support and encouragement provided by these organizations, offices, and institutions.

Additionally, we thank the Society for the Social Studies of Science for supporting the initial Buenos Aires meeting with a 4S New Initiatives Grant and acknowledge Rick Duque in particular for taking a leadership role in organizing the Buenos Aires meeting and promoting work on Latin America within the 4S community. His hard work over the years has been fundamental in galvanizing conversations between scholars in the north and the south on the social study of science and technology in Latin America. We further appreciate the contributions of those who participated in the Buenos Aires meeting for sharing their knowledge of the state of the field and its possible futures. This includes Rosalba Casas-Guerrero, Donna DeGennaro, Rick Duque, Adriana Feld, Pablo Kreimer, Luciano Levin, Maria Fernanda Olarte Sierra, Pablo Pellegrini, Raoni Rajão, Olga Restreo Forero, Sam Smiley, Yuri Takhteyev, Manuel Tironi, Ana María Vara, Leá Velho, and Hebe Vessuri. Many of these individuals further helped by circulating the open call for papers for this volume and offering general input on the project that led to its improvement.

We received constructive feedback on the introduction and framing of this volume from faculty and students at the MIT Program in Science, Technology, and Society and the University of Michigan Science, Technology, and Society Program. Participants at the 2012 meeting of the Tepotzlán Institute for Transnational History of the Americas in Mexico read and discussed an early draft of the introduction to this volume and improved our ability to bring together literatures from STS and Latin American studies. Fiona McDonald, Márcia Regina Barros da Silva, Joanna Radin, Gabriela Soto Laveaga, and the anonymous reviewers for the MIT Press provided additional feedback. Marcos Cueto deserves special thanks for writing the foreword

Preface

and for his pathbreaking work on the history of science and medicine in Latin America. Jeanne Barker-Nunn and Polly Kummel improved the book with their impeccable editing talents. We also thank Margy Avery and the MIT Press for backing this project from the outset and giving the book an ideal venue for publication.

Finally, this book would not have been possible without the love and support of our families. Corbin, Keeha, Russell, Gael, Cristian, and Márcia: muito obrigado, muchas gracias, thank you.

1 Introduction: Beyond Imported Magic

Eden Medina, Ivan da Costa Marques, and Christina Holmes

The essays in this collection employ critical frameworks from science and technology studies (STS) to formulate new ideas and knowledge about how Latin American peoples, countries, cultures, and environments create, adapt, and use science and technology. Two key themes run through the volume. First, its essays go beyond viewing science and technology in Latin America as imported from somewhere else and instead explore alternative views of how scientific ideas and technologies are created, move, change, and adapt. This may include travel from South to North; among Latin American regions, nations, and communities; and between different areas of the global South. Second, these essays examine the specificities of Latin American experiences to understand science and technology more broadly. They thereby augment our understanding of such categories as *global South*, *postcolonial*, and *developing* and reveal new dimensions of the relationships among science, technology, politics, and power.

The chapters in this book share a common goal of furthering conversations between scholars in the North and the South who study science and technology in Latin America. Although some Latin American academics find it necessary to access English-language scholarship published in the North, it is less common for academics in North America and Europe to seek out the Spanish- and Portuguese-language literature published by Latin American scholars. Consequently, the work of Latin American scholars who normally publish in those languages may not reach as large an audience, as English often functions as an unofficial lingua franca. Although this is typical of the unequal power relationships that influence global linguistic practices as well as technoscientific ones (Mar-Molinero 2006; Spivak 1987; Wright 2004), we hope that this collection will catalyze new conversations across continents and languages.

Moving beyond Imported Magic

In the 1970s, students in the engineering school of the Universidade Federal do Rio de Janeiro, Brazil, referred to computers as a form of "imported magic." This phrase, which came to be used even among members of the Brazilian technical elite, cast computer technology as highly effective, universal, sometimes mysterious, and always as coming from somewhere else (Marques 2005). Indeed, there is a widely held perception that science and technology necessarily come to Latin America from elsewhere, a notion fostered by ideas of modernization and development that originated outside Latin America and encouraged the transfer and diffusion of machinery and knowledge from more industrialized nations to less industrialized ones (Basalla 1967; Rostow 1991).[1] Such ideas have held considerable sway, shaped understandings of how science and technology should move, and guided decades of development policies. They have also propagated value systems that relegate Latin American nations to secondary or peripheral status. These perceptions endure, as seen in Latin American press coverage of Apple iPads, Nokia telephones, and forms of social media such as Facebook or Twitter that assumes that all design decisions about and the supply of such increasingly vital technology will come from elsewhere.[2] These perceptions also appear in government programs intended to replicate in Latin America imagined ideas about how US cultures of innovation and scientific expertise work.[3] Such views emphasize the transfer of supposedly superior technologies and ideas from North to South and do not acknowledge that innovation, invention, and discovery take many forms, occur in multiple contexts, and travel in many directions, nor do they acknowledge that diverse communities use scientific ideas and technologies in different ways. Framing science and technology as forms of imported magic overlooks processes of reinvention, adaptation, and use. It may also suggest that innovations such as a scientific way of thinking or an industrial technology should be adopted uniformly, rapidly, and unmediated, without considering the potentially negative or mixed effects of epistemic and technological change. All the essays in this volume contest this simplistic notion of a unidirectional movement of science and technology from the supposed center in the North to the assumed periphery in the South. The essayists' emphasis on science and technology, moreover, challenges past portrayals of Latin American life, which have stressed the magical, the exotic, and the primitive (Moya 2011). These elements, present, for example, in literary studies of magical realism, have drawn attention to important forms of Latin American culture. However, emphasizing them has also furthered the conceptual separation of the region from the West and, by extension, from science and technology.[4]

The essays' serious engagement with science and technology in Latin American contexts shines a light on this less studied aspect of Latin American life while illustrating the limitations of common diffusion narratives.[5] Such narratives give institutions and individuals in the most industrialized regions of the world disproportionate credit for scientific and technological creation, erasing the contributions of all other participants in this process and presenting those in other areas of the world, such as Latin America, as passive recipients or followers. The essays here offer alternative narratives that move the story of invention and innovation southward; study forms of local innovation and use; analyze the circulation of ideas, people, and artifacts in local and global networks; and investigate the creation of hybrid technologies and forms of knowledge production. Take, for example, the case of Argentine prosecutors who in 1892 used a bloody fingerprint left at a violent crime scene to secure the world's first criminal conviction based on fingerprint evidence, a form of forensic evidence with an accompanying science of dactyloscopy that would not appear in London or Paris courts for another decade. By the end of the nineteenth century, Argentina had become a center of advanced research on criminology, and ideas about fingerprint science traveled regularly across the Atlantic. Yet as Julia Rodriguez demonstrates in this volume, the early adoption and acceptance of this new science in the Argentine courts was rooted in the specificities of Argentine history, including the desire by the Argentine government to control its growing immigrant population. This example, like many provided in this volume, turns this idea of imported magic on its head.

A second example is the Soviet Union's 1963 donation of a factory to Cuba to produce large concrete panels for Soviet-style standardized housing, a supposedly humanitarian act also intended to further Soviet influence in Latin America during the Cold War. The story, however, looks quite different from the Cuban perspective. Modifications made by Cuban engineers to accommodate the Caribbean climate not only traveled back to the Soviet Union and shaped construction there but also created a local aesthetic that defined the architectural landscape of the Cuban revolution (Palmarola and Alonso, this volume). The story of the concrete panel factory is therefore more than a story of imported magic and Soviet hegemony. It is also a story of Cuban innovation, aesthetics, and resilience.

Through such cases, we see the benefits of adopting a more global view of science and technology that includes Latin America. Understanding the interrelationships among different communities and different parts of the world, however, requires broadening the study of the history and social aspects of science and technology to include scholarship on and by academics around the globe. It may entail encouraging STS and its constituent fields to look for producers or users of science and technology in places less conventional than, say, in the lab or among elites. The contributors to this book bring together literatures from varied fields and geographical areas to draw attention to a wide range of actors in the production of science and technology, including forms of invisible labor, neoliberal policies, and law, as well as the contributions of indigenous peoples, educators, children, activists, women, hackers, and shop floor workers.

In so doing, this book builds on and contributes to important previous work in the area of postcolonial science studies (e.g., Harding 2008; Harding 2011; Anderson 2002; Anderson 2009; Anderson and Adams 2008). As Anderson writes, in the context of science studies the term *postcolonial* "refers both to new configurations of technoscience and the critical modes of analysis that identify them" (2002, 643). These modes include studying the situated nature of technoscience; the connection and reconfiguration of the local and the global; the coproduction of identities, technologies, and cultural formations; and the transnational movement of people, practices, and technologies. Previous work in this area has raised questions about the role of science and technology outside elite institutions and pushed members of the STS community to look beyond US and European scientific discoveries and inventions. Indeed, Anderson (2008) has shown that cultural exchange can lead to discoveries, as seen in the case of the kuru brain disease and the discovery of prions. Other scholars, such as Harding, have harnessed feminist and postcolonial perspectives on science to highlight their shared interests in inequality, subjectivity, power, and the experiences of those whom Harding calls "modernity's Others" (Harding 2008). Many of the contributors extend concepts from the literature of postcolonial science studies and insights from feminist scholarship to generate new insights about the experience of science and technology in Latin America.

At the same time, it is important to note that the applicability of the postcolonial studies label to Latin America is a subject of ongoing debate. Postcolonial studies emerged largely from Western academic institutions, especially in the United States, where it gained currency in the 1980s, and is often used to refer to the study of people and places that were formerly under the control of a foreign power. The field has focused primarily on issues of personal, cultural, and political identities and possibilities in both colonized and decolonized environments. Postcolonial studies surely shares many concerns that are central to Latin American studies, including studies of empire and its legacy, the exertion of imperial power through representations of knowledge and culture, and the economic exploitation of land, natural resources, and peoples. Indeed, many members of the Latin American studies community have productively engaged with the concepts, findings, and frameworks of postcolonial studies (e.g., Seed 1991; Mignolo 1993; Vidal 1993; Adorno 1993; Moraña, Dussel, and Jáuregui 2008).

Nonetheless, as Fernando Coronil notes, "there is no corpus of work on Latin America commonly recognized as 'postcolonial'" (Coronil 2008, 396). Latin America, a region characterized by colonial ties to Spain and Portugal and comprising nations that typically gained independence in the first part of the nineteenth century, does not fit the historical mold of much of the postcolonial literature.[6] For this reason some scholars of Latin America have argued that the term *postcolonial* unnecessarily homogenizes experiences of colonization and decolonization (Adorno 1993; Klor de Alva 1992; Klor de Alva 1995; Moya 2011).[7] Adorno, for example, writes that while the sweeping expanse of colonial and postcolonial discourse may allow us to bring together the Indians of South Asia and the *indios* of Latin America, doing so "risks offering too much too easily and at too great a cost" (Adorno 1993, 141).

Europe's colonization of Latin America differs significantly from its colonization of Asia or Africa. It begins in 1492 with the arrival of Columbus in the New World, more than three centuries before the expansion of British colonial rule in India or the establishment of French colonial rule in Algeria, and four centuries before Belgium's colonial rule in the Congo. Latin American nations were also among the first to achieve independence from their colonizers and become modern nation-states. As Moya writes, "Spanish and Portuguese colonialism in the Americas began much earlier than other European colonial incursions elsewhere, lasted much longer, ended a century and a half earlier, and left a much deeper impact" (Moya 2011, 5). This can be seen in the overwhelming dominance of the colonial languages (Spanish and Portuguese) and the pervasive presence of the colonizing religion (Roman Catholicism). Colonization, moreover, altered the region at a microscopic level—70 to 90 percent of the indigenous population in the Americas succumbed to imported diseases such as smallpox, measles, and plague, a level of mass extermination that has not been seen since.

Additionally, the Spanish and Portuguese crowns created centralized, hierarchical systems of viceroyalties to manage their empires across the Atlantic; these systems differed in notable ways from British practices of colonial management in the north. Brazil even became the seat of the Portuguese crown in 1808 and thus served as the governing center of the Portuguese kingdom. Spanish and Portuguese colonization of the Americas, moreover, has been linked to the origins of European modernity in the sixteenth century. Latin America provided not only the other that Europeans could use to measure their achievements; it also provided a source of extractable resources that helped Europe gain a competitive edge at the beginnings of global capitalism. Europe, for example, paid for almost all its imports with silver mined from such places as Potosí in Bolivia. Overall, Latin America produced approximately 85 percent of the world's silver between 1500 and 1800 (Pomeranz 2000, 159). The colonies also contributed to European knowledge of the world through the collection of flora and fauna and the mapping of uncharted territories. This, in turn, allowed Spain to exert power over knowledge production by means of its supposed discovery and subsequent naming of plants and places (Cañizares-Esguerra 2001; Cañizares-Esguerra 2006).

Scholars of Latin America such as Mignolo (2000), Moraña, Dussel, and Jáuregui (2008), and Klor de Alva (1995) have called for greater recognition of colonial difference and greater attention to the historical specificities of how nations fit within the "world-system of colonial domination" (Moraña, Dussel, and Jáuregui 2008, 6). These scholars also reject the idea that the region has undone colonial relationships or moved beyond them, as the term *post*colonial might suggest. Instead, they argue that colonialism has changed over time as power has shifted from European colonizers to creole elites to those with connections to the metropolitan centers and global markets. Indeed, this shift in power from European colonizers to creole elites in the period following independence is yet another distinguishing factor of the Latin American colonial experience (Klor de Alva 1995). Thus, even as this book aims to foster conversation among the literatures of STS, Latin American studies, and postcolonial studies, it acknowledges these ongoing discussions within the Latin American studies community.

Through this volume we hope to enrich understandings of colonial and postcolonial science and technology by making Latin American history more visible within the field of STS. Many of the contributors to these pages use science and technology to illustrate how structures of power have shaped both knowledge production and material life in Latin America from its colonial past to its neoliberal present. In doing so, they explore how science and technology have played a role in perpetuating structural inequalities among nations and facilitating the spread of European, US, or Soviet ideas of modernity, even as the legacies of such ideas mutate when they are reused (Kowal, Radin, and Reardon 2013). At the same time, as the essays that follow demonstrate, Latin America provides scholars a space to study alternative epistemologies that challenge Western "scientific" ways of knowing or call into question science and technology policies that travel from North to South but are ill suited to Latin American realities or socially unjust. This book therefore also aims to offer science and technology as a lens through which to better understand Latin America's history, its connections to other parts of the world, and its ongoing struggle for intellectual, technological, and economic sovereignty.

STS in Latin America

Latin America is a region with a distinctive geography and history that have shaped its collective experience with science and technology. This includes a colonial history predominantly tied to the Iberian Peninsula, economies of colonial extraction that were later replaced by conditions of economic dependency on more industrialized nations, a history of centrality in economic globalization and understandings of modernity, and sustained US intervention in its political and economic life during the nineteenth and twentieth centuries. Scholars such as Saldaña (2006) have identified further commonalities within this area that pertain to the development of Latin American scientific and technological capabilities. For example, characteristic geographical features such as the Amazonian basin, Mesoamerican lowlands, and highlands in Mexico and the Andean cordillera, and the flora, fauna, and mineral deposits within the region, have contributed to the rise of particular areas of science and engineering expertise (e.g., mining, botany, zoology). In more recent times, the idea of Latin America as a region has led to the growth of networks of scientific collaboration and exchange, including the formation of professional societies or conferences such as the Latin American Conference on Informatics (CLEI). Regional analysis allows for researchers working in different disciplines to enter into conversation with one another and produce comparative knowledge about this part of the world.

However, although Latin America has a strong regional identity, it is not a self-contained, bounded unit but rather an open one, linked to global flows of labor, capital, and cultural hybridization. Thus, the regional focus of the book is intended as a basis for future transdisciplinary conversations using science and technology as a way to probe the global reality that is Latin America, past and present.[8]

That said, we do not mean to overlook the tremendous diversity of histories, cultures, and languages within the region. Although Spanish and Portuguese are the dominant languages in Latin America today, they are but two of many languages spoken historically in this part of the world. It is estimated that immediately before Columbus arrived in the New World, approximately 350 distinct languages were spoken in the areas that became known as Mexico and Central America and that 1,450 distinct languages were spoken in the areas that became known as South America (Sherzer 1991). African languages and cultures also arrived in Latin America with the transatlantic slave trade, which peaked in the eighteenth century.

We further recognize that while the term *Latin American* conveys a shared identity, it is politically fraught and always has been. The label was arguably first imposed on the region in 1836 in the writings of the French intellectual Michel Chevalier, who divided Europe into Latin (Roman) Europe and Teutonic (German) Europe and argued that such divisions carried over to the New World (Mignolo 2005). The French later used this construction of *Latinidad* to promote the linguistic root of Latin shared by Spanish, Portuguese, and French and thereby justify France's greater influence in the region, including its invasion of Mexico in 1862 and Napoleon III's installation of Maximilian as the short-lived emperor of Mexico (Chasteen 2001). Mignolo and others have also argued that the term *Latin America* is a geohistoric category that has since become naturalized and that the idea of Latin America has historically served as a necessary other for European modernity, claiming that "the Americas exist today only as a consequence of European colonial expansion and the narrative of that expansion from the European perspective" (Mignolo 2005, xi). More recently, Gobat (2013) has argued that Central and South Americans invoked the term *Latin America* in the mid-nineteenth century as a way to unify the region in protest against US expansion and European intervention. He notes that the term *Latin* also emphasized the region's connection to France, which was then viewed as one of the most modern European nations (in contrast to Spain), and provided the basis for a regional claim to whiteness. Sorensen, in contrast, has dated the construction of Latin American identity to the cultural and political imagination of the 1960s; "only then," she writes, "did a transnational cultural identity become rooted in the hemispheric imagination" (Sorensen 2007, 1). Thus, Latin America has historically functioned as a political category and contested form of regional identity. This book acknowledges this history but uses the term in a different way. It uses *Latin America* to refer to nations in the Americas that are south of the United States and where Spanish and Portuguese are predominantly spoken, including areas of the Caribbean. We opted to use *Latin America* instead of *Ibero-America* because it is familiar to a broader readership.

Although it is not possible within the scope of this introduction to do full justice to the existing interdisciplinary body of STS literature in Latin America, a brief overview of the multiple points of origin and its relationship to the development of related fields such as the history of science provides some useful context.[9] Our overview focuses on the development of STS within Latin America, as this history may be less known to readers of this volume.[10] It shows, moreover, that STS as a field has a diverse history that is interwoven with the historical experiences and intellectual currents found in different parts of the world.

Arellano Hernández and Kreimer (2011) date the precursors of the social studies of science and technology in Latin America to the 1950s, including work by the Brazilian scholar Azevedo (1955), the Cuban scholar López Sánchez (1967), and the Mexican scholar Gortari (1963). These early works limited the scope of inquiry by arguing that Latin America, and by extension Latin Americans, lacked science and attributed this deficiency to the history and culture of the region—perhaps contributing to the articulation of science and technology as forms of imported magic (López Sánchez 1967; Gortari 1963). Azevedo, for example, proposed that Brazil's colonial experience (including the tendency to focus on literary works and subjectivism) and the absence of a Protestant work ethic impeded its level of scientific activity. For this reason, Saldaña describes this early historical work as having "a limited horizon and distorted understanding of scientific activity" (Saldaña 2006, 5).

From the 1950s to the beginning of the 1980s, research on the social dimensions of science and technology did not have a strong institutional presence in Latin American universities or government offices. Instead, it coalesced as a current of thought or movement sometimes referred to as *pensamiento latinoamericano en ciencia, tecnología y sociedad* that connected science and technology to economic and political issues such as development and dependency. Much of this thought grew out of research activities at the Chilean offices of the United Nations Economic Commission on Latin America (ECLA) during the 1960s and 1970s (Arellano Hernández, Arvanitis, and Vinck 2012; Dagnino, Thomas, and Davyt 1996). Among those involved in this research were the ECLA economists Raúl Prebisch and Osvaldo Sunkel, and scientists and technologists such as Jorge Sábato, Amílcar Herrera, Oscar Varsavsky, and Marcel Roche, many of whom became social thinkers and ideologues who connected science, technology, and politics in their writings. It was in this way, according to Vaccarezza, that "Latin American thought on political science and technology was shaped without forming a conscious community identified as STS" (Vaccarezza 2011, 47).

Introduction: Beyond Imported Magic

This thinking on Latin American science and technology was influenced by dependency theory, one of the most important Latin American contributions to economic thought.[11] It arose at a moment when economists firmly cast countries as either developed or developing and argued that the unequal power relationship between developed and developing nations was a necessary part of industrialization. Dependency theory challenged the idea that Latin American nations could move from developing to developed by following linear models of innovation that included adopting Western models for scientific practice and industrial technologies for economic growth and increased consumption. Instead, dependency theory framed underdevelopment as necessary for the developed nations to amass their wealth and argued that Latin American nations could not follow the same trajectory as nations such as the United States. The condition of Latin American nations could be improved, however, with increased national autonomy and less reliance on foreign capital, foreign goods, and foreign priorities.

Latin American thought on science and technology during this period similarly argued that Latin Americans should develop science policies and technological practices that reflected the specific conditions of Latin American life. In Cuba, Chile, and elsewhere, this resulted in national policies that viewed science and technology as mechanisms for social change and activities that could be oriented toward addressing national problems (Núñez Jover and López Cerezo 2008; Medina 2011). The ideas of Varsavsky, Herrera, and others were embodied in government reference texts that shaped science policy. In some countries, such as Chile, these ideas merged with platforms for socialist change (Medina 2011) and Marxist critiques of capitalism. But the national programs for science and technology development proposed in these texts also appealed to nationalistic wings of the military, especially in countries such as Brazil and Argentina (Adler 1987; Evans 1995; Marques 2003).

One of the most influential of these early thinkers about the role of science and technology in Latin America was the Argentine mathematician Oscar Varsavsky, who wrote a number of scholarly texts on the structure and practice of Latin American science. In his 1969 *Ciencia, política y cientificismo,* which exemplifies the critical stance he termed *cientificismo,* Varsavsky argued that Argentine scientific practices, norms, and values were following those of larger research centers in Europe and especially the United States, perpetuating a form of scientific colonialism that prioritized foreign business interests and encouraged young scientists to study problems that were of little relevance in their own countries, a phenomenon more recently described as an "internal brain drain" (Polanco 1987). Varsavsky argued that it was both possible and necessary for Latin American countries to develop scientific styles adapted to their own national projects (Núñez Jover 2002; Schoijet 2002). His later works critiqued scientific objectivity, cast science as an ideological extension of existing power relations, and accused it of being overly focused on problems that would not bring about revolutionary change (Varsavsky 1971). Varsavsky's writings, and those of his contemporaries, connected science and technology to political and economic structures and unequal power relations between Latin America and more industrialized parts of the world.[12] This attention to social justice and public policy remains a hallmark of Latin American STS scholarship, and the acronym STS, or CTS in Spanish and Portuguese, is often used to denote scholarship on the subject of science, technology, and public policy.

The 1980s to the second half of the 1990s witnessed a shift in STS research, as well as research in the related field of the history of science, and reflected the growing maturity of both. Beginning in the 1980s, the history of science in Latin America moved away from Eurocentric understandings of science and began to focus on how the specificities of Latin American history created a different context for science, which was often referred to as "science of the periphery." The field also increased its professional identity with the founding of the Sociedad Latinoamericana de Historia de las Ciencias y la Tecnología in 1982 and the creation of its journal *Quipu: Una Revista Latinoamericana de la Historia de las Ciencias y la Tecnología* in 1984. The journal was published in Spanish and based in Mexico from 1984 to 2000 (producing thirteen volumes of three numbers each), with a recent resurgence of the journal and publication of a fourteenth volume in 2012 (Quipu 2012). *Quipu* was a disciplinary landmark for the history of science in Latin America because it opened a space for local historical narratives and abandoned earlier views of Latin America as an empty space for the production of scientific and technological knowledge.

Like the history of science, STS also developed a professional identity during the 1980s. Whereas previous work on science, technology, and society had been conducted primarily by scientists who were reflexive about their practices, the field increasingly became a domain that included social scientists and humanists (sociologists, psychologists, anthropologists, historians, philosophers) who considered STS their specific area of expertise and who gave greater attention to theory. Within Latin America, the pressure to shift from economic models of socialism and developmentalism to those of neoliberalism also affected the content of the field and motivated scholars to abandon such influential frameworks as dependency theory. This included challenging the center-periphery dichotomy and studying how scientific and technological exchanges transcended the rigidity of these categories. As Cueto has noted, although the concept of the periphery that had emerged from dependency theory had been "useful for locating scientific communities that are considered to be on the 'outskirts' of the traditional centers of knowledge," the theory's "lack of temporality, the linearity, [and] the passivity assigned to the periphery" must be rejected by scholars if the field is to continue and mature (Cueto 2006, 231).

Writing in the journal *Social Studies of Science* in 1987, Hebe Vessuri observed that social studies of science were still fragile in Latin America, despite increasing research activity and a growing number of publications in Spanish and Portuguese. Although she credited the emerging field with fostering a deeper understanding of the role of science and technology in development (which was in keeping with the early interests of the field in the 1960s and 1970s), she also expressed discomfort with the conceptual dependency of Latin American scholarship on intellectual production in the United States and Europe.[13] This concern remains, given that many of the theoretical frameworks found in the STS literature on Latin America have originated in France, Great Britain, or the United States (although, as this book illustrates, theories change as they are invoked in different historical, cultural, and geographical contexts and often become more nuanced in the process).[14]

The field has continued to grow within Latin America since the mid-1990s. For example, the biannual Latin American Conference on the Social Study of Science and Technology (ESOCITE) has held nine meetings in different locations in South America and Mexico since 1994 and attracted an increasing number of participants.[15] Nevertheless, certain areas of Latin America have been more engaged in STS research than others. For example, a historical analysis of participation in STS conferences reveals a strong concentration of scholarship in such countries as Brazil, Argentina, and Mexico; fewer contributions from such nations as Colombia, Costa Rica, and Chile; and a dearth of work from many areas of Central America and the Caribbean (Arellano Hernández 2012).

This growth has led scholars to reflect upon where the field should go in the future. Arellano Hernández and Kreimer (2011), for instance, posit that one of the key contemporary challenges facing the field is to understand how Latin American technoscience functions in the global context. Velho (2011), for her part, argues that a key paradigm of Latin American science studies in the twenty-first century is studying the relationship of science to social good. And Vessuri (2011) situates the future direction of the field in a larger political context to wonder whether the return to power of various leftist governments in Latin America will result in a resurgence of earlier topics of importance, including issues of sovereignty, legitimacy, and power. The essays in this volume also consider other potential directions, such as studying science and technology in transnational frameworks (Mateos and Suárez-Díaz), moving beyond the North-South dichotomy to study exchanges among areas of the global South (Fressoli, Dias, and Thomas), examining forms of internal colonialism (Cukierman), as well as challenging what we view as legitimate innovation and ingenuity (Marques).

The Book's Structure and Conceptual Themes

The rest of this volume is divided into three parts: Latin American Perspectives on Science, Technology, and Society; Local and Global Networks of Innovation; and

Science, Technology, and Latin American Politics. The first part examines the politics of knowledge and representation in specific Latin American contexts and possible frames of analysis for studying science and technology in the region. The second traces the circulation of scientific ideas within community, national, and transnational networks. The final part addresses the mechanisms through which scientific projects and technologies are linked to Latin American politics and political will. This includes how Latin American politics have shaped the historical development of science and technology in the region and how science and technology have shaped the contours of Latin American politics and history.

Part I: Latin American Perspectives on Science, Technology, and Society

The chapters in this part use multiple frames of analysis to help us better understand the development and practice of science and technology in Latin America. Although the authors employ concepts drawn from different analytical frames and geographic regions, all are concerned with the politics of knowledge and representation in and about Latin America. They also examine what is made newly visible by these approaches and illustrate their broader applicability to our understanding of Latin American science and technology.

In the first of these essays, Henrique Cukierman examines a scientific expedition to inland Brazil in 1911–1913 that he ties to nation building and the colonization of both the scientists involved (through their desire to emulate European science) and the residents of an inland northern tropical region of Brazil. Cukierman explores the use, perception, and adaptation of European medical science in inland Brazil and how local material conditions and cultures come into contact with Western science and technology to shape the production of scientific knowledge.

Mariano Fressoli, Rafael Dias, and Hernán Thomas highlight the tensions between market logic and local participation in innovation by describing how ideas for technological development and social inclusion that were originally developed in India are being taken up in Latin America and by such global institutions as the World Bank. Combining analyses of cognitive praxis and of the sociotechnical alliances involved in these movements allows the authors to consider how ideas circulate among areas of the global South, their relevance to a new local context (in this case, Latin America), and how these ideas intersect with the market, market thinking, and concerns about the well-being of the poor.

In their investigation, Tania Pérez-Bustos, María Fernanda Olarte Sierra, and Adriana Díaz del Castillo deploy feminist theory on the politics of care to examine how gender figures in the ways various scientific practices and practitioners are made visible or invisible and their value increased or diminished. In particular, they examine how women forensic geneticists in Colombia have framed their work within a broader concern for serving justice. They demonstrate that giving care can be at odds with

Introduction: Beyond Imported Magic

academic concerns, such as the desire to publish results, and thus diminish the status of the work but not its technical difficulties.

Ivan da Costa Marques uses the concept of ontological politics to examine various frames of reference used by scientists and activists in Brazil to determine the legitimacy of a local nutritional innovation, *multimistura*. Ontological politics is the sum of choices and decisions made to establish or stabilize what constitutes a reality—a frame of reference for understanding the world. By granting an ontological role to narratives, Marques uses the case of *multimistura* to explore the different worlds that surrounded the development, adoption, and use of this technology, including those of nutritionists, social scientists, and activists/users.

Finally, Michael Lemon and Eden Medina analyze how two different academic communities, Latin American studies and the history of technology, have approached the telling of Latin American history of technology, based on a review of the English-language literature on the history of technology in Latin America in two of the top journals on the history of technology and three of the top journals on Latin America. Through an analysis of the number of articles published, they identify important differences in how each community addresses the topic of technology, and the authors highlight what is made visible and invisible in each field. They conclude by suggesting ways to broaden how scholars in these fields write technology history in the Latin American context.

Part II: Local and Global Networks of Innovation

The essays in this part investigate the flows of technology and scientific knowledge across communities, regions, institutions, and nations, examining the formation of networks of exchange and how ideas and technologies are transformed or contested as they move within these networks. These studies address translations, hybrids, adaptations, and the creation of charismatic objects; uniformly reject the idea that technoscience travels unchanged from a central to a peripheral context; and examine how politics shape the movement, selection, and use of ideas, practices, and artifacts.

Julia Rodriguez's historical study of the development of the science of fingerprinting demonstrates that the system for forensically filing and categorizing fingerprints to allow their easy retrieval for use in criminal cases was first developed in Argentina and then spread to other parts of the world. Yet her story also details the transatlantic exchanges that contributed to both the initial development of forensic fingerprinting, including circulation among members of shared language groups within the scientific community, and the rapid movement of immigrants into Argentina that first prompted the Argentine state to develop new sciences for social control.

Hugo Palmarola and Pedro Ignacio Alonso's essay examines the confluence of science, technology, and ideology in one of the largest social housing projects in

Latin America, the large-concrete-panel housing projects in Cuba. The authors demonstrate that Cuba's imagination of revolutionary change was simultaneously local and imported, arguing that the Cuban slogan "socialism is building" furthered the Cuban political project and that the Cuban government's framing of prefabricated housing paralleled its framing of the Cuban New Man. Large-panel construction, they argue, is a hybrid technology that is both Soviet and Cuban and made possible through local innovation and transnational exchange.

In their chapters Anita Say Chan and Morgan G. Ames study the design, use, and meaning of the XO laptops created by the One Laptop per Child program. Chan first traces the process from the differing perspectives of a programmer affiliated with the program in the United States and an engineer working on the deployment of the laptops in Puno, Peru. Chan delineates two models for laptop design—an engineering-centric vision that minimizes engagement with local users and contexts, and a participative model that accepts input from multiple members of the local community—and argues that the adoption of new technologies necessarily depends on processes of local translation that can open new technological possibilities. Ames describes the appeal of the XO laptop in Paraguay in terms of the concept of the charismatic object, which exerts power not because of what it is but because of what it promises to do. Tracing the everyday challenges related to the use of the XO laptop in the classroom and the networks of NGOs, teachers, and government officials involved in the program, Ames explains why the Paraguayan program has been more successful than most and retains its appeal despite those challenges. Both Chan and Ames describe how the actions and networks of globally dispersed actors, including students, rural teachers, technicians, activists, software programmers, and engineers, create unique sociotechnical ecosystems.

Noela Invernizzi, Matthieu Hubert, and Dominique Vinck study the development of nanoscience policy in Brazil, Mexico, and Argentina to examine how the scientific priorities of nations outside Latin America shape those in Latin America. Acknowledging the strong policy influence of the United States and Europe, they nonetheless challenge the narrative of technological imitation by showing how the distinctive character of national and international scientific networks in different Latin American countries, and the scientific traditions within each country, shape the specific character of these policies despite their common rhetoric. They thus argue that local factors contribute to the ways in which scientific development is materially organized and scientific policies travel and change.

Amy Cox Hall details the origins of the first photo of Machu Picchu that appeared in *National Geographic* and that has since become one of the most iconic depictions of Peru. The photograph was taken as part of an anthropological expedition sponsored by Yale University during a period in which the camera was viewed as a scientific instrument that could capture objective data. Cox Hall, however, details how the use

Introduction: Beyond Imported Magic

of the camera during the expedition was a negotiated and disciplined practice. Presenting Machu Picchu as a lost civilization that had been found again by the expedition involved a particular configuration of human and nonhuman actors and practices that presumably stabilized the facts created by the expedition. Borrowing a term from Ames, Cox Hall suggests that cameras and their photographs thereby also served as charismatic objects as well as inscription devices.

Part III: Science, Technology, and Latin American Politics

The chapters in this part examine more closely the relationship of science, technology, and politics in Latin American settings, including how the coproduction of technology and politics shapes bureaucratic decision making and how science and technology have been used to achieve specific political ends. They also include analyses of the ways in which international relations have shaped the character of national research programs and how technologies can interrupt or change legal practices, regulations, and classification systems and intersect with such state structures as policy making and legislation.

Three accounts in this part deal with cases of nuclear power. Jonathan Hagood uses the history of nuclear energy in Argentina to challenge overly simple and often decontextualized understandings of technological success and failure in Latin American contexts. Examining how the political world of 1950s Argentina made it possible for Argentine president Juan Perón and the German émigré scientist Ronald Richter to envision creating a "bottle of atomic energy," Hagood argues that although these efforts are often viewed as a failure (or an example of a president's being duped by a savvy scientist), the project helped Perón achieve his short-term political goals and thus can be seen as a political, if not technological, success.

Gisela Mateos and Edna Suárez-Díaz use a transnational framework to study the history of Mexican nuclearity during the Cold War. As they show, the development of national nuclear capabilities required the international circulation of physicists, nuclear materials, standards, and practices. Yet the authors further argue that the history of the Mexican nuclear program cannot be understood outside the context of Mexican political history in the twentieth century, as Mexico's dual goals of political independence and national scientific capability resulted in a nuclear program that was dedicated exclusively to peaceful applications—such as energy—and eschewed the building of bombs.

Manuel Tironi and Javiera Barandiarán examine Chilean energy policy in the context of neoliberalism, which they argue functions as a "political technology"—a form of applied knowledge that can be used in pragmatic and intentional ways to transform the state and society. To demonstrate how neoliberalism and its use as a political technology have changed over time, they examine two cases. The first is the dismantling of Chilean nuclear energy policy in the 1970s, when the Chilean government's instantiation of the neoliberal model devalued the expertise of Chilean nuclear engineers while giving greater power to Chilean economists. The second case examines the more recent hydropower generation to argue that although current forms of neoliberalism permit multiple and noneconomic forms of expertise to contribute to environmental regulations, they do so only in ways that do not threaten the structure of the neoliberal state.

Shifting topics, Ana Delgado and Israel Rodríguez-Giralt detail the attempts of Brazilian peasant movements to include creole seeds in the national agrarian insurance system. Creole seeds, which have been developed, adapted, or produced by farmers and landless and indigenous peoples in situ over long periods of time using local methods and management systems, are by definition not standardized. Yet if they cannot be standardized or classified, they cannot be entered into the national registry, which provides insurance in times of crop failure. The authors draw on Moreira's discussion of interferences, or copresences through which the social is reenacted, to describe how creole seeds disrupt what is considered a legally valid entity in the Brazilian context.

Finally, João Biehl examines how "right-to-health" legislation in Brazil is changing the relationship among technology, medicine, and the law. Right-to-health legislation requires the state to pay for health care in the form of certain pharmaceuticals, which has spurred a phenomenon referred to as the "judicialization of health." Biehl chronicles how some Brazilian poor have turned to suing the government to claim their right to access newer medical technologies, often expensive proprietary pharmaceuticals that otherwise would be beyond their reach. This in turn shifts public understanding of the state's role in health care toward the acquisition of and payment for pharmaceutical treatments that are often manufactured by multinational pharmaceutical companies. Biehl examines the practices of citizenship and governance that are emerging from these struggles and how the market, acting through the acquisition of pharmaceuticals, is influencing the way Brazilians understand their relationship with the state.

Collectively, then, this book features a range of voices from Latin American, US, and European institutions. Although we cannot hope to cover the geographic, cultural, and historical richness of Latin America in a single volume, our intention is to provide a snapshot of some of the promising, provocative, and groundbreaking research taking place in this growing field and to suggest areas for future work and interdisciplinary engagement. As such, the collection deepens our knowledge of Latin American experiences, extends the geography of science and technology and makes it more inclusive, and provides new conceptual approaches for STS-related research. Freed from the framework of imported magic, Latin American science and technology, as well as their study, take on new relevance in how we understand the history of the region and its global connectedness, diverse practices of knowledge production, and the many forms of innovation that create and maintain our worlds.

Notes

All translations in this chapter from Spanish and Portuguese sources were made by the authors.

1. One of the more frequently cited models of scientific diffusion is George Basalla's (1967), which tracked the spread of modern science from a handful of European countries to the rest of the world through military conquest, colonization, missionary work, and political and commercial relationships. Rostow's theory of economic growth (1991), which charted the progression of societies from traditional to modern, with modernity defined as an industrial economy and the mass consumption of high-value goods, also exerted considerable influence in Latin America and guided decades of foreign aid policies, Latin American economic planning, and discourse regarding development (Escobar 1994). Rostow and Basalla have rightly been criticized for idealizing the historical experiences of the United States and Western Europe and for insisting on the global imposition of these experiences.

2. For example, in 2012 *O Globo*, a daily newspaper in Rio de Janeiro, published an article on trends in cell phone manufacturing that lamented the decline in "the techno-diversity" of cell phone models while also praising a new Nokia model and expressing disappointment that this new phone "has no foreseen date of arrival in Brazil." Instead of arguing that Brazilian technologists should change cell phone design, the underlying assumption of the article is that technology necessarily comes from somewhere else (Rónai 2012).

3. Start Up Chile, for example, is a program run by the Chilean government to replicate in Chile the start-up culture of Silicon Valley and increase Chilean economic competitiveness. The Brazilian government is also trying to place 100,000 Brazilian students and researchers in top universities worldwide as part of its Science without Borders Program, based on the premise that training Brazilian students and researchers abroad in science, engineering, and innovation and having them subsequently return to Brazil will provide a mechanism to import knowledge and expertise and strengthen the Brazilian economy. We point to these programs not as a criticism, but rather to demonstrate the enduring belief that science and technology are only and should only be imported to Latin America from elsewhere.

4. This does not mean that we wish to establish a dichotomy with magic on one side and science and technology on the other. Indeed, technoscientific objects can take on some of the mysterious connotations of magic in the sense that magic represents an emotion-laden experience of the unknown. Furthermore, as this book shows, literary works such as Gabriel García Márquez's *One Hundred Years of Solitude* can suggest new and productive directions for the study of technology history (Lemon and Medina, this volume). Our point, rather, is that scholarly attention to the magical has propagated a particular image of the region that is not associated with scientific practice or technological creation.

5. Indeed, the analyses developed in this book illuminate the limitations of diffusion narratives in national and international contexts outside of Latin America. *Diffusion* can often be used interchangeably with *technology transfer*, although the two terms are sometimes distinguished by the movement they describe. *Technology transfer* is often used to describe the movement of a technology from one country to another, whereas *diffusion* is used to describe movement within

a community (Staudenmaier 1989). We do not draw this distinction here, and our assessment of diffusion would apply to many discussions of technology transfer.

6. Perhaps this is why it has been noticeably absent from major works of postcolonial scholarship. For example, Ashcroft, Griffiths, and Tiffin's 1989 *The Empire Writes Back*, an important early book on postcolonial theory, did not include the Latin American experience, with the exception of the Caribbean. The second edition of the book, however, recognized the omission of "the oldest, and second largest and most complex modern European empire—that of Spain" and suggested that the distinct character of coloniality in Latin America "may fundamentally change our view of the post-colonial" (Ashcroft, Griffiths, and Tiffin 2002, 202). Although the authors did not include additional material on Latin America in the text, their comments support our contention that broadening what is considered postcolonial to account for Latin American experiences is imperative.

7. This argument does not imply that the analytic frameworks of postcolonialism are not relevant, but rather that they have not been sufficiently expanded to include the Latin American context.

8. Drawing from the transnational turn in historical scholarship, we know that we cannot understand the histories of colonizing nations without understanding those of the colonized and vice versa. Nor can we understand the centers of scientific and technological production without understanding processes of imports and exports, particularly asymmetries in resource exchanges (e.g., in raw materials, geographically specific knowledge, the external and internal brain drain of Latin American technical experts), and the ways that ideas, artifacts, and techniques travel. This is even truer today, as contemporary science and industrial production become increasingly globalized and distributed, a phenomenon made possible by the increasing speed of transportation and communication and the resulting time-space compression (Harvey 1990; Harvey 2000). The presence of Latin America in the social studies of science makes visible the inseparability of the histories of various regions and demonstrates how the production of colonialism, modernity, and Western hegemony is intertwined with the production of technoscience.

9. While the history of science can be considered a constituent or related field of science and technology studies, the two fields have different historical paths of development in the Latin American context and overlap only occasionally. One goal of this book is to increase the degree of conversation among these epistemic communities.

10. This review, for instance, does not detail the rich and growing literature on the history and social study of science and technology being written by scholars outside Latin America (e.g., Soto Laveaga 2009; Rodriguez 2006; Wolfe 2010; Gallo 2005; Grandin 2009; Hayden 2003; Chan 2013; and Takhteyev 2012), nor does it trace the development of all the varied areas of the field within Latin America.

11. For example, Herrera's views on Latin American science and technology were intellectually tied to the work on dependency theory coming out of ECLA (the United Nations Economic Commission on Latin America) during the 1960s and 1970s. His 1970 *Latin America: Science and Technology in the Development of Society* was edited by the ECLA economists Fernando Henrique

Cardoso, Aníbal Pinto, and Osvaldo Sunkel, leading figures in the development of dependency analysis.

12. For example, Herrera (1971) argued that actual scientific practice was based on the implicit policies of elites whose interests were often aligned with foreign powers. Roche (1972) extended the framework of dependency theory to science and technology research and worried that the increasing levels of foreign investment in Latin American manufacturing and the importation of capital technologies from elsewhere would have long-term effects on the development of Latin American science and technology research capabilities.

13. This evocative observation extends Xavier Polanco's 1987 critique of internal brain drain to the production of STS scholars and STS scholarship.

14. For example, Varsavsky and his contemporaries are relatively absent in the bibliographies of present-day ESOCITE papers, while distinguished STS scholars from the United States and Europe appear frequently (Silva and Marques 2012).

15. For example, for the 2008 ESOCITE meeting in Rio de Janeiro, scholars submitted 470 extended abstracts; 313 abstracts were selected, and 228 complete texts were distributed in 43 sessions (available at http://www.necso.ufrj.br/esocite2008/index.html).

References

Adler, Emmanuel. 1987. *The Power of Ideology: The Quest for Technological Autonomy in Argentina and Brazil*. Berkeley: University of California Press.

Adorno, Rolena. 1993. Reconsidering colonial discourse for sixteenth- and seventeenth-century Spanish America. *Latin American Research Review* 28 (3):135–145.

Anderson, Warwick. 2002. Introduction: Postcolonial technoscience. *Social Studies of Science* 32 (5–6):643–658.

Anderson, Warwick. 2008. *The Collectors of Lost Souls: Turning Kuru Scientists into Whitemen*. Baltimore: John Hopkins University Press.

Anderson, Warwick. 2009. From subjugated knowledge to conjugated subjects: Science and globalisation, or postcolonial studies of science? *Postcolonial Studies* 12 (4):389–400.

Anderson, Warwick, and Vincanne Adams. 2008. Pramoedya's chickens: Postcolonial studies of technoscience. In *The Handbook of Science and Technology Studies*. 3rd ed., ed. Edward J. Hackett, Olga Amsterdamska, Michael Lynch, and Judy Wajcman, 181–204. Cambridge, Mass.: MIT Press.

Arellano Hernández, Antonio. 2012. Mapeo de los estudios sociales de la ciencia y la tecnología en América Latina. Paper presented at the IX Journadas Latinoamericanas de Estudios de la Ciencia y la Tecnología, Ciudad de México, June 5–8.

Arellano Hernández, Antonio, Rigas Arvanitis, and Dominique Vinck. 2012. Global connexity and circulation of knowledge. *Revue d'Anthropologie des Connaissances* 6 (2).

Arellano Hernández, Antonio, and Pablo Kreimer. 2011. Estudio social de la ciencia y la tecnología desde América Latina: Introducción general. In *Estudio social de la ciencia y la tecnología desde América Latina*, ed. Antonio Arellano Hernández and Pablo Kreimer, 9–19. Bogotá: Siglo del Hombre Editores.

Ashcroft, Bill, Gareth Griffiths, and Helen Tiffin. 2002. *The Empire Writes Back*. 2nd ed. New York: Routledge.

Azevedo, Fernando de, ed. 1955. *As ciências no Brasil*. São Paulo: Edições Melhoramentos.

Basalla, George. 1967. The spread of Western science. *Science* 167 (3775):611–622.

Cañizares-Esguerra, Jorge. 2001. *How to Write the History of the New World: Histories, Epistemologies, and Identities in the Eighteenth-Century Atlantic World*. Stanford: Stanford University Press.

Cañizares-Esguerra, Jorge. 2006. *Nature, Empire, and Nation: Exploration of the History of Science in the Iberian World*. Stanford: Stanford University Press.

Chan, Anita Say. 2013. *Networking Peripheries: Technological Futures and the Myth of Digital Universalism*. Cambridge, MA: MIT Press.

Chasteen, John Charles. 2001. *Born in Blood and Fire: A Concise History of Latin America*. New York: W. W. Norton.

Coronil, Fernando. 2008. Elephants in the Americas? Latin American postcolonial studies and global decolonization. In *Coloniality at Large: Latin America and the Postcolonial Debate*, ed. Mabel Moraña, Enrique Dussel, and Carlos A. Jáuregui, 396–416. Durham: Duke University Press.

Cueto, Marcos. 2006. Excellence in twentieth-century biomedical science. In *Science in Latin America: A History*, ed. Juan José Saldaña, 231–239. Austin: University of Texas Press.

Dagnino, Renato, Hernán Thomas, and Amílcar Davyt. 1996. El pensamiento en ciencia, tecnología y sociedad en Latinoamérica: Una interpretación política de su trayectoria. *Redes* 3 (7):3–52.

Escobar, Arturo. 1994. *Encountering Development: The Making and Unmaking of the Third World*. Princeton: Princeton University Press.

Evans, Peter. 1995. *Embedded Autonomy: States and Industrial Transformation*. Princeton: Princeton University Press.

Gallo, Rubén. 2005. *Mexican Modernity: The Avant-Garde and the Technological Revolution*. Cambridge, MA: MIT Press.

Gobat, Michel. 2013. The invention of Latin America: A transnational history of anti-imperialism, democracy, and race. *American Historical Review* 118 (5):1345–1375.

Gortari, Eli. 1963. *La ciencia en la historia de México*. Mexico City: FCE.

Grandin, Greg. 2009. *Fordlandia: The Rise and Fall of Henry Ford's Forgotten Jungle City*. New York: Metropolitan Books.

Harding, Sandra. 2008. *Sciences from Below: Feminisms, Postcolonialities, and Modernities*. Durham: Duke University Press.

Harding, Sandra. 2011. Introduction: Beyond postcolonial theory: Two undertheorized perspectives on science and technology. In *The Postcolonial Science and Technology Studies Reader*, ed. Sandra Harding, 1–31. Durham: Duke University Press.

Harvey, David. 1990. *The Condition of Postmodernity: An Enquiry into the Origins of Cultural Change*. Cambridge, MA: Blackwell.

Harvey, David. 2000. *Spaces of Hope*. Berkeley: University of California Press.

Hayden, Cori. 2003. *When Nature Goes Public: The Making and Unmaking of Bioprospecting in Mexico*. Princeton: Princeton University Press.

Herrera, Amílcar. 1971. *Ciencia y política en América Latina*. Mexico City: Siglo Veintiuno Editores.

Klor de Alva, Jorge. 1992. Colonialism and postcolonialism as (Latin) American mirages. *Colonial Latin American Review* 1 (1–2):3–23.

Klor de Alva, J. Jorge. 1995. The postcolonization of the (Latin) American experience: A reconsideration of "colonialism," "postcolonialism," and "mestizaje." In *After Colonialism: Imperial Histories and Postcolonial Displacements*, ed. Gyan Prakash, 241–275. Princeton: Princeton University Press.

Kowal, Emma, Joanna Radin, and Jenny Reardon. 2013. Indigenous body parts, mutating temporalities, and the half-lives of postcolonial technoscience. *Social Studies of Science* 43 (4):465–483.

López Sánchez, José. 1967. *Tomás Romay et l'origine de la science à Cuba*. Havana: Institut du Livre.

Mar-Molinero, Clare. 2006. The European linguistic legacy in a global era: Linguistic imperialism, Spanish and the Instituto Cervantes. In *Language Ideologies, Policies and Practices: Language and the Future of Europe*, ed. Clare Mar-Molinero and Patrick Stevenson, 76–88. New York: Palgrave Macmillan.

Marques, Ivan da Costa. 2003. Minicomputadores brasileiros nos anos 1970: Uma reserva de mercado democrática em meio ao autoritarismo. *História, Ciências, Saúde—Manguinhos* 10 (2): 657–681.

Marques, Ivan da Costa. 2005. Cloning computers: From rights of possession to rights of creation. *Science as Culture* 14 (2):139–160.

Medina, Eden. 2011. *Cybernetic Revolutionaries: Technology and Politics in Allende's Chile*. Cambridge, MA: MIT Press.

Mignolo, Walter. 1993. Colonial and postcolonial discourse: Cultural critique or academic colonialism? *Latin American Research Review* 28 (3):120–134.

Mignolo, Walter. 2000. *Local Histories/Global Designs: Coloniality, Subaltern Knowledges, and Border Thinking.* Princeton: Princeton University Press.

Mignolo, Walter. 2005. *The Idea of Latin America.* Malden, MA: Blackwell.

Moraña, Mabel, Enrique Dussel, and Carlos A. Jáuregui. 2008. Colonialism and its replicants. In *Coloniality at Large: Latin America and the Postcolonial Debate*, ed. Mabel Moraña, Enrique Dussel, and Carlos A. Jáuregui, 1–20. Durham: Duke University Press.

Moya, José C. 2011. Introduction: Latin America—the limitations and meaning of a historical category. In *The Oxford Handbook of Latin American History*, ed. José C. Moya, 1–24. Oxford: Oxford University Press.

Núñez Jover, Jorge. 2002. Ética, ciencia y tecnología: Sobre la función social de la tecnociencia. *LLULL: Revista de la Sociedad Española de Historia de las Ciencias y de las Técnicas* 25:459–484.

Núñez Jover, Jorge, and José Antonio López Cerezo. 2008. Technological innovation as social innovation: Science, technology and the rise of STS studies in Cuba. *Science, Technology and Human Values* 33 (6):707–729.

Polanco, Xavier. 1987. La ciencia como ficción. Historia y contexto. In *El perfil de la ciencia en América*, ed. J. J. Saldaña, 41–56. Mexico City: Cuadernos de Quipu, Sociedad Latinoamericana de Historia de la Ciencia y la Tecnología.

Pomeranz, Kenneth. 2000. *The Great Divergence: China, Europe, and the Making of the Modern World Economy.* Princeton: Princeton University Press.

Quipu. 2012. Inicio. *Quipu, Revista Latinoamericana de Historia de las Ciencias y la Tecnología.* http://www.revistaquipu.com (accessed January 24, 2013).

Roche, Marcel. 1972. Dependence and the development of science in Latin America. *Cambridge Review* 93 (2207):115–119.

Rodriguez, Julia. 2006. *Civilizing Argentina: Science, Medicine, and the Modern State.* Chapel Hill: University of North Carolina Press.

Rónai, Cora. 2012. Notícias do ecosistema. *O Globo*, March 3, 30.

Rostow, W. W. 1991. *The Stages of Economic Growth: A Non-Communist Manifesto.* New York: Cambridge University Press.

Saldaña, Juan José. 2006. *Science in Latin America: A History.* Austin: University of Texas Press.

Schoijet, Mauricio. 2002. Ultra-left science policy and anti-modernization in Argentina: Oscar Varsavsky. *Science and Public Policy* 29 (1):69–75.

Seed, Patricia. 1991. Colonial and postcolonial discourse. *Latin American Research Review* 26 (3):181–200.

Sherzer, Joel. 1991. A richness of voices. In *America in 1492: The World of the Indian Peoples before the Arrival of Columbus*, ed. Alvin M. Josephy, 251–275. New York: Alfred A. Knopf.

Silva, Márcia Regina Barros da, and Ivan da Costa Marques. 2012. Rio e conhecimentos situados. Paper presented at the IX ESOCITE—Jornadas Latinoamericanas de Estudios Sociales de la Ciencia y la Tecnología, June 5–8, UNAM, Mexico City.

Sorensen, Diana. 2007. *A Turbulent Decade Remembered: Scenes from the Latin American Sixties.* Stanford: Stanford University Press.

Soto Laveaga, Gabriela. 2009. *Jungle Laboratories: Mexican Peasants, National Projects, and the Making of the Pill.* Durham: Duke University Press.

Spivak, Gayatri. 1987. *Other Worlds: Essays in Cultural Politics.* New York: Methuen.

Staudenmaier, John. 1989. *Technology's Storytellers: Reweaving the Human Fabric.* Cambridge, MA: MIT Press.

Takhteyev, Yuri. 2012. *Coding Places: Software Practice in a South American City.* Cambridge, MA: MIT Press.

Vaccarezza, Leonardo Silvio. 2011. Ciencia, tecnología y sociedad: El estado de la cuestión en América Latina. *Ciência e Tecnologia Social* 1 (1):42–64.

Varsavsky, Oscar. 1969. *Ciencia, política y cientificismo.* Buenos Aires: CEAL.

Varsavsky, Oscar. 1971. Ideología y verdad. *Ciencia Nueva* 12 (September):44–47.

Velho, Lea. 2011. Conceitos de ciência e a política de ciência, tecnologia e inovação. *Sociologias* 13 (26):128–153.

Vessuri, Hebe. 1987. The social study of science in Latin America. *Social Studies of Science* 17 (3):519–554.

Vessuri, Hebe. 2011. La actual internacionalización de las ciencias sociales en América Latina: ¿vino viejo en barricas nuevas? In *Estudio social de la ciencia y la tecnología desde América Latina*, ed. Antonio Arellano Hernández and Pablo Kreimer, 21–55. Bogotá: Siglo del Hombre Editores.

Vidal, Hernan. 1993. The concept of colonial and postcolonial discourse: A perspective from literary criticism. *Latin American Research Review* 28 (3):113–119.

Wolfe, Joel. 2010. *Autos and Progress: The Brazilian Search for Modernity.* Oxford: Oxford University Press.

Wright, Sue. 2004. *Language Policy and Language Planning: From Nationalism to Globalization.* New York: Palgrave Macmillan.

Part I Latin American Perspectives on Science, Technology, and Society

2 Who Invented Brazil?

Henrique Cukierman

The Manguinhos Institute, popularly known as Manguinhos, was founded in 1900 in Rio de Janeiro and continues to be a leading Brazilian institution for research, teaching, and the production of medicine in the public health area. During the first years of its history, Maguinhos was involved in efforts to build the Brazilian nation and national identity through creating a "modern" Brazilian science (Cukierman 2007, Benchimol, 1990).[1] The scientific expeditions to Brazil's vast hinterland undertaken by the institute between 1911 and 1913 are particularly important for understanding the role of science in constructing Brazil as a modern nation. These expeditions allow us to see the interactions between local and global forms of knowledge in the locations in which they came into contact and the tensions this contact produced.

In this chapter, I sketch several stories from these early scientific expeditions, which together make three main points. First, they portray Brazilian science as an example of scientific practices inscribed in the universe of modern science. Although these expeditions took place outside the "civilized centers" of the time, it was modern science that was being enacted. Second, the construction of sanitation science in the "frontier" regions of Brazil was part of a larger project to build the new Westernized, Europeanized nation-state of Brazil under the auspices of modern science, a project that I call "the invention of Brazil." Expeditions to "discover" Brazil were an integral part of this invention, which included not only the symbolic aspects of categorizing "unknown" territory through science, but also economically motivated attempts to expand Brazilian markets into the interior and to improve worker health in these areas by combating tropical diseases. This was accompanied by a desire to foster the racial "improvement" of the Brazilian population through creating conditions that would promote European immigration to inland Brazil. Third, these stories of scientific expeditions feature an oscillation between colonized and colonizers on the part of the scientists involved. On the one hand, Brazilian scientists had a colonized fascination for modern science, which was introduced into the country together with the cosmopolitan life existing in splendid isolation on the coast. On the other hand, in the anguish of the traveler-scientist confronted with the misery and unhealthiness of the inland regions, these scientists also appear as colonizers who aim to correct those ills through modern science. These scientists therefore were caught up in questioning where and what Brazil was. In their attempts to find an answer, the expeditionaries revealed their struggle against the inadequacy of a cultural environment that they deemed hostile to modernity.

In a certain sense, this was a struggle of those scientists against themselves, fractured between two characters, the Brazilian scientist colonized by European science and the Brazilian scientific colonizer of the country, entrusted with the civilizing mission of guiding it toward modernity. This fracture was sometimes expressed by a mixture of shame and pride at being Brazilian. This is shown in a letter sent by Rocha Lima, one of the Manguinhos pioneers, from Germany, where he was working with Hermann Dürck, the renowned German researcher, to Oswaldo Cruz, the founder and first director of Manguinhos: "I was therefore able to establish relationships with everybody and show them that we are not *monkeys*, always taking part in discussions and always making objections … seeing that with our studies we are not backward" (letter dated July 7, 1906, emphasis mine).[2] On other occasions, this fracture corresponded to the difference between the Iberian and the northern European world, as could be seen from articles published in the press when Manguinhos won first prize at the Hygiene and Demographics Exhibition held in Berlin in 1907. As one put it, "A Latin country, with an imagination that is more than ardent—one could even say boiling hot—Brazil has become accustomed to considering oratorical triumphs as one of its greater glories. … Whereas the work of Dr. Oswaldo Cruz constitutes a positive conquest, without the exaggerations of fantasy, the real and pure conquest of the scientific victory" (quoted in Guerra 1940, 383–384).

As such documents suggest, the scientists at Manguinhos behaved as if the constitution of Brazilian citizenship had to wait for the invention of a country of which they could at last be proud. The dictum *ultra aequinoxialem non peccari* (there is no sin below the equator) had been widely used in Europe during the seventeenth century, as if the line that divided the world into two hemispheres also separated virtue from vice (Holanda 1996, 198). These scientists wanted to invent Brazil under the auspices of science, to undo this implied separation between North and South and to include Brazil on the political and cultural map of the planet. A century later, I, too, insist on demolishing this wall between the hemispheres, but now to restore vice and virtue everywhere. My approach is built on the possibility of rethinking old polarities—Iberian versus Anglo-Saxon, coast versus interior, urban versus rural, intellectual versus lay people, native versus foreigner, center versus periphery, colonized versus colonizer—so that they are no longer based on separations, but on the contact between communities and cultures (Pratt 1992). My task thus consists of understanding "how meanings locally produced are multiplied in other sites and how representations circulate and acquire global legitimacy" (Lenoir 1997, 18), or, in other words, of developing a kind of history that does not repeat the Eurocentric pattern Sandra Harding calls *isolationist*, which merely retells "the histories of Europe, Africa, China, the Americas, and the societies of other parts of the world as largely separate and self-contained chronologies, more or less isolated from each other except for the one-way diffusion of the achievements of European societies to the others" (1998, 7).

The Manguinhos Scientific Expeditions: Brazilian Science in Action

What is Brazil other than an *a posteriori* invention? This is the taunting provocation made by Lamartine Babo, the beloved carnival songwriter, in his well-known song "História do Brasil" (History of Brazil):

Who was it who invented Brazil?
It was Mr Cabral! It was Mr Cabral!
On the 21st of April
Two months after the carnival.[3]

In this version of Brazilian history, in the beginning there was the carnival, a rite of songs and dancing bodies, Dionysiacally dissolved in a formless multitude. Afterward, and only afterward, did the need arise to fit those carnival revelers into the frame of a narrative that would subordinate them into a greater history, an epic history of the nation's foundation and its heroes. Beginning with the fabrication of a European "discoverer" of the country, Pedro Álvares Cabral, who arrived, as the carnival march claims, long after the carnival had already testified to the existence of the people who would later emerge as the "Brazilian people," wrapped in the splendor of a great nation.

Lamartine Babo's witty lyrics suggest the existence of two conflicting narratives, one dealing with the "symbolic" existence of the Brazilian people, the other with "embodied" beings, thus revealing the degree of polarization in the debate over the country's national identity. At the dawn of the twentieth century, soon after the proclamation of the Republic in 1889, narratives about the discovery of Brazil constituted two distinct histories. One was that of a glorious discovery, a narrative made of patriotic and conservative fervor. The other was a very different tale of a late, and thus not so glorious, "discovery" of already occupied territory, one that was more willing to include the perplexities and questionings regarding an already "embodied" country that was complex and problematic. This more anguished gaze of sectors of the elite who were attuned to northern Europe's modernizing trends considered arbitrating between these distinct narratives their primary task.

For these elites, such as Manguinhos scientists, it was urgent to invent the "true" Brazil once and for all and thus catch up with a carnival (as a metaphor of those populations already living in "Brazil") that was already far ahead of them. It was in

Figure 2.1
Carlos Chagas (in the middle) in São Gabriel, Rio Negro, Amazonas, 1913. Archive Casa de Oswaldo Cruz, Archive and Documentation Dept. (BR RJCOC 02.10.20.35.005.232 Acervo da Casa de Oswaldo Cruz, Departamento de Arquivo e Documentação). Used with permission.

the context of this zeal to engage in the late foundation of nationality that Manguinhos organized its scientific expeditions to the country's interior, which were aimed at inventing Brazil under the sign of science, more specifically of medical science. This was an invention based on the laboratory, the microscope, the "neutral" scientific report. According to Carlos Chagas's report from his expedition to the Amazon in 1913, this national project involved abandoning the "thankless terrain of more or less arbitrary doctrines, in order to enter scientific reality, in the light of modern knowledge, taking advantage of the determinism of today's research methods that have made medicine an exact science." Figure 2.1 clearly captures how Chagas's

Figure 2.2
Belisário Penna and Arthur Neiva encamped with their expedition in Bebe Mijo (Piauí), 1912. Arthur Neiva is the bald one sitting in the middle; Belisário Penna sits at his right. Archive Casa de Oswaldo Cruz, Archive and Documentation Dept. (BR RJCOC BP.06.TP.03.V2.019 Acervo da Casa de Oswaldo Cruz, Departamento de Arquivo e Documentação). Used with permission.

absolute confidence in science empowers him as the triumphant colonizer of Amazonian nature.

According to Arthur Neiva and Belisário Penna, leaders of the scientific expedition undertaken in 1912 (figure 2.2) to the states of Bahia, Pernambuco, Piauí, and Goiás, the "carnival" they found consisted of a sick and maimed group of "revelers." Their report records that it was "usually said that the *sertanejo* people are slothful and without initiative. The truth, however, is quite different. This poor folk's lack of effort and initiative is due to the state of abandonment in which they live and the physical and intellectual incapacity caused by depressing and annihilating diseases" (Neiva and Penna 1984, 221).[4] The central question of the carnival march—that of the period of time that had elapsed between the "carnival" (or inland settlement) and the arrival of the men from Manguinhos in the country's interior, bringing their scientific eyes to "save" the ill—could be measured in centuries as they encountered "populations that in the course of three centuries had assimilated almost nothing of the great transformations that had taken place throughout the universe" (183). Implicit in this lag was a spatial demarcation between the coast and the interior, an isolation between the two worlds that had been broken by the supposedly impartial eyes of the enlightened men of the coast, a gaze problematized by Mary Louise Pratt in her *Imperial Eyes* (1992). In that work, Pratt proposes the concept of anti-conquest, "a way of taking possession without subjugation and violence" that refers to the strategies of representation through which this type of observer, the bourgeois and civilized subject, the scientist, the white European male, seeks to assure his innocence while at the same time affirming his hegemony (57). For her, this protagonist of anti-conquest is the subject "whose imperial eyes passively look out and possess" (7).

The displacement of Manguinhos's imperial gaze toward the interior of the country would have to be matched by a displacement of the rhetoric that described it and conducted it to modernity. As Neiva and Penna noted, they sought to produce a rhetoric that would be able to articulate a discourse "in which the impartiality of the language could be very useful" (1984, 179), which they understood to be the discourse of science. Thus, under its command, Brazil could finally be invented, three centuries after the ragged carnival of those folk who had been "iniquitously forgotten by the public authorities" and "living in a total state of abandonment" (174, 179).

The scientists from Manguinhos were so dumbfounded by what they saw that it seemed to them that this invention would have to be literal. An awareness of the federation and its states was completely absent among the populations visited, a problem the Neiva-Penna expedition's report describes with appalled clarity: "A woman we spoke to … was not able to tell us whether she was from Bahia or Pernambuco—'I'm from over there.' … Very few people know what is Brazil. Piauí is a land, Ceará another land, Pernambuco another one, and the same goes for the other states. … Asked if these lands … are linked to each other, constituting a nation, a country, they say they don't understand anything about this" (Neiva and Penna 1984, 187, 191).[5]

A Brazil without edges or margins was the response of the inhabitants to the scientists' questions about frontiers. The notion of frontiers was exclusively a part of the scientists' national landscape—that is, the frontier existed solely from the coast's expansionist perspective. This is why Pratt avoids using the concept of colonial frontiers and proposes instead the more potent idea of contact zones in "an attempt to invoke the spatial and temporal co-presence of subjects previously separated by geographic and historical disjunctures, and whose trajectories now intersect" (1992, 7). In this particular intersection of trajectories, the foreigners of the coast represented the side of "civilization," formulated in the first person plural ("to them, we were gringos, *lordaços* [superlative form of *lords*]"), while the other side, described in the third person plural, was "a race that is cretinized, owing for the most part to a cruel and avoidable disease, incapable and useless" (Neiva and Penna 1984, 191, 212). Although characterized by enormous inequalities, the contact between these "lords" and "cretins" ended up blurring these frontiers significantly. In the case of the

Figure 2.3
The Manguinhos expeditionaries in the "contact zone" in Caldeirão (Pernambuco, 1912). Artur Neiva is sitting on the right. Archive Casa de Oswaldo Cruz, Archive and Documentation Dept. (BR RJCOC 02.10.20.35.003.022 Acervo da Casa de Oswaldo Cruz, Departamento de Arquivo e Documentação). Used with permission.

Neiva-Penna expedition, this can be seen in the way the records kept by "their lordships" oscillate between proximity and distance in relation to the "cretinized." This oscillation is certainly related to the debates that, according to Roberto Ventura, reveal a tension whose origins can be traced back to Enlightenment philosophy itself, more specifically to the ambivalence of the European discourse regarding "the positive image of natural happiness and innocence of the inhabitants of fertile climates and the condemnation of their barbarian customs" (1991, 22), a discourse in which Montesquieu and Buffon provided a negative view of American Man and nature and Rousseau's noble savage supported a positive image based on "natural Man's" superiority over the civilized European.

This oscillation also expresses the multiplicity of the contact zone itself, based on a cultural coexistence that resulted in heterogeneous and unforeseeable results (figure 2.3).[6] There are plenty of examples of this oscillation within the Neiva-Penna report, beginning with its examination of folk healing. The report, in discussion of treatments of conjunctivitis, makes a point of highlighting that "they are usually barbarically treated with pipe nicotine and in some places also with iron filings and lemon" (Neiva and Penna 1984, 163). In the case of worms, however, the expeditionaries were obliged to recognize that the inhabitants were treated in a "rational" way, because "they use wormseed[7] as an anti-helminth ... which is in fact effective" (163). The polemical issue of the people's supposed slothfulness and unwillingness to work constitutes another good example of such oscillation. Whereas in some passages the scientists emphasize that the population is "slothful, as indeed all over Brazil" (198), in others we can read that it is above all "a resilient, useful and vigorous race and worthy of a better fate" (200). As pointed out in the study by Lemon and Medina (this volume), the oscillation between "civilization and barbarism" is commonly found in descriptions of Latin American modernization projects.

An amalgam of languages was another very common phenomenon in that contact zone. It should be remembered that there was very little unity in spoken language, another of the fundamental symbols of nationality, in Brazil during this period. Thus the invention of Brazil would have to include a process involving the translation and incorporation of syntactic constructions and vocabularies that the colonizer was encountering for the first time. In this process, the master of the "official language" was obliged to give up his colonial grammar and come to grips with those people's descriptive resources so that he could gain access to the symptomatology of the various diseases encountered during the journey. This task constituted a scientific challenge: completely unknown clinical conditions, or unknown variations of diseases that were already widely cataloged, appeared before the members of the expedition, indicating the existence of epidemics of diseases that had never before been observed. One of them was *vexame* (literally, shame), or *vexame do coração* (*vexame* of the heart), apparently a disease of the nervous system that had "a truly epidemic character in the dry zones we crossed." It was often observed among women but could not be classified as hysteria or epilepsy, or as "any other known nervous disorder" (Neiva and Penna 1984, 139).

To research something with which they were not familiar, like *vexame*, they had to listen to the sick (figure 2.4 shows Belisário Penna in one of these moments at his expedition's open-air medical office). However, the sick had their own language, which was quite different from that spoken on the colonizing coast. The report includes a scene that provides a perfect illustration of this, in which the authors attempt to translate the language with which a man describes his wife's strange symptoms:

The wife has made the mother of the body (uterus) *angry because she washed her body* (had a shower) *when she was oxed*[8] (menstruating). The thing *rose up* (suspension) and the *month* (period) did not occur again. Every *return of the moon* her belly *swells up* (becomes tympanic) and she does not *deseste*[9] (defecate). She has already taken two *purgatives,* one made of olive oil and the other from pinhão[10] and a whole lot of *mezinhas*[11]—and nothing's changed. People *have already recommended blessings* because it seems to be *something that was ordered by someone* (spell). (Neiva and Penna 1984, 211–212, italics in original)

Figure 2.4
The doctor and his patients: immediate encounters in the "contact zone." Belisário Penna is sitting in the hammock on the right. Archive Casa de Oswaldo Cruz, Archive and Documentation Dept. (BR RJCOC BP.06.TP.03.V2.007 Acervo da Casa de Oswaldo Cruz, Departamento de Arquivo e Documentação). Used with permission.

For the expeditionaries, these differences in language were ample proof of "the complete lack of exchange of ideas, facts and things between the coast and central Brazil" (Neiva and Penna 1984, 178). Nonetheless, it was their duty to observe *vexame*'s symptoms, and their only option was to use the method at hand, the people's own words, obliging the scientists to resort to a description that was typical of the contact zone: "the attack manifests itself as a *baticum* in the heart (palpitations), darkening of vision, and fainting, with an absence of muscular contraction, convulsions, profuse sweating, screams, or moans" (139, italics in original). This passage shows that *vexame* could not be described without referring to the *baticum* in the heart, an onomatopoeic reference to a drumbeat or beating on a percussion instrument used by that population. Moreover, it constituted the most important symptom of the illness, as it was the first and practically the only one actually present, with all other symptoms characterized by their absence.

Indeed, the list of absent symptoms showed how mystified the expeditionaries were by the disease, as if they were commenting between the lines on the great challenge of understanding it without the symptoms' being "logically" explained by the vast amount of medical knowledge that had already been cataloged. The logic of their medical paradigm (Kuhn 1995) determined the scientists' surprise and their characterization of the disease as "an extremely curious nervous manifestation" (Neiva and Penna 1984, 139). That they found this so curious displays the serious problems they faced in describing the illness in "refined" language, and they thus had to adopt, even if only temporarily, the "uncultured" naming of the "other," which was why they reported that "the sertanejos call [the disease] *vexame* or *vexame do coração*" (139, italics in original), a term borrowed to fill a void that their language could not cope with and which their science had not yet colonized.

One does not borrow a word lightly, however. One needs to have a certain sympathy for the foreign language, a sympathy that can easily develop in a contact zone. In the case of these expeditionaries, this was an almost compulsory sympathy for their own folk, the Brazilian people, for one must remember that their purpose was to invent Brazil and they would therefore have to give the nation its people. Thus, the confession they make almost between the lines is very significant: "Today, what moves us forward is a deep sympathy for those folk" (Neiva and Penna 1984, 179). This is a *moving* sympathy, one that necessarily produces displacements, thus inviting the question: displacements from where to where? It may be an unceasing displacement between the positions of colonized and colonizer that fills the foreign gaze with hesitations, as its quest for the other, the Brazilian people, is also, after all, the gaze of the native seeking himself.[12] This is one of the reasons Sérgio Buarque de Holanda's description (1996, 31) of Brazilian intellectuals as "exiled people in their own land" has become a reference. This sense of exile challenged these intellectual elites to propose a political, economic, and social project to which imperial Europe had little to contribute, as it had not experienced certain institutions that were deeply rooted in the daily life of Brazilian society, such as slavery, the large plantation and extensive monoculture regime, and the colonial yoke itself. Thus, these elites, besides fostering an autonomous national order that had not yet been clearly established, would also have to conceive of themselves as Brazilian intellectuals (cf. Pratt 1992, 175–176). This phenomenon was revealed intensely during the Neiva-Penna expedition, whose report inadvertently showed that the contact zone was not only a laboratory of invention but also—and perhaps mainly—one of self-invention.

This self-invention was marked by an identity crisis, as if the contact zone had penetrated deeply into their innermost selves, torn by the conflict between the erudite scientist, the prince of "universal knowledge," and the perplexed Brazilian, a lost soul searching for its nation. If they were scientists, what was the purpose of this science? To uphold the ideals of the "universal good"? What could be done for the good of their own folk, which, after all, had to be the overriding purpose of a science that deserved to be called "Brazilian"? It was not easy for the scientists to look on "those folk" as their countrymen. That this demonstrated a torn innermost self is revealed with the utmost clarity by the Neiva-Penna report itself: "The whole region is very backward ... we can now understand the painful opinion of a notable jurist and eminent politician of the monarchy, that for any form of government to drive the nation forward we had to have a people, and what we had 'was not a people but the manure of a people that was still to come'" (Neiva and Penna 1984, 198).

The scientists' *moving* sympathy forcefully enacted a displacement between two languages, thus raising the contemporaneous question of translation. As Bruno Latour points out, translation in "its linguistic and material connotations, refers to all the displacements through other actors whose mediation is indispensable for any action to occur. In place of a rigid opposition between context and content, chains of translation refer to the work through which actors modify, displace, and translate their various and contradictory interests" (1999, 311). The case of *vexame* demonstrates the scientific interest of the Manguinhos scientists in identifying and describing the disease, and it is precisely the displacements between languages that help them sketch the first outlines of a disease that still had to be "discovered." Not taking these displacements into account would lead to a rigid opposition between content and context, between the "illness in itself"—the content of the investigation—and a multitude of sick people speaking of a certain *baticum*—the context of the investigation. We already know that this false opposition is adequately sustained *a posteriori* when science has already taken possession of the object and the results of its investigation, transforming the disease into a black box and thus making all the work undertaken to attain it invisible. Such is not the case here because the "disease in itself" does not yet exist and therefore was not solemnly described in some textbook of tropical medicine. On the contrary, it was necessary to begin the work of science, and the quotation from the Neiva-Penna report—"the attack manifests itself as a *baticum* in the heart (palpitations)"—demonstrates the literalness of the translation, which, in this case, occurs in one of its most evident forms, that of a translation from a language of origin to another of destination, the first visually identified by italics and the second by parentheses.

It is noteworthy that the scientist who wrote the report did not choose to make his translation invisible but preferred to indicate the presence of two languages instead of simply affirming that the "attack manifests itself as heart palpitations." When a "lord" accepts the name proposed by a "cretinized person," I would argue, the effort to provide the country that has still to be invented with a people allows a respectful bow to their customs and traditions. It also demonstrates, I would posit, that in the actual process of producing science, not a science that is already established but a science in action, the emphasis necessarily falls on its location and materiality, revealing an image of the scientific process as one of precariousness and uncertainty in its interactions with the world.

The report appears to want to make us believe that the Mother Nature to which *vexame* belongs had not yet been revealed, surreptitiously insinuating a happy ending in which Science would inevitably decipher the enigma and discover Nature, offering its shameless nakedness to the patrimony of "universal knowledge." Yet it is in this gap between ignorance and discovery that the scientific process, observed through the various mediations involved in the translation effort, enables a considerably more plausible explanation for the hesitations of our expeditionaries. Their reports provide us with a clear confirmation of what Michel Callon summarizes so succinctly: "Science is the product of a fabrication process, in which the selection of relevant problems, signs, and events plays an essential role. In concrete terms this means that the scientist never engages with actual nature itself but always with more or less faithful and more or less numerous and distant representatives of nature" (1989, 10).

Expeditionary Science and the Expansion of Markets

The next step in this investigation of the Manguinhos expeditions is to examine the close links between expeditionary science and capitalist expansion in the country in its quest for new products and markets in regions far from the Atlantic coast. This association between science and capital could be seen in the sponsors of expeditions, which included the Departamento de Obras Contra as Secas (Drought Works Department) and the Superintendência da Borracha (Rubber Superintendency). The interests of those sponsors were reflected in two broad purposes: to foster the expansion of capital toward the interior by identifying good business opportunities and infrastructure projects to support economic growth, especially those related to transportation networks and the supply of power and water; and to prevent tropical diseases from hampering the exploitation of the vast Brazilian hinterland by improving workers' health conditions and thus making the work involved more productive. This is reflected in a query in Carlos Chagas's report on his 1913 expedition to the Amazon regarding the enormous loss of life on the rubber plantations caused by malaria: "How can one reduce the cost of production in an extractive industry without first normalizing the coefficient of human work that has been reduced to a minimum in those areas?" An answer, as Manguinhos scientists were proposing, should come from the unrestricted adoption of medical science. The message that "production must be increased" was embedded in the Manguinhos reports in the form of diagnoses and advice regarding opportunities to be grasped, resources to be developed, surpluses to be commercialized, infrastructure to be built, values to be transformed, and races to be "perfected" (as discussed later). The scientists scanned the land looking for profitable projects to be developed while at the same time naturalizing the coastal colonizer's interests and intentions in the format of a "civilizing" mission aimed at redeeming those backlands that they conveniently characterized as being materially and spiritually destitute. Their reports served to prove that "those folk" were incapable of exploiting the surrounding nature, not only because of their slothfulness, when this was the case, but also because of their inability to exploit nature methodically and thus expand production.

Underlying the scientific rationality of the reports was the economic rationality of capital, with the scientists playing the role of capitalism's spearhead, their mission to reveal all the opportunities potentially available for development and progress, which they insisted were "universally necessary." It is not that the expeditionaries failed to record a series of problems at odds with the modern spirit that they sought to embody: the concentration of rural ownership in the hands of only a few families, the then-common practice of using slave and even child labor. Rather, these were viewed as merely small bumpy detours from the oneiric road that would lead to modernity under the banner of a science at the service of capital. They thus sought to convey an unshakeable faith in the power of science at the cost of instating a generic, abstract, and therefore depoliticized Man, showing that they were faithful worshippers of their leader Oswaldo Cruz's motto of "eternal faith in science"—a strange faith in the idea of a purely scientific solution for a country that, despite the social ills that could be seen with the naked eye, would miraculously emerge renewed at the hands and microscopes of science. This echoes Marques's description (this volume) of a Westernized faith in science as the unique and perfect mirror of reality. The West has knowledge while other people have interpretations and beliefs about reality.

To the imperial gaze of the coast, a way of life whose organization violated industrial logic could only be seen as uncivilized. The logic of growth and market expansion was completely incompatible with an economic regime based on mere subsistence and local, nonmonetary exchanges. Aghast at what they observed, the expeditionaries were forced to record scenes that testified to the total lack of a certain "civility" devoted to the accumulation of goods and merchandise:

Since we entered the state of Goiás, the main currency used to persuade the inhabitants to supply us with eggs, chicken, manioc, sweet potatoes, etc., has been spools of thread, needles, pins and trinkets ... no one cares for money, and the relatively large amounts offered for a dozen eggs ... are rejected disdainfully. The small lots where they plant things are always located far from their house ... [they take] home only what they will consume that day. This was one of the reasons why they refused to give us anything solely in exchange for money: the fact that it was worth little to them and they could not be bothered to go to their small lots to get what they wanted. ... Their needs are so small that they can satisfy them with local resources. ... They barter for or exchange food. (Neiva and Penna 1984, 215)

To the expeditionaries, their refusal to glorify money was itself a sufficient manifestation of laziness. This atmosphere of condemnation prevented the expeditionaries from noticing any inconsistency between the systematic accusations of native slothfulness and their own observations regarding the overexploitation of labor by the local regimes, not to mention the exhausting work performed by their native bearers, whose unceasing labor ensured the survival of the scientific excursion itself.[13] The scientists had only critical eyes for what seemed to them the people's totally senseless choice of a demonetarized economy.

The Manguinhos scientists' annotations perfectly demonstrate Pratt's observation regarding European colonizers. "The bottom line in the discourse of the capitalist vanguard was clear: America must be transformed into a scene of industry and efficiency," she notes, with the task of this vanguard being to conduct the civilizing mission constituted through a language with which they sought to "produce other peoples (for themselves) as 'natives,' reductive, incomplete beings suffering from the inability to have become what Europeans already are, or to have made themselves into what Europeans intend them to be. So did the capitalist vanguard read themselves into the futures of those they sought to exploit, as a kind of moral and historical inevitability" (1992, 155, 152–153). More than seeing themselves as a historical inevitability, what was perhaps at stake for this vanguard was the imperial idea of giving a history to those who did not have one. Yet in fact, it can be argued, those people were being expelled from history, after which their territory would be devoid of any local meaning and thus at the mercy of the meanings that the more "advanced" would bring from outside, or, better, from history itself (cf. Ventura 1991, 28). All the vocabularies, customs, and knowledge of that local culture would have to disappear and make way for a scenario of capitalist prosperity, transforming that slothful, ignorant, superstitious, destitute, and unambitious population into salaried workers and consumers of the goods and values of those distant cities by the sea.

The report of the expedition undertaken in 1912 to the São Francisco River, led by Adolpho Lutz and Astrogildo Machado, formulated a proposal for the industrial exploitation of fishing: "Surubim comes from the São Francisco and is dried and consumed inside the state. This fish—*Pseudoplatystoma corruscans*—could be canned and exported to more distant places, competing with the best imported fish" (Lutz and Machado 1915, 10). The Neiva-Penna report contains various suggestions relating especially to cotton cultivation: "From what we observed during our whole journey ... these parts have a potential that we are sure will wrench them from their current state of poverty. We are referring to cotton cultivation. ... The day people decide to take it seriously, studying the species and varieties that are best suited to the soil, a great transformation will take place and prosperity will arrive" (1984, 176). As we have already seen, the Carlos Chagas expedition report indicated what measures should be adopted to combat malaria and thus prepare the way for the production of rubber in the Amazon.

By cataloguing the tropics to ensure the position of Brazilian science in the vanguard of the nation's invention, the reports also reaffirm Manguinhos's defense of the country's healthiness, which they no longer considered a tellurian problem but one of the availability of scientific resources. The defense of the tropics included the act of scientific naming itself, which they disputed with international science in favor of the advancement of the Brazilian nation. This is illustrated by the following example from the Neiva-Penna report when it describes spasmodic dysphagia, also popularly known as *entalação* (tight squeeze), *mal de engasgo* (choking disease), *entalo* (squeeze), and *engasgue* (choking), which was another of the endemic diseases whose extraordinary frequency surprised the expeditionaries. They deemed it to be the male version of *vexame*, as it tended to occur in the same places and families, and argued that it should be so named: "The name given by Paranhos—'tropical dysphagia'—has the disadvantage of limiting the disease to a specific geographical region where it may not exclusively occur, thus further worsening the already negative view of the tropical regions. ... It is high time that we reacted against these incorrect designations that only serve to increase the disrepute in which all tropical zones are held" (1984, 133).

This dispute evokes Pratt's observations regarding the ordering of the world undertaken by natural history, especially during the eighteenth century, when the discipline considered it its task to scour the planet in order to locate all animal and vegetal species. In so doing, its purpose was to rescue all species from arbitrary and particular locations in order to insert them, by giving them a new name, in an appropriate place in a classification system, in an effort to reestablish order where chaos had previously reigned: "Here the naming, the representing, and the claiming are all one; the naming brings the reality of order into being" (Pratt 1992, 33). Heirs to the classification systems that preceded them, the Manguinhos scientists were aware that, in this reality of the order, a name like *tropical dysphagia* would only reinforce the location of tropical countries in the pigeonhole reserved for places that were intrinsically unhealthy and thus unsuited for "civilized" life.

The "Perfecting" of the Races

The Neiva-Penna report defended emphatically the healthiness of the tropics: "The accusations made against the climate in order to discourage foreign immigration are absurd; the climate on the banks of the big rivers, where there is always water and which constitute practically the only fertile land in the region, is perfectly compatible with the human life of foreigners of any race" (1984, 175). It is noteworthy, however, that their defense of tropical life also exploited a racial theme: that the impossibility of "civilized" life in the tropics could be attributed not only to the diseases that were peculiar to tropical nature but also to the inferiority of nonwhite races. The scientists were there to vouch for the tropics but did not feel capable of doing the same for the problem of race, whose solution depended on the incorporation of "superior" ethnic groups composed of immigrants, who were expected to speed up this ethnic reinforcement, or whitening, of the Brazilian population. This is attested by the report of Lutz and Machado's expedition to the São Francisco River, which graced the annals of science with a rather pessimistic description, to say the least, of the races and mixtures of races living in the Brazilian interior: "The São Francisco river valley seems to be a poor and backward region. As for the population, one must say that the Indian element is almost entirely absent. However, the black race is very numerous and often predominates. The white element is often absent among the natives. This naturally influences the character of the population, which usually lives in a very primitive way. ... Of course there can be no progress where the people just vegetate without improving their living conditions" (Lutz and Machado 1915, 8–9). More discreetly, the Neiva-Penna report also proclaimed the need to "perfect the races," its prescription being "the intelligent exploitation of the land, its settlement by capable and conscious men, giving them the means for rapid and cheap communication with consuming centers, education and exact notions and practices for the prophylaxis of regional ailments, which can all be avoided with rational and continuous assistance and by wise laws relating to the cautioning and perfecting of races" (Neiva and Penna 1984, 221).

Thus, "the disrepute in which all tropical zones are held" was reduced in the reports to the "single" condition of racial disrepute. Progress would be the prerogative of the white race, as one can read in the Neiva-Penna report: "In the south [of Goiás] the white element already predominates and the inhabitants are more vigorous" (167); "in short, the southern region we traveled through, inhabited for the most part by healthy people, with developed agriculture, fields containing sizeable herds of livestock, decent housing with relative comfort, offered a stark contrast to the northern region of the state" (223). There was thus a solution available. Nature could be redeemed by science and race by immigration, which, given the context of the reports, appeared to constitute an equally scientific solution: "progress in Brazil is, in large measure, due to foreigners. ... We are convinced that one of the main causes, and in our opinion the most important, of the backwardness of the northeastern regions is the lack of immigrants. ... Without the cooperation of immigration it will be difficult to galvanize populations that are stuck in their ways" (175, 181). The expeditionaries imagined that combating the legend of "the North's unsuitability for the European immigrant" would suffice to inundate those lands with white Europeans, without whom it would be useless "to expect the miracle of the transformation of the northeastern backlands into the oft-announced land of promise" (181).

The idea of immigration, associated with the discrimination reflected in the claims arising from racial disrepute, evidently found support in all these supposed findings we have seen of slothfulness and irrationality, a veiled denial of the possibility that any form of knowledge could be produced by those populations. Situated at a great distance from the "truth," or rather, from the truth instituted by modern science in the service of capital, the people of the interior were the object of the scientists' compassionate consideration, which, despite the latter's good intentions, could serve as an appropriate epitaph: "they are relatively happy because they are unaware of the real situation they live in" (Neiva and Penna 1984, 179).

Conclusion

Perplexed by much of what they saw during their expedition, Arthur Neiva and Belisário Penna confessed that, if they had been able to express what they saw through poetry, "we would write a tragic poem, describing the poverty, the misfortune of our abandoned backland inhabitants. The poetry of the landscapes and panoramas would be blotted out by the tragedy, by the desolation and by the misery of our unfortunate backland inhabitants, our fellow countrymen, … these regions are more akin to a picture of hell which could only be described in masterly fashion by the immortal Dante" (Neiva and Penna 1984, 222).

What convinced the scientists that their report did not constitute a tragic poem? After all, where was the thin line separating the scientific report from the literary work? The answer is that, considering the scientific scope of their reports, their purpose was to fulfill their commitment to scientific truth, presumably immune to the distortions imposed by a literary style. They had convinced themselves of the existence of a very special truth that lies beyond the language that establishes it, a truth that, in the case of these reports, would have to affirm itself beyond their authors' oscillations between colonized and colonizers. Thus persuaded, perhaps they lived happily ever after. However, in the role I have chosen to play more than a century afterward, I hope I have convinced the reader that the work of science is concomitant not only with the artifices of rhetoric but also with the political and ideological project of inventing modern Brazil under the aegis of science and capital. Following Marques (this volume), I also hope to have articulated a different ontological-political perspective that moves Western science from its privileged place of universality and neutrality and places it among other valuable knowledges.

After all, what Brazil did the Manguinhos expeditionaries invent, if they indeed managed to do so? Today, we know that science did not fulfill the purpose attributed to it by its representatives and heralds in Manguinhos, which was to eliminate the gap between the discovery of Brazil recorded in the annals of history and the popular carnival that had got there sooner. If the invention of Brazil by expeditionary science lost its way due to an excessive confidence in the power of salvation in science's supposed universal truth, I have sought to avoid this same fate by arguing against this blind faith in the existence of "pure" science with its redeeming capacity. Having freed science from the myth of a science without a myth (Michel Serres, cited in Latour 1994, 93), it becomes possible to think of it as an undertaking that, through the same movement in which it constructs facts and artifacts, not only constructs societies, cultures, and nations but is also constructed by them. If so recognized as "impure," the technoscience so necessary to the world in which we live will be able to resume—now in a stronger position—its work of replying to Lamartine Babo's question, Who invented Brazil? This is a question that remains open in a society that longs to overcome racial discrimination, social inequality, the ineptitude of governments, poverty, and disease—a Brazil, in other words, that still has to be invented.

Notes

1. This chapter draws on a chapter of the same name published in Cukierman (2007). Another article that also draws on this same material, "History of Science in Brazil: The Case of Instituto de Manguinhos," was presented at the meeting of the Latin American Studies Association in Rio de Janeiro, Brazil, in June 2009.

2. These pioneers were all doctors trained in the School of Medicine of Rio de Janeiro, founded in 1808. The letter can be found in the Archives of the Casa de Oswaldo Cruz/Fiocruz, Rio de Janeiro.

3. Lamartine Babo (1904–1963) was a Brazilian composer especially known for his carnival songs, including this famous one composed in 1934. "Mr. Cabral" refers to Pedro Álvares Cabral, the Portuguese navigator to whom the official Brazilian historiography attributes the discovery of Brazil on April, 22, 1500. Copyright © 1933 by Mangione, Filhos & Cia Ltda. All rights reserved. Used by permission.

4. The word *sertanejo* refers to an inhabitant of the *sertão*, itself an abbreviated form of *desertão* (big desert), which refers to the region's semiarid tropical climate. The *sertão* covers a large part of Brazil's northeast, including the interior of Bahia, Pernambuco, Paraíba, Rio Grande do Norte, Ceará, and Piauí, as well as a small part of Sergipe and Alagoas. It also stretches to the northern part of Minas Gerais and the Jequitinhonha Valley, located in the same state. In its most generalized meaning, *sertanejo* refers to a poor person living in the interior of the country far from the coast.

5. Bahia, Pernambuco, Piauí, and Ceará are northeastern states in Brazil.

6. This was reinforced, in the specific case of the scientists from Manguinhos who had ventured deep into the country's interior, by the fact that the contact zone consisted of a broad swath of time and space, a long and ample sojourn in a quotidian existence that was very distant from usual references. The Neiva-Penna expedition, for example, lasted around seven months, three of them with no news from the coast.

7. A small herb (*Senebiera pinnatifida*) known for its medicinal properties.

8. Translator's note: "Oxed" is my coinage in English in hopes of conveying the meaning of the original Portuguese expression—*tava de boi* (*boi* means ox)—as closely as possible.

9. This word is difficult to translate even in Portuguese. Its meaning is similar to the verb "to give up."

10. The pinhão-barbados nut (*Jatropha curcas*), with yellow-green flowers and three- to five-lobed leaves on trees six meters tall from Mexico and Central America, produces seeds from which cooking oil, soap, and a strong purgative are obtained.

11. Homemade remedies.

12. There is no record of any women participating in the expeditions.

13. Not to mention that a gang of "loafers" used litters to carry "ladies of the most reputable kind" in the region that stretched from the capital of Goiás to Anhanguera (Neiva and Penna 1984, 171).

References

Benchimol, Jaime, ed. 1990. *Manguinhos, do sonho à vida: A ciência na belle époque*. Rio de Janeiro: Fiocruz/Casa de Oswaldo Cruz.

Callon, Michel. 1989. *La science et sés reseaux*. Paris: La Découverte.

Chagas, Carlos. 1913. Notas sobre a epidemiologia do Amazonas. *Brazil-Medico* 27:450–456.

Cukierman, Henrique. 2007. *Yes, nós temos Pasteur—Manguinhos, Oswaldo Cruz e a história da ciência no Brasil*. Rio de Janeiro: Relume Dumará/FAPERJ.

Guerra, E. Sales. 1940. *Osvaldo [sic] Cruz*. Rio de Janeiro: Editora Vecchi.

Harding, Sandra. 1998. *Is Science Multicultural? Postcolonialisms, Feminisms, and Epistemologies*. Bloomington: Indiana University Press.

Holanda, Sérgio Buarque de. 1996. *Raízes do Brasil*. 26th ed. Rio de Janeiro: José Olympio.

Kuhn, Thomas. 1995. *A estrutura das revoluções científicas*. São Paulo: Perspectiva.

Latour, Bruno. 1994. *Jamais fomos modernos*. Rio de Janeiro: Editora 34.

Latour, Bruno. 1999. *Pandora's Hope: Essays on the Reality of Science Studies*. Cambridge, MA: Harvard University Press.

Lenoir, Timothy. 1997. *Instituting Science: The Cultural Production of Scientific Disciplines*. Stanford: Stanford University Press.

Lutz, Adolpho, and Astrogildo Machado. 1915. Viagem pelo Rio São Francisco e por alguns de seus afluentes entre Pirapora e Juazeiro. *Memorias do Instituto Oswaldo Cruz* 7:5–62.

Neiva, Arthur, and Belisário Penna. 1984. Viagem científica pelo norte da Bahia, sudoeste de Pernambuco, sul do Piauí e de norte a Sul de Goiás. Brasília: Academia Brasiliense de Letras. Facsimile of first publication in *Memórias do Instituto Oswaldo Cruz* 8 (1916):74–224.

Pratt, Mary Louise. 1992. *Imperial Eyes: Travel Writing and Transculturation*. New York: Routledge.

Ventura, Roberto. 1991. *Estilo tropical: História cultural e polêmicas literárias no Brasil, 1870–1914*. São Paulo: Companhias das Letras.

3 Innovation and Inclusive Development in the South: A Critical Perspective

Mariano Fressoli, Rafael Dias, and Hernán Thomas

As Medina, Marques, and Holmes argue in the introduction to this book, science and technology studies (STS) scholars in Latin America have a strong tradition of critiquing foreign science and technology (S&T) models of development and their unidirectional model of technology transfer. One of the most salient issues in this trend has been a critique of the failure to relate "universal" research agendas to the pressing needs of the local population, a critique that has also been instrumental in producing alternative, autonomous agendas for the development of S&T (Dagnino, Thomas, and Davyt 1996).

In addition to this critical STS stance on "universal" S&T models, a more marginal and less studied source of critique has been social movements and networks that are oriented toward the use of local S&T capabilities to construct solutions for greater social inclusion. Scholars in innovation and development have hailed these movements and networks, which have been recognized as experimental spaces offering alternative pathways of cognitive production and technological change (Jamison 2001; Hess 2007). Reciprocally, by constantly highlighting the sociopolitical, situated, and nonautonomous character of S&T agendas, these innovation movements and activists sometimes also push STS scholars to engage in the design of new visions, approaches, and policies (Waks 1993).

Latin America has a long though not very well known history of alternative innovation and technologies for social inclusion, mainly based on the work of centers of appropriate technology in the late 1960s and 1970s (Whitecombe and Carr 1982; Brandão 2001). Amílcar Herrera, a key early STS scholar in Latin America, was a regular contributor to the appropriate technologies movement in the region and became an international consultant on the theme. He also wrote several papers on it and proposed a methodology for the development of appropriate technologies for the rural sector, which strongly encouraged local participation (see Herrera 1981).

More recently, a new wave of activists, scholars, and development institutions has begun to combine a critique of mainstream S&T models with a search for solutions to social problems in South America and other parts of the world. Since the beginning of the twenty-first century, new social innovation movements have appeared in Latin America: the Social Technology Network in Brazil and the Technologies for Social Inclusion Network in Argentina. Other new innovation movements, like Grassroots Innovation and Bottom of the Pyramid, have been developed in India since the mid-1990s and early 2000s and have also rapidly become highly influential worldwide.

At the same time, many international organizations working in Latin America, such as the World Bank, the Inter-American Development Bank, the Organization for Economic Co-operation and Development, and the United Nations Development Program, are designing and advocating specific programs, workshops, and studies on what has become labeled inclusive innovation.[1] As a result, the old issue of which S&T approaches are best for the region has reemerged, reinvigorating the persistent question of how to best combine local S&T capabilities with the local needs of the population and introducing issues like social agenda setting in S&T, the participation of users, and the development of a broader politics of knowledge. Thus, as the issue of innovation and inclusive development is increasingly being discussed in STS forums and adopted as policy by national governments in countries like Argentina, Brazil, and India, it has become a contested space within which several actors advocate their own visions, strategies, and networks (Smith et al. 2012).

In this chapter, we examine two competing—though not necessarily contradictory—approaches to inclusive innovation in South America and India. Our aim is to understand how different approaches in South America and India imply different politics of knowledge regarding participation in the innovation process, empowerment, and thus distinctive ways of fostering social inclusion. By comparing the cognitive praxis of these initiatives we seek to analyze their lessons and constraints for policy making on innovation and social inclusion in the region. Regarding these differences, we want to argue that despite the recent interest in inclusive innovation, adopting foreign pro-poor innovation models for South America might jar existing experiences and alternative pathways in the region.

To understand these differences, we analyze the movements' cognitive praxis (Jamison 2001), focusing on their visions and strategies, their conception of technological solutions, and their networks. We also refer to the concept of sociotechnical alliances (Maclaine Pont and Thomas 2012)[2] to analyze how asymmetries of power and participation affect the process of technical change.

In the first section, based on a survey of initiatives and experiences of technologies for social inclusion and case studies, we look at two innovation movements in South America: the Social Technology Network in Brazil and the Technologies for Social Inclusion Network in Argentina.[3] In the second section, we analyze innovation movements from India that have influenced, to different extents, the emergent movement toward innovation and social inclusion in South America: the Grassroots Innovation approach and the Bottom of the Pyramid approach. In the third section, we analyze what we can learn from these Indian movements that may be applicable to South American networks. In the fourth section, we explain why, despite what we can learn from foreign models, there is still a need to create autonomous and situated approaches to inclusive innovation for South America.

Innovation and Social Inclusion in South America

Since the early 2000s, the issue of technology and development has received renewed attention among South American public institutions, research and development (R&D) laboratories, cooperatives, and nongovernmental organizations (NGOs). Fueled by new center-left populist governments in the region, different strands of ideas about innovation and development began to take shape in Brazil and Argentina (as well as in other countries such as Uruguay). More recently, this emerging activity has caught the attention of mainstream S&T institutions. Argentina, for instance, has created a new National Program of Social Innovation in the Ministry of Science and Technology, and Brazil has created the position of Secretary for Science, Technology, and Social Inclusion at the Ministry of Science, Technology, and Innovation.[4]

Vision and Strategies

That older appropriate technology approaches still overlap with new initiatives in the region has led to an ongoing debate about what concepts and methodologies are most suitable here. While some actors remain committed to earlier terms and visions related to appropriate technologies and the transfer of technologies, others explore new strategies, including a dialogue between scientific and local knowledge. At the same time, other actors encourage new, more participatory dynamics within the innovation movement to strengthen cooperatives and solidarity economic initiatives (see Singer and Portella Kruppa 2004).[5]

Undoubtedly, the main new movement is the Brazilian Social Technology Network (STN), which includes public R&D groups, universities, federal and regional institutions, public companies, social movements, and especially NGOs. The STN's vision and principles include the need for social commitment and collaboration of S&T institutions; according to Miranda, Lopez, and Couto Soares, "It is built and reapplied from proactive, collective, solidarity and democratic means. Knowledge production occurs alongside the deepening of community awareness, with scientific knowledge coupled with local and traditional knowledge" (2011, 151). In Argentina, the appropriate technologies tradition reaches back to the 1960s and later found a place in academic extension activities at universities and public R&D labs. In 2011, the Technologies for Social Inclusion Network (TSIN) was launched with support from the Ministry of Science and Technology and a few other public institutions.[6] Its main aims are empowering NGOs, cooperatives, and social movements by linking them to public R&D

institutions and translating available R&D capabilities into solutions to problems of poverty, lack of basic services, and environmental risks.

Most of these South American initiatives not only seek public funding and public support for research and development, but also have originated within public R&D labs. Therefore, they cannot be regarded as strictly bottom-up innovation movements, but rather as a mixture of different initiatives and approaches, including several state-driven innovations generated by public research institutions. In most cases, commercial innovations are not considered the ultimate goal of these movements, which have strong sympathies with and bridges to the social movement of the solidarity economy. Thus, innovation and development sociotechnical alliances in Brazil and Argentina involve a more prominent state presence in R&D and funding than similar movements elsewhere. By trying to forge networks between universities, public agencies, NGOs, and social moments, the STN and TSIN, as we shall see, are learning how to provide services, how to co-construct knowledge and technologies, and (what has proved harder) how to influence local, national, and regional agendas of innovation in order to redirect the attention of S&T efforts to the basic needs of the population.

Networks and Artifacts

The STN has also played an important role in creating a new innovation and social inclusion agenda in Brazil, promoting the creation of new public policies and social awareness on the subject. Funded by, among others, the Bank of Brazil Foundation and Petrobras, the STN connects more than 850 institutions from Brazil and other countries and has spent approximately US $175 million on projects from 2005 to 2010 in areas like agroecology, recycling, water sanitation, and support for new social entrepreneurs (RTS 2010). Its first activities included a survey of social technologies and the creation of a digital databank of local innovation experiences, and later it started to promote the reapplication of selected technologies at a larger scale. A good example of these experiences is the One Million Cisterns Program funded by the National Minister of Social Development and implemented by several different local NGOs in Brazil's semiarid region. It is an important program that supported the construction of over 550,000 water collectors in ten years. These cisterns were built and operated by local inhabitants with the help of the STN and another local network, the Semi-Arid Association. The cistern system was originally designed by a brick mason from the region and later modified in collaboration with public R&D labs. The program's main contribution is its self-built aspect, which encourages creating bonds within the community, learning to use and modify the technology, and empowering people (Dias 2012).

In Argentina, our survey of such programs has found over one hundred institutions and programs committed to creating technological solutions and development in the areas of agroecology, renewable energy, health, and social housing.[7] These are largely public institutions, ranging from small public R&D laboratories to large national programs like the Pro-Huerta program at the National Institute of Agricultural Technologies (INTA), an agroecology initiative that serves more than three million people (Montaña 2010). An example of the challenges faced by these initiatives is the case of the Social Habitat group based at the University of Córdoba. This R&D group of architects advocates the co-construction of social technologies in a process that attempts to link technicians and scientists, local authorities, social cooperatives, NGOs, and beneficiaries. Their goal is to procure solutions for social housing by promoting the use and production of local materials like wood instead of bricks or concrete, thus fostering local employment and industry. They also promote the codesign of the houses with the beneficiaries and train local technicians in production systems. They avoid patenting their designs and try to work with cooperatives instead of private companies. The result is an attempt to provide self-sustaining technological and economical solutions involving a negotiation of knowledge between academics and the local population. However, this program still has some shortcomings: the scale of implementation is small, and its process requires a lot of negotiation and redesign. In this sense, the flexibility of the approach can be considered a strength when adapting a design to local needs, but also a problem when trying to scale up the intervention to a wider population.

Preliminary Balance

Although innovation is being recognized as an important driving force of social inclusion initiatives in South America, slowly gaining recognition in the formation of public policy, it is also confronting new challenges and problems. As these movements grow, they start to face new forms of resistance from traditional S&T institutions, especially those that lean toward entrepreneurship and business innovation. For example, the STN in Brazil has recently been suspended, and its main funder, Banco do Brasil, seems to be shifting its policy toward the construction of centers for the demonstration of social technology.[8] In Argentina, a novel initiative in innovation and inclusion in the National Institute of Industrial Technology (INTI) also has been suspended. These setbacks show the fragility of such movements in the region, their dependence on state funding, and the resistance they face from established policy makers and mainstream S&T actors in their attempt to go beyond poverty eradication programs.

Apart from political challenges, these South American initiatives also suffer some of the same limitations as the old appropriate technologies programs and the new inclusive innovation movement (discussed in the next section). These shortcomings are related to the approaches and knowledge strategies that the movements apply (Fressoli, Smith, and Thomas 2011). One of the deepest problems involving innovation movements in Argentina and Brazil is their tendency to propose "technological-fix" solutions (Weinberg 1991) for the problem of poverty. In this case, technological fixes focus on single issues, like the lack of drinkable water or food, which they tackle individually without looking for more systematic solutions. In this sense, most of these organizations and R&D units in Argentina and Brazil tend to focus on the development of "the right artifact" to solve specific social problems, such as using recycled materials for social housing or improving water sanitation systems. Therefore, they tend to rely on in-house innovations with subsequent technology transfers to the community without providing other mechanisms of participation and learning.

In recent years, a few of these initiatives have started to recognize the need to negotiate knowledge between lay people and experts in the co-construction of technologies.[9] These changes in the politics of knowledge, however, are still very novel and have been largely restricted to low-tech solutions, thus generally excluding technological capabilities in high-tech (and better-funded) areas such biotechnology or the development of new materials.[10]

Inclusive Innovation and Social Entrepreneurship in India

India has a long history of social movements aimed at solving the problems of poverty through the use and development of technologies, tracing back to the debates about social development in the 1930s and 1940s (Abrol 2012). These ideas and experiences, especially Gandhian ideas about technology, influenced the shape of the appropriate technology movement founded by E. F. Schumacher in the United Kingdom in the mid-1960s (Willoughby 1990). As a result, approaches and movements conceived in India continue to be regarded as models for new policies of innovation around the world. During the last two decades, new initiatives broadly described as inclusive innovation and social entrepreneurship programs have been created by several scientists, activists, and entrepreneurs in India. These include movements like Grassroots Innovation, the Honey Bee Network, and the Bottom of the Pyramid that have had a wide impact worldwide and have already been tried in South America.

Grassroots Innovation and Bottom of the Pyramid: Vision and Strategies

The Grassroots Innovation (GI) movement created by Anil Gupta in the early 1990s and the Bottom of the Pyramid (BoP) movement launched by C. K. Prahalad in the 2000s share some common characteristics, namely: (a) reconsidering traditional knowledge and local ingenuity, (b) regarding entrepreneurship as a way to solve the shortcomings of aid provided by the state, and (c) attempting to solve local problems through pro-poor innovation,[11] transforming local knowledge into commercial patents and commercial ventures. By focusing on business-based approaches, such approaches differ from more politically engaged movements like the leftist People's

Science Movement or the nationalistic Patriotic and People-Oriented Science and Technology movement (see Abrol 2012).

One of the main differences among business-oriented approaches like GI and BoP involves how they envision which actors should be the leading knowledge producers during the pro-poor innovation process and how. The GI perspective is mainly concerned with the asymmetries of knowledge production between S&T institutions and local communities. This concern also implies a critique of the imposition of industrial development models on local communities and of the risks of economic exploitation and environmental damage that they pose (Gupta et al. 2003). Thus, a central goal of the GI perspective is to properly recognize and add value to the traditional knowledge and natural resources held by local communities and poor people (Gupta 1997, 37). The BoP approach differs in that it regards the poorest population as a dynamic force for innovation, but also posits multinational and national companies instead of grassroots movements as the locus and main beneficiary of pro-poor innovation.

The Honey Bee Network (HBN), one of the best examples of GI in India (Utz and Dahlman 2007; Smith et al. 2012), has expanded rapidly since its founding in the mid-1990s.[12] The HBN's main activity is the scouting and documentation of innovations and traditional knowledge through such methods as community visits, interviews, awards, and competitions.[13] It also explores the commercial potential of the products and processes identified during this scouting, which involves supporting local grassroots innovators in the process of obtaining patents and of prototyping, incubating, and seed funding to ensure commercial viability (Sone 2012).

The BoP approach was created by management expert C. K. Prahalad, who proposed that multinational companies should regard the poorest population (those living on less than US $2.50 per day, largely outside formal markets) as a new space for innovation and economic development (Prahalad 2010). Thus, it assumes that innovation in products and services will help pro-poor development and increase commercial profits at the same time, apparently ensuring mutual benefits. To serve this market, the BoP approach suggests, firms should produce innovative products, services, and business strategies that will be suitable for the poor. This implies new ways to design products, new distribution channels, and new funding strategies. According to Prahalad, "the basic economies of BoP markets are based on small unit packages, low margins per unit, high volume and high return on the capital employed" (2010, 48). In order to sell products, the BoP strategy therefore is to create hybrid networks between firms, NGOs, and public institutions by funding capacity-building initiatives or developing microfunding schemes to gain access to these new markets.

As London and Davidson (2011, 3) recently noted, the creation of new distribution and organizational schemes by early BoP initiatives was neither simple nor straightforward. Accordingly, in the last few years BoP advocates have changed their discourse and strategy from "fortune finding at the BoP" toward the creation of new markets, and from the mere provision of cheaper products toward the creation of new funding and evaluation mechanisms. This change of strategy is intended to overcome the difficulties of earning profits from consumers within the informal economy.

Networks and Artifacts

In recent years, pro-poor innovation initiatives like GI and BoP gained support from governmental institutions, multinationals, and international development institutions. New coalitions with these actors have been crucial for expanding their activities, securing new funding, and influencing inclusive innovation debates at the international level.

The Honey Bee Network, for example, has managed to secure the support of several institutions such as the Grassroots Innovation Augmentation Network (GIAN) and the Society for Research and Initiatives for Sustainable Technological Institutions (SRISTI), as well as the Indian government. The GIAN itself is the result of the efforts of the emerging GI coalition in India. It was created by the Gujarati state government in 1997 through the joint efforts of HBN, SRISTI, and other organizations with the goal of creating a channel to connect local innovations, funding mechanisms, and firms to provide new goods and services (Gupta 1997).

In 2000, India's Department of Science and Technology established the National Innovation Foundation (NIF) to help foster grassroots innovations. Apart from linking informal and formal knowledge and skills by providing technical assistance and credit, the NIF has supported the HBN's recent expansion in India and to China and other developing countries (see Smith et al. 2012).

A representative case of how the HBN's approach works in translating local knowledge into commercial innovations is that of the groundnut digger and separator. Originally designed by Yusuf Khan, a Rajasthani farmer, this technology employs a small tractor that makes the separation and extraction of nut waste from the soil, usually done manually, significantly cheaper. The HBN identified this innovation in 2003 and put the innovator in touch with the Rajasthan Agriculture University, which tested the design of the technology and made incremental changes to it. In 2004, a patent request was filed, and in 2005 the technology was awarded a prize from the NIF. The result was the licensing of this technology to the ARDEE group of companies from Visakhapatnam, India (Singhal n.d.).

According to the BoP model, commercial companies willing to engage in pro-poor markets should learn to operate in informal economies and spaces that lack basic infrastructure (Prahalad 2010). Therefore, these companies seek to partner with NGOs that already have social networks and platforms in place. A good example of how BoP tries to set up networks is the Unilever Shakti initiative, an alliance between Unilever and CARE, an NGO with long experience in capacity building and women's empowerment in poor communities. In this case, CARE provided the platform for training women in the rural sector, while Unilever funded courses in health, infant care, and entrepreneurship along with microfunding for business development. As a result, Unilever acquired a platform for selling its products in new markets in India (London and Davidson 2011; see also Jaiswal 2008).

Since 2005, several international development institutions like the United Nations Development Program (UNDP), the Inter-American Development Bank (IDB), and the World Bank have adopted part of BoP's approach. For example, the Inclusive Markets Program, launched by the UNDP in 2006, surveyed and analyzed fifty successful cases of inclusive business around the world (UNDP 2008). In South America, the IDB (IDB n.d.) claims that the BoP market represents 70% of the population, or almost 360 million people. In 2007, the IDB also launched the Opportunities for the Majority Program, with US $250 million available for funding. Opportunities for the Majority has already funded fourteen projects, most of them focused on seed funding schemes for small producers, new financial services and infrastructure for the development of local markets, and, to a lesser extent, innovation in new products.

Both the IDB's and the UNDP's programs differ slightly from the original BoP approach. In particular, they emphasize the creation of new incomes and the strengthening of the capabilities of the poor. At the same time, the issue of the innovation of new products does not play a central role here as it does in BoP. Instead, the focus is on organizational innovations. The inclusive business programs also seem to pay more attention to small local firms and NGOs than to multinational companies (see Lizuka and SadreGhazi 2011).

Preliminary Balance
The achievements of inclusive innovation initiatives in India and their usefulness for South American science and technology institutions and social movements have been mixed. On one hand, the alliance formed by HBN, SRISTI, and NIF has made it possible to map about one million ideas or forms of traditional knowledge and innovations. In terms of patenting, however, the results have been significantly more modest: the NIF has filed five hundred patent requests in India and seven in the United States, of which only thirty-five and four were approved, respectively (NIF n.d.). Among these, some of the more common types involved the exploration and use of medicinal herbs, artifacts and methods for pest control in agriculture, techniques for handling farm animals, technologies for water and soil preservation, and small tractors, among others.

Despite the large number of ideas and projects in the process of development, very few of these innovations have been patented and only sixty-five products have been licensed to commercial companies, according to the NIF. At the same time, microfunding schemes associated with the HBN have had difficulties being implemented (Utz and Dahlman 2007). More importantly, many of the innovations that the HBN promotes can be characterized as incremental improvements on existing traditional knowledge rather than new scientific knowledge. This may be related to the fact that the HBN strategy is focused on commercial ventures and technological fixes rather than on democratizing access to science and technology capabilities. Therefore, the engagement with local innovators is limited to the process of patenting, incubation, and scaling up, and does not necessarily include the co-construction of knowledge with scientists and engineers. This raises questions about how fairly local innovators are treated in terms of knowledge creation and empowerment (see, e.g., Visvanathan 2005).

In the case of BoP, the divide between local innovators, local knowledge, and companies is clearly worse. After a ten-year history, the BoP approach still has to demonstrate its capacity to provide a development alternative for social inclusion (Abrol 2012). Many of Prahalad's general considerations were based on assumptions that have been widely questioned. For instance, Karnani (2009) and Jaiswal (2008) have criticized his overestimation of the market potential for BoP consumption. Jaiswal (2008) also questioned the success of some of the cases studied by Prahalad, and Halme, Lindeman, and Linna (2012) have argued that short-term profit-maximizing strategies and risk avoidance tendencies actually hamper inclusive innovation experiments in commercial companies. Furthermore, following Christenson, Kaplinsky (2011) has suggested that the patterns of innovation and accumulation within multinational companies, especially regarding the value of their brands in central markets, will not be changed without difficulty. More importantly, BoP posits a sociotechnical alliance in which innovation is concentrated within firms and NGOs, and social actors participate only tangentially in the process of creation. This is reinforced by the kind of simple, deskilled artifacts and services promoted. By these means, the logic of BoP captures and profits from the social networks of the poor without really changing the politic of knowledge typical of mainstream R&D strategies.

Lessons from the HBN and BoP for South America

Inclusive innovation initiatives and programs are being adopted and promoted in South America by international institutions like the World Bank, IDB, and UNDP. But can these initiatives help to improve existing capabilities and social innovation movements in the region? The experiences of the HBN and BoP offer interesting lessons in the role that local knowledge and capabilities might play in the production of new forms of innovation for social inclusion.

Heterogeneous Network Building

Instead of pursuing top-down strategies of technological development that generally disregard local idiosyncrasies and social asymmetries, BoP and GI are keen to recognize and use local capabilities and existing informal networks with NGOs and other institutions. Expanding these networks with local actors may trigger new kinds of negotiations on what is problematic and what is not, which kinds of solutions will work and which will not. In that sense, the creation of heterogeneous networks like those pursued by BoP and GI is an interesting strategy for reflecting new kinds of public policies for knowledge development and innovation. This attempt to create new networks can be taken as an example of heterogeneous network building.

In South America, it is still difficult to find examples of such collaboration between private companies, social movements, and NGOs in the creation of technologies and knowledge. In Brazil, funding from public and private firms allowed the STN to flourish. Yet private actors can play a bigger role in innovation for social inclusion. For instance, private and public companies and more public universities in South America could also offer their important pool of knowledge and capabilities to foster innovation for social inclusion.

Retrieving Traditional Knowledge and Promoting the Coconstruction of Knowledge
Both GI and BoP approaches consider retrieving traditional knowledge to be a crucial aspect of solving environmental and poverty problems. Yet bottom-up strategies of innovation by themselves do not seem enough to encourage a dialogue between such knowledge and scientific and technological knowledge. It is difficult to reconcile this bottom-up vision with the inequalities that arise during the development of products and the distribution of benefits.

In the case of BoP, this problem is straightforward: this approach regards innovation as the realm of private companies, displacing NGOs and other social actors from R&D activities and limiting them to the commercialization and diffusion of innovations. In addition, BoP innovations may be regarded as the simplification of already available products and processes (see Lizuka and SadreGhazi 2011). To a lesser extent, the creation of knowledge inequalities is also a problem for the GI movement. For example, despite HBN attempts to revalue and promote local knowledge through its networks, this approach ultimately relies on knowledge transfer to public R&D units for its development, scale-up, and patenting. This strategy generates problems of power asymmetries between experts and lay people (largely disregarding local capacity to participate in the design of the technological solution). In this respect, lessons from the appropriate technology movement show how low-tech solutions can in fact worsen inequalities and enlarge the divide between the high technology available to the wealthy and the technologies accessible by the poor (see Fressoli, Smith, and Thomas 2011; Willoughby 1990).

Both the STN and TSIN also reflect problems of knowledge inequality between R&D labs that control innovation and local beneficiaries. For example, these networks suffer from some of the limitations of pro-poor approaches such as relying on low-tech solutions and adopting paternalist approaches to the design and implementation of technologies (e.g., Garrido and Lalouf 2012; Dias 2012). In this sense, S&T scholars and policy makers should be cautious about the benefit of adopting initiatives like inclusive innovation without questioning their shortcomings, as such approaches could reinforce local problems instead of solving then.

From a South American point of view, innovation for social inclusion movements may need not only to revalue traditional knowledge (as Gupta argues), but also to help reorient the mainstream scientific agenda to foster the co-construction of knowledge between local actors and scientists. Thus, more than transforming local knowledge into commercial ventures, the challenge for the STN and TSIN seems to be how to tap existing capabilities in science and technology and translate them for the resolution of local problems (Thomas, Fressoli, and Becerra 2012). Innovation for social inclusion could thus be considered a new way to generate learning and new technological developments for everyone (and not just for the poor). But to do this will require developing a new cognitive praxis, including, for example, new forms of participation in agenda setting and fair interactions between local and/or traditional knowledge and scientific knowledge.

Fighting Poverty or Translating It into a Business Opportunity?
Inclusive innovation supporters propose building new markets to create new wealth. But what kind of social inclusion do these strategies provide? Even though the explicit aim of both BoP and GI is to alleviate poverty, there is some tension between the goals and interests of commercial firms, NGOs, and social actors involved in innovation movements. Both approaches encourage the inclusion of the poor in formal, commercial markets as entrepreneurs (in HBN terms) or consumers (in Prahalad's terms). But this is not easily achieved, especially given the deeply embedded motive of profit maximization in commercial companies (Kaplinsky 2011; Halme, Lindeman, and Linna 2012). Indeed, BoP and GI seek to combat poverty through intellectual property arrangements and strong entrepreneurship without necessarily providing broader public participation and the democratization of technologies.

Although this might empower people as users or commercial innovators, it could hardly equalize the long-term inequality in the distribution of knowledge, resources, and capabilities. In their current form, these approaches can be thought of, at the most, as a remedy for the market's failures or a way to ameliorate market economies' unintended effects. On the other hand, while the discourse against poverty generates immediate sympathy, it may also open the gate for opportunistic behaviors or simplistic state interventions. Where some actors see a need to develop new strategies against poverty, others may see the potential to increase their benefits. In particular, some private companies view these schemes only as a business opportunity (see Prahalad 2010).

The Brazilian experience of funding the building of cisterns discussed earlier might help us understand this risk. Despite the fact that the water collectors are rather simple, one of the key aspects of the program is that it generates a sense of identity and cooperation and empowers local people in the building process while providing autonomy from local governments (which previously distributed water in water tanks). This means that the process of building a cistern is more important in generating inclusion than the cistern itself. However, some private companies saw a business opportunity in the scheme, disregarding the process of participation and empowerment that was central to the One Million Cisterns Program. Urged by the need to show fast results, the Brazilian government recently announced the purchase of 300,000 plastic water collectors at almost twice the price of the original scheme (Dias 2012). Situations like this can not only produce setbacks such as higher costs, but may well reverse the creation of capabilities and values associated with the process of technological inclusion.[14]

Thus, a significant but understated problem for these approaches seems to be the question of who decides the direction, scope, and pace of technological developments. This includes the possibility of generating a more democratic politics of knowledge. In this sense, we assert that if inclusive innovation (as well as SNT and TSIN) approaches do not tackle the issue of citizen participation in the process of innovation and development, they run the risk of leading to further inequalities (Thomas 2011), especially cognitive inequalities (Visvanathan 2005).

Final Remarks

In this chapter we have examined how innovation movements and networks in South America and India have developed different cognitive praxes based on their visions, technical artifacts, and sociotechnical alliances.

We tried to show how GI and BoP offer interesting lessons about how to create heterogeneous networks of innovation and how to retrieve and co-construct knowledge with local communities. At the same time we noticed a tendency toward proprietary innovations shared by GI and BoP that does not necessarily meet (or even goes against) the goals and visions of networks in Brazil and Argentina. Thus, while GI and especially BoP aim to empower people as entrepreneurs and consumers of technology, the STN and TSIN networks would like to empower them as active participants in the decision-making process of technological change. In spite of their common goal of social inclusion, then, these movements differ crucially in their politics of knowledge and empowerment.

This distinction is important, since the theme of inclusive innovation is increasingly attracting attention in international and local forums, so that policy makers in South America might be tempted to adopt foreign ideas and models (as the cistern experience suggested) without really considering whether they meet the local innovation movement's visions and strategies. As in other cases (see Cukierman in this volume), there is a risk that the linear adoption of foreign S&T models and ideas could disregard or displace local values and practices, thus undermining autonomous processes of R&D (see Varsavsky 1969; Valderrama and Jiménez 2008). Although those alternative knowledge spaces, like the Social Technology Network in Brazil and Technology for Social Inclusion Network in Argentina, have flourished in recent years, most of them might face serious difficulties at the stage of negotiating incorporation of ideas, visions, and artifacts into mainstream policies of innovation and development (see Hess, 2007).

Rather than seeking to impose brand-new ideas on inclusive innovation, we should take into account that local innovation movements are able to create and carry out their own ideas on technological change, participation, and social inclusion. In this sense, as sources of alternative sites of knowledge production, technology for social inclusion movements can offer us a twist on the classic issue of autonomy. Since the 1960s, scholars in the STS field have helped create a tradition of analysis concerned with the need to foster endogenous technological development that, with notable exceptions (see Herrera 1983), was mainly based on high-tech areas.[15] Nowadays, as novel movements and networks of technology for social inclusion attempt to develop an alternative politics of knowledge and more participatory science and technology practices, they posit an interesting challenge for STS scholars in the region: the task of going beyond accepted models of high-tech investment and technology transfer in order to achieve social development. But that will require the creation of new, situated frames of S&T that can escape the asymmetrical dichotomy between innovation for economic growth and competitiveness and innovation for poverty alleviation.

Notes

This paper is based on the research projects Social Technologies and Public Policies in Latin America, funded by the International Development Research Centre, and Grassroots Innovation in Historical and Comparative Perspective, funded by the STEPS Centre, University of Sussex.

1. Inter-American Development Bank and United Nations Development Program initiatives will be discussed in the third section of this chapter. The World Bank has incubation programs for pro-poor innovation in India (Sone 2012), and the OECD did a special workshop on the issue in November 2012 (OECD 2012).

2. Alliances are dynamically constituted by movements of coordination between artifacts, ideologies, norms, knowledge, institutions, social actors, economic resources, environmental conditions, etc. Different alliances imply different strategies for developing and implementing certain technological approaches (Maclaine Pont and Thomas 2012).

3. This analysis is based on a survey of experiences and institutions funded by the International Development Research Centre. For preliminary results, see www.redtisa.org. Case studies are mentioned in the following section.

4. The authors would like to thank Adrian Smith for his contribution to the ideas of this section.

5. The concept of social economy includes issues in areas such as economic and solidarity relations, work economy, and alternative economic arrangements in civil society. There is some consensus among groups and movements that a solidarity economy entails the search for economic alternatives to a full-fledged capitalist market economy.

6. See www.redtisa.org.

7. These include twenty-four institutions and programs involved in social housing, fifty-one in public production of drugs, sixty in agroecology, and thirty in renewable energy. (For further information see www.redtisa.org.)

8. It is still not clear whether this change implies leaving the challenge to mainstream R&D agendas in order to focus on pro-poor and low-tech innovation approaches.

9. See Garrido and Lalouf (2012) for an account of examples in solar energy, and Fenoglio, Fressoli, and Picabea (2011) for social housing.

10. Public investment allocated by the ministries of science and technology in response to social problems in Argentina and Brazil is quite limited. For example, in 2008 Brazil allocated only 1.1% of government spending to R&D for social development (Brito Cruz and Chaimovich 2010). In the case of Argentina, the results of PROCODAS (the only area in the Ministry of Science, Technology, and Productive Innovation explicitly focused on social problems) are not mentioned among the administration's achievements for 2007–2011.

11. Pro-poor innovation usually refers to entrepreneurial business innovation in services and goods that can be provided to the poor at lower cost (see Sone 2012; OECD 2012).

12. The Honey Bee Network's image of a bee pollinating flowers metaphorically shows the need to draw elements from and to improve local knowledge through cooperation with other actors, which does not weaken but strengthens the communities and their practices.

13. One of their main scouting activities is the Shodh Yatra or journey of scouting (but also dissemination of previous ideas and information) through Indian villages in search of traditional knowledge and local innovations.

14. In this case, the commercial water collector costs approximately US $1,500, more than double the cost of the cisterns under the previous scheme (Dias 2012).

15. For a historical account of this tradition, see Dagnino, Thomas, and Davyt (1996) and Kreimer and Thomas (2004).

References

Abrol, Dinesh. 2012. Pro-poor innovation making: Lessons from India. Paper presented at the II International Conference on Science, Technology and Society Studies: Technology, Innovation and Inclusive Development, Buenos Aires, October 1–2.

Brandão, Flávio Cruvinel. 2001. Programa de Apoio às Tecnologias Apropriadas—PTA: avaliação de um programa de desenvolvimento tecnológico induzido pelo CNPq. MA thesis in management of science and technology, Universidade de Brasília.

Brito Cruz, Carlos H., and Henrique Chaimovich. 2010. Brazil. In *The Current State of Science around the World*, UNESCO World Science Report, 77–103. Paris: UNESCO. Available at http://www.unesco.org/new/en/natural-sciences/science-technology/prospective-studies/unesco-science-report/unesco-science-report-2010/download-report/.

Dagnino, Renato, Hernán Thomas, and Amilcar Davyt. 1996. El pensamiento en ciencia, tecnología y sociedad en América Latina: Una interpretación política de su trayectoria. *Redes* 3 (7): 13–52.

Dias, Rafael. B. 2012. Uma análise sociotécnica do Programa Um Milhão de Cisternas (P1MC). Paper presented at the IX ESOCITE—Latin American Congress of Social Studies of Science and Technology, Mexico City, June 8.

Fenoglio, Valeria, Mariano Fressoli, and Facundo Picabea. 2011. Soluciones puntales vs. soluciones integrales en el campo del hábitat. Desafíos y aprendizajes en la construcción. La experiencia Paranacito. Paper presented at the 1er Congreso Latinoamericano de Estudios Urbanos, Universidad Nacional de General Sarmiento, August 24–26.

Fressoli, Mariano, Adrian Smith, and Hernán Thomas. 2011. From appropriate to social technologies: Some enduring dilemmas in grassroots innovation movements for socially just futures. Paper presented at Globelics 2011, Buenos Aires, November 15–17.

Garrido, Santiago, and Alberto Lalouf. 2012. The socio-technical alliance: Bringing new tools to the design of policies aimed to promote social inclusion. *Review of Policy Research* 29 (6):733–751.

Gupta, Anil K. 1997. The honey bee network: Linking knowledge-rich grassroots innovations. *Development* 40 (4):36–40.

Gupta, Anil K., R. Sinha, D. Koradia, R. Patel, M. Parmar, P. Rohit, H. Patel, K. Patel, V. S. Chand, T. J. James, A. Chandan, M. Patel, T. N. Prakash, and P. Vivekanandan. 2003. Mobilizing grassroots' technological innovations and traditional knowledge, values and institutions: Articulating social and ethical capital. *Futures* 35:975–987.

Halme, Minna, Sara Lindeman, and Paula Linna. 2012. Innovation for inclusive business: Intrapreuneurial bricolage in multinational companies. *Journal of Management Studies* 49 (4): 743–784.

Herrera, Amílcar. 1981. The generation of technologies in rural areas. *World Development* 9 (1):21–34.

Herrera, Amílcar. 1983. Transferencia de tecnología y tecnologías apropiadas: Contribución a una visión prospectiva a largo plazo. Mimeo. Campinas: UNICAMP.

Hess, David. 2007. *Alternative Pathways in Science and Industry: Activism, Innovation and the Environment in an Era of Globalization*. Cambridge, Mass.: MIT Press.

Inter-American Development Bank (IDB). n.d. Opportunities for the majority: Bringing market-based solutions to Latin America and the Caribbean to promote social change. Available at http://www.iadb.org/ (accessed June 18, 2012).

Jaiswal, Anand Kumar. 2008. The fortune at the bottom or the middle of the pyramid? *Innovations: Technology, Governance, Globalization* 3 (1):85–100.

Jamison, Andrew. 2001. *The Making of Green Knowledge: Environmental Politics and Cultural Transformation*. Cambridge: Cambridge University Press.

Kaplinsky, Raphael. 2011. Schumacher meets Schumpeter: Appropriate technology below the radar. *Research Policy* 40 (2):193–203.

Karnani, Aneel. 2009. The Bottom of the Pyramid strategy for reducing poverty: A failed promise. DESA Working Paper No. 80, United Nations. Available at http://www.un.org/esa/desa/papers/2009/wp80_2009.pdf (accessed May 5, 2012).

Kreimer, Pablo, and Hernán Thomas. 2004. Un poco de reflexidad o ¿de dónde venimos? Estudios sociales de la ciencia y la tecnología en América Latina. In *Producción y uso de conocimientos. Estudios de sociología de la ciencia y la tecnología en América Latina*, ed. Pablo Kreimer and Hernán Thomas, 11–90. Bernal: Universidad Nacional de Quilmes Editorial.

Lizuka, Michiko, and Shuan SadreGhazi. 2011. Understanding dynamics of pro-poor innovation: Mapping the disputed areas. Paper presented at DIME—Dynamics of Institutions and Markets in Europe, Maastricht, April 6–8.

London, Ted, and William Davidson. 2011. Creating a fortune with the base of the pyramid. In *Next Generation Business Strategies for the Base of the Pyramid: New Approaches for Building Mutual Value*, ed. T. London and S. Hart, 1–18. Upper Saddle River, NJ: Pearson Education.

Maclaine Pont, Polly C. A., and Hernán Thomas. 2012. The sociotechnical alliance of Argentine quality wine: How Mendoza's viticulture functions between the local and the global. *Science, Technology and Human Values* 37 (6):627–652.

Miranda, Isabelle, Michelle Lopez, and María Clara Couto Soares. 2011. Social technology network: Paths for sustainability. *Innovation and Development* 1:151–152.

Montaña, Sebastián. 2010. Procesos de co-construcción de usuarios del programa de autoproducción de alimentos Pro huerta de INTA. Análisis de dos manuales de usuario. Paper presented at VIII Jornadas Latinoamericanas de Estudios Sociales de la Ciencia y laTecnología, Ciencia y tecnología para la inclusión social, Buenos Aires, July 20–23.

National Innovation Foundation (NIF). n.d. Africa calling. Available at http://nif.org.in/dwn_files/africa%20calling.pdf (accessed May 5, 2012).

Organization for Economic Co-Operation and Development (OECD). 2012. Innovation and inclusive development report. Available at http://oecd.org (accessed January 10, 2012).

Prahalad, Coimbatore Krishnarao. 2010. *The Fortune at the Bottom of the Pyramid: Eradicating Poverty through Profits*. Upper Saddle River, NJ: Pearson Education.

Rede de Tecnologia Social (RTS). 2010. Relatório dos 5 anos da RTS. Available at http://www.rts.org.br/ (accessed May 10, 2011).

Singer, Paul, and María Sonia Portella Kruppa. 2004. Senaes e a economia solidária—democracia e participação ampliando as exigências de novas tecnologias sociais. In *Tecnología social. Uma estrategia para o desenvolvimento*, ed. Renato Dagnigo, 89–103. Rio de Janeiro: Fundación Banco do Brasil.

Singhal, Rajeed. n.d. Groundnut digger cum separator: A case study. Grassroots Innovations Augmentation Network-North (GIAN-North), India. Available at http://www.gian.org/north/files/Groundnut%20Digger-Case%20Study.pdf (accessed, May 5, 2012).

Smith, Adrian, Elisa Arond, Mariano Fressoli, Hernán Thomas, and Dinesh Abrol. 2012. Supporting grassroots innovations: Facts and figures. Sci-Dev Net—Science and Development Network. Available at http://www.scidev.net/ (accessed November 15, 2012).

Sone, Lina. 2012. Innovative initiatives supporting inclusive innovation in India: Social business incubation and micro venture capital. *Technological Forecasting and Social Change* 79:638–647.

Thomas, Hernán. 2011. Sistemas tecnológicos sociales y ciudadanía socio-técnica. Innovación, desarrollo, democracia. In *Culturas científicas y alternativas tecnológicas*, ed. Fernando Tula Molina and Gustavo Giuliano, 65–86. Buenos Aires: Ministério de Ciencia, Tecnología e Innovación Productiva.

Thomas, Hernán, Mariano Fressoli, and Lucas Becerra. 2012. Science and technology policy and social ex/inclusion. Analysing opportunities and constraints; Problems and solutions in Latin America. The cases of Brazil and Argentina. *Science and Public Policy* 39 (5):539–571.

United Nations Development Programme (UNDP). 2008. Creating value for all: Strategies for doing business with the poor. Available at http://web.undp.org/ (accessed May 10, 2012).

Utz, Anuja, and Carl Dahlman. 2007. Promoting inclusive innovation (India). In *Unleashing India's Innovation: Toward Sustainable and Inclusive Growth*, ed. Mark A. Dutz. World Bank. Available at http://web.worldbank.org/.

Valderrama, Andrés, and Javier Jiménez. 2008. Desarrollos tecnológicos en Colombia. Superando categorías de oposición. *Revista Redes* 14 (27):97–115.

Varsavsky, Oscar. 1969. *Ciencia, política y cientificismo*. Buenos Aires: Centro Editor de América Latina.

Visvanathan, Shiv. 2005. Knowledge, justice and democracy. In *Science and Citizens*, ed. M. Leach, I. Scoones, and B. Wynne, 83–94. London: Zed Books.

Waks, Leonard J. 1993. STS as an academic field and a social movement. *Technology in Society* 15:399–408.

Weinberg, Alvin. 1991. Can technology replace social engineering? In *Controlling Technology: Contemporary Issues*, ed. E. B. Thompson, 131–141. Buffalo, NY: Prometheus Books.

Whitecombe, Richard, and Marilyn Carr. 1982. *Appropriate Technology Institutions: A Review*. London: Intermediate Technology Development Group.

Willoughby, Kelvin W. 1990. *Technology Choice: Critique of the Appropriate Technology Movement*. London: ITDG.

4 Working with Care: Narratives of Invisible Women Scientists Practicing Forensic Genetics in Colombia

Tania Pérez-Bustos, María Fernanda Olarte Sierra, and Adriana Díaz del Castillo H.

In this paper we focus on the case of women geneticists working in the field of forensic genetics in Colombia.[1] These scientists, who contribute to the identification of victims of the armed conflict in the country and in civil or criminal cases, as well as in the adjudication of paternity, are usually trained as microbiologists or bacteriologists, scientific fields that have been largely feminized in the country.[2] We argue that this feminization situates the scientific practice of these geneticists within a paradox, as they are simultaneously admired and respected by the general public—due to their work in identifying victims—and yet, as women scientists, are invisible in the media. Furthermore, the practice of forensic genetics—regardless of whether it is performed by women or men—has traditionally been undervalued vis-à-vis academic genetic research, with those working in the field regarded by their scientific peers as mere technicians.

By analyzing this particular experience in a Latin American country, we aim to provide new theoretical tools and perspectives with which to perform situated studies of science and technology (STS) in dialogue with feminist discussions. In Latin America, gender studies and STS have not followed the same processes of institutionalization as in Europe or North America (Cutcliffe 2003; Wotipka, Ramírez, and Díaz Martínez 2007); thus feminist discussions of scientific and technological knowledge production are only just beginning to take place. In this sense, although in general there has been increasing consolidation of these fields as two interdisciplinary academic projects (Vessuri 2007; Bartra 1997), in Colombia the dialogues between them have not been systematic, and in some cases are nonexistent. This contrasts with longstanding reflections about feminism in STS in hegemonic Anglo-Saxon contexts, where the intersections with STS posed by feminists have responded to the different waves of thought that have characterized these fields.

From our perspective, this inconsistent or nonexistent dialogue in Latin America has had two consequences. On the one hand, it explains the questions that are now emerging about the epistemological implications of gender, sexuality, and power in the shaping of science and technology in the global South. One area to examine, for example, would be the STS research related to war and internal conflict in Latin America. Studies undertaken in Argentina (Lalouf 2004; Blinder 2011) and Brazil (Dagnino 1988), and some of the research proposals currently being developed in Colombia (i.e., Gómez 2011), which are mainly focused on the study of the technology of war and illegal innovations in the context of internal conflict, do not take into consideration how those technologies might be gendered. Following Wajcman (2006), we can observe how these studies assume that the identities and interests of those who design technologies and routinely use them do not shape the technologies themselves, but merely shape the hierarchies between subjects.

The second consequence of this limited dialogue between STS and feminist and gender studies in regions like ours is that most of the research produced, which seeks to articulate the category of gender in relation to the study of science and technology, does so without explicit nourishment from the reflections of feminist theories that have flourished in Latin America. Furthermore, such an approach overlooks the reflections that the feminist critique of hegemonic science and technology have produced since the early 1980s in Western Europe and North America.

This is the case with most of the research related to the underrepresentation of women in scientific fields in the region (for example Fernández Rius 2006; Pérez Sedeño 2011). The main goal of these studies has been to discuss the socioinstitutional barriers that explain the low presence of women in science and technology, an approach which underscores the general idea that these fields are predominantly masculinized.[3] Though this emphasis might be considered by some as strategic, in the sense that it helps to address gender inequalities within national systems of science and technology, paradoxically it also contributes to the nonrecognition of other types of diversities and power relations within these systems. Moreover, it contributes little or nothing to the comprehension of how procedures, methodologies, innovations, technologies, and social impacts embedded in scientific knowledge production and application are situated, embodied, and crosscut by gender.

We want to argue that studying science and technology from a feminist perspective in Latin America implies more than studying the differences between men and women scientists, or merely stressing women's roles and contributions to particular fields. As a perspective, it suggests ways of observing and understanding the politics of science and technology. Among other things, it questions the practices of doing and thinking science that take place in everyday life, including within labs. These practices are performed by particular people, for example women forensic geneticists, and build up a diversity of utopias. Such a perspective advocates for an understanding of the value of issues and things that are taken for granted or made invisible in the composition and possibility of situated sciences and technologies.

In line with this and based on ethnographic data, we argue that the scientific practice of forensic genetics in Colombia can be framed by particular politics of care (Puig de la Bellacasa 2011). Care has been one of those ignored issues in STS in Latin America. In Colombia, feminists have reflected upon care as a feminized practice related to the sustainability of life (Arango Gaviria 2011). It is a practice that involves a way of doing, and so an ethos, embodied in those who do something for someone and produce work that contributes directly to the maintenance or preservation of another's life. Such work is feminized in the sense that it is mainly developed by women—though not restricted to them—but also in the sense that it is devalued by society. These studies have focused on the politics of care embedded in invisible jobs within the sexual, aesthetic, and domestic arenas (Arango Gaviria and Molinier 2011). Following the work of Puig de la Bellacasa (2011; 2009), Haraway (2007), Singleton (2011), and Mol (2008), we argue that this ethos of care has special characteristics when situated within the field of science and technology, and we find it of particular relevance in contexts of internal conflict, such as that which persists in Colombia.

With this in mind, we propose that women geneticists who perform this practice over many years find ways to situate themselves, and their work, in a context marked by social injustices of different kinds and in relation to a scientific field that hierarchically stratifies knowledge production. We understand care as a performative practice, a way of doing and intervening in everyday life from particular ethical, political, and affective standpoints (Puig de la Bellacasa 2011; Puig de la Bellacasa 2009; Haraway 2007; Singleton 2011; Mol 2008). Within science and technology, matters of care are ambivalent and highly political. We see them as referring to services or practices of everyday life that sustain life itself and are embedded in a variety of scientific fields. These practices are situated and context-specific and, like domestic labor in macroeconomics, are in general rendered invisible. This is the case with the modest work of geneticists working in the field of forensic genetics in labs such as the ones we studied, who also happen to be women. However, when we say that matters of care are highly political, we refer not only to the moral necessity to make visible the role of caring practices in the shaping of science and technology, but also to the epistemological need to comprehend how these practices perform situated science (Haraway 1988). In the case of the women geneticists we discuss here, we describe various dimensions of this notion of care. It highlights certain modes of articulating *emotional*, *rational*, and *practical* work, both at the lab and with regard to the victims, their families, or those awaiting paternity tests. It also refers to the ways in which they position their work toward the future.

In order to better understand how the politics of care operates specifically in the case of forensic genetics in Colombia, in the following section we present the context of this scientific field, outlining the emergence of forensic genetics in Colombia and its public image in the media. We link the latter with the circulation of globalized representations of science and argue that in this context there is a particular invisibility of women geneticists and their work. In order to problematize this, in the second part of the paper we address the experiences of six women geneticists in relation to the above-mentioned dimensions of care.

Forensic Genetics in Context

The study of forensic genetics is relevant in the Colombian context as it is one of the main activities of geneticists in the country. Forensic genetics includes all cases of human identification for paternity adjudication, identification and reparation of victims of the armed conflict, and resolution of crimes and civil cases. In this sense, the field of forensic genetics has contributed to the development of genetics in general, not only through the import of technologies and the development of associated knowledge, but also through the constant provision of samples and related genetic profiles (in paternity testing and civil cases). Furthermore, the technical requirements of paternity testing were the driving force behind the construction of a table of reference of Colombian genetic markers (Usaquén 2012).

The technology and the related knowledge to perform DNA tests to resolve cases of dubious paternity were first imported to Colombia in 1997. Nevertheless, DNA technology was not implemented straightaway, as geneticists needed to be trained in it and other technical requirements were pending (Usaquén 2012). Such training took place at the National Institute of Legal Medicine and Forensic Sciences (NILM) in Bogotá, where biologists, bacteriologists, and microbiologists worked—together with forensic anthropologists, forensic dentists, and forensic physicians—in the area of identification. Those who were trained as geneticists thus came from the varied disciplines working in the NILM in the identification of bodies and crime scenes. In terms of the composition of the forensic labs of NILM and the Attorney General's Office (where there is a department for forensic genetics), women bacteriologists and some women microbiologists have been a majority in their fields since these labs were created.

This predominance of women can be explained by several factors, in particular the fact that many forensic geneticists begin their careers in bacteriology. The job of forensic genetics requires the highest standards of care for the samples; thus the training that bacteriologists receive is ideal, and bacteriology is a field predominantly made up of women (as all six of the interviewed women geneticists pointed out). Additionally, many bacteriologists migrate to the field of forensic genetics because it offers mostly better working (and salary) conditions than does clinical bacteriology. However, there is no single standard with regard to working conditions, which vary according to the type of forensic genetics that is practiced and the lab in which the geneticists work. For example, Sandra expressed:[4]

When I was working in clinical bacteriology I was the professional who earned the least; this job is not valued, and that discouraged me and I decided not to work in clinical bacteriology ... then I decided to do the master's in human genetics. When I finished studying, I started working as head of the forensic lab; we work with paternity cases. It is easier to find a job if you focus on paternity tests ... the pay is much better now.

In contrast, Margarita commented:

I do not do this for the money, it is a vocation. There are several disincentives, one is the monetary compensation. People believe that it pays a lot, but it doesn't. There is little financial reward.

As the armed conflict in Colombia continued to worsen and genetics grew as a science for identification, forensic genetics became recognized as important in the identification of both victims and perpetrators. In the 1990s a new forensic lab was created at the Attorney General's Office, owing to an increase in criminal investigation that took place at the time. In the frame of Law 975 of 2005, also commonly known as *De Justicia y Paz* (Of Justice and Peace), the identification of victims of the conflict is currently conducted with the goal of returning the bodies to their families, as part of what is considered a reparation process (Usaquén 2012).[5] Forensic geneticists working at the NILM and the Attorney General's Office deal today with an overwhelming number of cases pertaining to Law 975/2005, while simultaneously working on unrelated criminal and paternity cases.

The work of forensic geneticists is considered a matter of public interest, and has permeated the public sphere by means of information media. In Colombian newspapers, for example, it is not unusual to find news about criminal investigations that mention the role of forensics in solving cases. In such news reports, genetic identification is presented as an all-powerful tool to provide justice and peace through the identification of victims and their return to relatives. Such representations have been studied in other contexts by Kruse (2010), who called them the "*CSI* effect." She argues that the US TV series *CSI* presents an idealized vision of science capable of reestablishing order, justice, and peace based on evidence that speaks for itself—and always tells the truth. This evidence is revealed by passive scientists who are presented as sacrificial, dedicated workers, capable of keeping their personal lives and emotions at bay in their research. While genetic identification is celebrated, in Colombian news the scientists performing this identification are either seldom mentioned or are presented anonymously using the Spanish masculine form of *expertos*, *forenses*, or *investigadores* (Díaz del Castillo, Olarte Sierra, and Pérez-Bustos 2012).[6] Our previous research into genetics in Colombia (Olarte Sierra and Díaz del Castillo, forthcoming; Olarte Sierra and Díaz del Castillo 2013) called our attention to the fact that these *expertos* were, more often than not, women. This contrast between media reports and practice inspired us to further explore an area of work (and the scientists performing it) that is simultaneously paramount and invisible.

Visible Scientists, Invisible Women

There were two main factors influencing our decision to conduct and present this research in the way we have. First, we were interested in making visible a type of work that is usually concealed in the informative media, and that is commonly undervalued vis-à-vis academic research (which is considered more scientific as opposed to technical, and thus more respectable), yet is nonetheless highly relevant in the context of Colombia's past and current history. Second, we considered it more than a mere coincidence that this work is performed predominantly by women scientists. The inclusion solely of women in our study thus had to do both with the characteristics of the field and with our interest in bringing their work to light as a situated science.

It was with these aims that we looked for women geneticists currently working in forensic genetics at NILM and the Attorney General's Office; later, as a source of contrast, we decided to include one woman geneticist working at a paternity testing laboratory who is also part of academia. We contacted the scientists through common acquaintances (the first ones contacted subsequently led us to others), informed them about the purpose of our research, and invited them to an interview either in person or by teleconference. All three authors conducted all the interviews together (except one in which only two authors were present) based on a previously designed guide. Through the interviews we aimed to trace these women scientists' experiences as forensic geneticists through their personal histories as well as their practical work in the labs, and to investigate their positioning regarding the public image circulated about them in the media.

The six interviewed women geneticists involved in this study have worked in genetics for most of their professional lives; in spite of previous training in bacteriology (five of the six) or microbiology, all of them consider themselves geneticists. For nearly all, their interest in genetics surfaced early in their careers and led them to occupy important managerial positions in the labs where they work or have worked. Most of them have also participated in the development or consolidation of such labs and in the training of new colleagues.

We initiated the conversations by explaining how the contrast between media representations and practice inspired our interest in them as scientists and their work. Interestingly, however, for these women the media portrayals of them (or the complete lack thereof) were not striking. One woman geneticist explained that those in her field are not interested in media visibility because their work requires discretion; it could be risky to be publicly exposed. Another ascribed their absence in the media to the modesty of their work.[7] Their personal invisibility has thus been normalized, and is actually considered desirable.

In contrast, the women were eloquent regarding representations of their work, especially its portrayal in TV series. They recalled images that show it as easy, effortless, and performed by high-tech all-purpose equipment, work that is even magical. "I want one of those centrifuges and techniques that they use in *CSI*," Azucena commented ironically. Consequently, they are often expected by the public to deliver 100% reliable results in hours or minutes ("as during TV commercials," as Margarita put it), or to provide unfeasible information (i.e., genetic profiles that automatically match with photos, as is the general practice on TV). Victims' relatives, attorneys, or journalists have even accused them of negligence when they fail to comply with these imagined possibilities. Despite these incongruities, they were nonetheless ambivalent toward the TV series. As a teenager, one of the women had been inspired by similar shows of the time; several commented that these series got people interested in their work, and that they use *CSI* to explain their work to others.

In the following sections, we will present these women's work as "matters of care" using three categories: touching/thinking/feeling; handling with care; and thinking toward the future.

Caring through Touching, Thinking, and Feeling

We use this category following Rose's (1983) argument in favor of transcending the division of hand, brain, and heart in technology and knowledge production, and link it to Puig de la Bellacasa's (2011, 95) argument that *caring* is a situated hand, brain, and heart endeavor. In science and technology, it might be seen as an example of a particular type of ontological politics (see Marques, this volume). It is not a general moral disposition, but a linkage that is used to make everyday decisions. "How to care?" asks Puig de la Bellacasa (2011). How do the women scientists who work in genetic identification in Colombia deal with the sometimes divergent agendas of manipulating samples, empathizing with victims, and analyzing complex DNA profiles?

The combination of mental, manual, and emotional labor arose without prompting during the interviews. Most of the women geneticists emphasized the emotional dimension of their work, an aspect they considered inseparable from the technical and "rational." Like Margarita in the following account, they described in detail the cases that *touched* them, referring to samples, tests, and challenges but also to families, horror, and accomplishment.

There are many cases that impact you, but I remember very much a case that was very difficult to solve. An old lady, 82 years, from a town near Bogotá, who lived alone in the country. She heard noises at night, came out with her flashlight, and the only thing she remembers is that someone knocked it down. ... The next day the neighbors found her nearly dying. They had caused her horrifying lesions in the head with a machete, wounds longer than 11cm. Apart from all that, they raped her. ... The lady fell into a profound depression, a grandma who had been raped! She wanted to die. When that case was assigned to me, it was only a slide, nothing more. A very difficult sample; with only one spermatozoid it is very difficult to obtain results. We only had this slide and a blood sample from the man. I remember that the person who assigned me the case was even disparaging: "Take it, see if you can perform miracles." That's how s/he told me. And I thought: "Well, ok. I'll see if I can perform miracles." I worked on the case and eureka! I got a masculine genetic profile that coincided with that man [the main suspect] but I also obtained an additional profile. ... Those cases really give you a huge satisfaction, to be able to contribute to the delivery of justice.

As Margarita describes, genetic identification requires skill and knowledge. Yet it is not devoid of emotions, subjectivities (of victims and researchers), or commitment and gratification (professional and personal). Some of the women geneticists mentioned that they allow themselves to cry, to "feel" what the victims feel, to "see" beyond the bones. They are able to weave the victims' life histories and contexts into the samples. Nevertheless, they are not willing to completely rid themselves of their invisibility cloak, and they (all) emphatically claimed that these feelings do not affect their objectivity:

In a paternity case, you see the child or the mother, poor, with nothing. That does not mean that in a given moment we think "we have to assign paternity because she needs it"; never. But you do feel the mother's pain. ... We cry watching the news, talking to a victim. ... [Emotion and reason] are always linked. Without ever losing objectivity and impartiality, but feeling and justifying our work every day. (Antonia)

Although this modesty coincides with media representations of their work, the link between brain, hand, and heart, palpable in the women's descriptions, is muted in the representations that circulate in *CSI* (Kruse 2010) or local news (Díaz del Castillo, Olarte Sierra, and Pérez-Bustos 2012). The neutral, invisible scientist that these media put into circulation is far from "feeling the mothers' pain."

At times, the geneticists seem to romanticize this modest and objective way of presenting their work, even treating it as normative for their job. "Losing commitment can lead to things not being performed correctly," said Lorena. Yet this perspective is not shared by all. The inseparability of the three dimensions is not without tensions. Some of the women try to avoid emotional involvement by focusing on the *brain* (the "spirit of research") and the *hand*. Yet instead of denying the emotions connected with technical work, this actually reasserts the fact that the geneticists' subjectivities are so present at work that they need to be protected: "When we get to the lab, we have to perform a technique. If we cry all over it [the sample], it gets contaminated" (Azucena). Similarly, for some, the *heart* part of the equation has more to do with working with passion and commitment, with feeling satisfaction when training colleagues or obtaining results. Several mentioned that they are "in love" with their work. The opposite happens as well; when this love is over, so is the job.

The hardship of dealing with human remains can shake "rationality." This work can be scatological, as Margarita described: "You have to put your hand inside [the remains] and try to extract. ... That sensation is so terribly horrifying, that is when you lose all the scientific [objectivity], it is really nasty ... you never lose your human condition, [never stop] having those sensations." Likewise, although their work involves an intellectual endeavor, in the interviews the *hand* and the *heart* seemed unbalanced with the *brain*. Some women referred to their work as "not academic," since there is not enough time for research. Others explained that even though both the labs of the NILM and the Attorney General's Office produce constant DNA profiles, which could be used for further analysis and the writing of scientific papers (as is done in labs working on paternity testing), *caring* for the samples implies not publicizing the data in any way. Therefore, the writing of scientific papers is out of the question. The fact that these geneticists do not publish renders their work merely "technical" in the eyes of their academic peers. Sandra said: "Women geneticists at NILM and the Attorney General's Office just do the technical part, they don't go further, they don't do any research ... they are limited to the lab part, that's all they do. They don't know how to do [academic] research." In turn, Azucena explained:

Many fellow geneticists keep on telling us, "All those samples you have and don't share; let's do something to those samples [further molecular analysis]," and we can't. In the informed consent it is very clear that these samples are for exclusive use for identification, we can't do anything more to them ... well, we could ask donors to sign for molecular analysis allele frequencies and population genetics, but donors are usually so shaken up that asking them for more [which also implies explaining more to them] is not always the right thing to do.

(Please) Handle with Care

To work as a forensic geneticist in the context of reparation for victims (or when testing for paternity) is to work with what Latour has named matters of concern (1993). These geneticists do not merely produce facts; the facts they work with are political, and their nature is of public interest. However, such concern in itself does not suffice to understand their work. The dimensions of care become crucial, especially when trying to answer questions about how to care or for whom (Puig de la Bellacasa, 2011, 97).

Care, as was evident in our fieldwork, is central to forensic genetics in Colombia, and it is necessary for science to be able to respond to public concerns and to compensate victims in criminal cases. We found this dimension of care in the way in which geneticists handle the samples (all throughout the identification process). It was also evident in how they relate to and do their job, as a craft that requires care and discipline, mostly because it involves others (directly and indirectly) in vital ways (such as family members of victims, the victims themselves, and those awaiting paternity test results). In this sense, forensic geneticists have a commitment not only to science and scientific endeavor but also to others whom they try to help.

Understanding the job as central—though not the only factor—for the resolution of crimes, reparation processes, and paternity testing presupposes the handling of samples (blood, bone, or otherwise) with utmost care. One geneticist explained that "the samples that we work with are scarce; if you contaminate the sample you lose the whole case … you cannot mess it up here. It's a juridical investigation and it is the reparation of families that we are talking about" (Margarita).

But the need to care for the sample is not limited to the commitment they have to the victims (and/or their families), or those involved in paternity cases. There is another dimension to the careful handling of the samples that directly involves them as individuals. These geneticists are expert witnesses, and their analyses and results are demonstrative evidence. If a sample is damaged, and thus becomes useless for the identification process (or for the adjudication of paternities), the geneticist working on the case faces juridical and legal consequences. Thus, as Margarita further explained, "If you mess up you'll be investigated. Both disciplinarily and juridically … you cannot mess it up."

A way of avoiding "messing it up" is to work in what these geneticists describe as an extremely organized, precise, and clean environment. Laboratory training in how to perform a technique and sticking to protocol—that is, learning to take care of the samples—is seen as one of the most important parts of becoming an authorized forensic expert. The interviewed geneticists were emphatic in this regard. They also expressed how, once a person is socialized into this laboratory culture, routines and protocols become a sort of second skin. Lorena expressed how "it is gratifying to see assistants and analysts growing into their dexterities. How they do their job better every time until they can do it by themselves … with some it takes longer than with others."

Doing it well not only implies the technical skills required for not "messing up" and producing a satisfying result, but also—as stated above—requires the analytical sophistication to understand the work, to solve the cases, and to be able to deal with the potential contingencies of each case. Juliana explained it as follows:

I prefer to say "expert in forensic genetics" [instead of forensic geneticist] because it is not only about mastering the technique, of doing all the process well. You become an expert when you are able to analyze and solve problems, when you go beyond the first results and dig deeper. For doing that, you need to learn to recognize problems and to know where to look for answers.

Here again we find the collective and the personal intertwined in the job. In solving complex cases, it becomes a challenge to find an answer "for the family, who are expecting the news and are waiting to bury their loved one, but also it is a personal challenge of being able to solve a case that's difficult both technically and analytically" (Juliana).

The dimensions of care shape the work that these women geneticists do, yet care is neither singular nor fixed. It may take different forms as cases come in different shapes; sometimes it even costs them the scientific status in what they do. However, for them care is constitutive of their work, and a good geneticist needs to know how to respond to particular needs for care. If she fails to do so, analysis is compromised, quality cannot be guaranteed, and ultimately she is unable to produce the results required, results that are, in most cases, of national concern. In this sense, we could argue that to look at and unpack matters of concern in technoscience from the perspective of care, as we have done here, helps to respond to some of the questions posed by Marques (this volume) with regard to ontological politics and why they are relevant. Our reading of these issues makes evident how care is an ethos that configures ontological entities in a way that makes up forensic geneticists' individual and collective realities as they respond to particular questions pertaining to a nation.

Caring toward the Future

The work is rewarding, yet the situation is so bad that you wonder "Why?" What we can do is minimal ... it is contradictory. What is the real contribution if this person—the family, because they too are victims—has already suffered so much? ... This should not happen, but it is because it happens that what we do exists. (Azucena)

As we have seen, the work of these women geneticists is framed by a certain ethos of care that orients their thinking and doing; not just in relation to others—those who work with them and whom they work for—but also in relation to the work they do at the lab with the human samples. In this section we argue that caring in forensic genetics in Colombia is not simply a matter of *doing* and *intervening* in particular ways, but involves an issue of *hoping*. Here we want to take up Puig de la Bellacasa's invitation to answer the question of caring: *what for?* (2011, 97). In this sense, we aim to comprehend how these women geneticists imbue their work of identifying victims with the symbolic power of repairing a broken nation; a kind of speculative commitment (Puig de la Bellacasa 2009; Puig de la Bellacasa 2011) to participate in change, affecting—in a minor but significant way—an aching national reality. It also entails a commitment to understand and deal with the contingencies that that same reality imposes on their work at the lab, and which at the same time paradoxically qualifies it.

As Margarita puts it, in contrast to what the "*CSI* effect" makes people believe, results in forensic genetics do not take place between commercials. However, as in the

TV series, their work of identifying victims is marked by standards of high quality and perfection. As we have seen, results require detailed hard work and time, but this is also conditioned by the reality of the country's internal conflict. Thus, the ideal of perfection is shaped by the day-to-day casuistry, which brings with it a variety of contingencies.

A high-profile case, the child who was kidnapped in the National Hospital. It was almost a year and a half looking for Juan Diego, with a sample of umbilical cord. And in that search, more than 60 cases arrived that were not Juan Diego's, they were cases of other children who were kidnapped or were in the possession of persons other than their parents. (Antonia)

On some occasions, contingencies affect the quality of the sample. This can happen when remains are found in places where military confrontations are taking place, and they have to be taken out quickly with backhoes. In cases like these, the women geneticists' work becomes an analytical challenge, which is as much applied as it is technical. It situates them in the daily domain of casualties, which tinges the will to achieve perfection with the social responsibility of reparation. Reparation is an act of touching. It displaces these women scientists' work from *the view from nowhere* that is constructed by the media, to a concrete domain: the act of carefully touching the remains to achieve the analysis, and in doing so being touched by what the analysis implies. As Margarita said, "When I have many of these cases, I pray to God to make me an instrument of justice. To help me get the truth." Thus, in the daily reality of their work, the challenge is not only technical but painfully emotional, as well as highly gratifying. In their words:

The most rewarding is what we do in the routine of everyday life: the identification of Jane/John Does. Knowing that this person was a Jane/John Doe for many years, and today has a name, today is the son of Mrs. someone. ... Then to deliver that report ... so their families can grieve. (Antonia)

It is not just the family but the entire country that is demanding justice. (Margarita)

As we can see in these quotes, the geneticists' speculative commitment to reparation, which is technical, practical, and emotional, takes shape at different levels. First, with the victims' families, it gives them the right to grieve. This, in turn, is a small contribution to the reparation of a country that lacks justice. Finally, this ethos of care, which situates the work of these women scientists toward a particular future, is shaped by daily technical practices that have effects and are *useful* now. In this sense, we could argue that when analyzed from the perspective of care, the kind of forensic genetics that we present here is, in many ways, a social technology of the kind discussed by Fressoli et al. (this volume). This is so in the sense that what these women geneticists do—and how they do it—fosters an inclusive space in which individual victims, as well as the nation, can heal.

As we mentioned at the beginning, this ethos of care toward the future has also had implications for the development of genetics in general in Colombia. What is paradoxical is that this development has been possible not only thanks to grief but also to the daily practices and challenges that forensic geneticists face, challenges that, being mainly technical, are on many occasions devalued, especially in comparison with more academic approaches to genetics. This makes the work of these women not just a speculative commitment to reparation, but, in Puig de la Bellacasa's (2011) words, a speculative commitment to neglected things.

Final Remarks

In this chapter we have addressed women geneticists working in the field of forensic genetics in Colombia. Our purpose was to underscore that the work of these women scientists can be framed within particular politics of care, from which it is possible to question and reinforce persistent ideas about science and its practice, in particular in regions such as Latin America.

With the study of this particular case, we intend to highlight a particular contribution to the fields of STS and feminist studies. First, by analyzing a specific local scientific practice, we were able to unpack what takes place beneath globalized popular representations of science, such as the ones promoted by *CSI*. Instead of a science "from nowhere" and "by no one," we have described a situated science, performed by women geneticists, who carefully touch and are touched by their studied subjects in a context of war. Second, and in direct relation to the first point, we argue that this view of a situated science, a caring science, constitutes in itself a novel feminist approach to the social study of science and technology in Latin America, one that proposes a new angle from which to comprehend science and technology in the global South. This is a perspective that goes beyond justice and inclusion as matters of concern that shape the political agenda of STS in regions marked by inequalities and privileges, a perspective suggesting that a caring science exists in the quotidian realities interwoven in these public agendas. We also see this contribution as a possibility for geopolitically connecting feminist studies in the North and South. We see this particularly in terms of *thinking with* (Puig de la Bellacasa 2012) Northern feminist discussions on matters of care in technoscience, to propose new horizons for feminist sociologists studying the ethics and economics of care in Latin America, which till now have focused on labor fields such as the domestic, the sexual, and the aesthetic.

In this sense, we consider care to be crucial to the doings of science, especially in a context of internal conflict and daily crimes, such as is the case in Colombia. We have addressed care through three core dimensions: one that relates to feeling/thinking/doing with care; one that supposes careful daily relations both with organic samples and with humans in different conditions, such as victims of war; and one that faces the future with the speculative commitment to think about a science that can potentially repair. We found that although care is constitutive of forensic genetics and is required to fulfill its prescribed tasks of contributing to crime resolution, paternity adjudication, and victim reparation, care is also invisible and taken for granted. The absence of scientific research produced by these women geneticists, which is one of the tradeoffs of working with such a volume of (often sensitive) cases, is seen by fellow geneticists as displaying a lack of analytical abilities, and the forensic geneticists are consequently considered mere technicians in the field.

Despite this invisibility, we argue that, through the lens of care, the scientific practice of these women geneticists is twofold. First, it can be seen as contributing to the weaving of Colombia's web of social relations, which has become unknitted by crimes in daily life, enduring internal conflict, and dubious paternity. Second, care can be understood as a way of reimagining scientific practice in Latin America. In this regard, we consider that the analysis we have provided here enriches significantly Latin American discussions on STS, such as those presented in the introduction to this chapter. We could say that our approach helps in *feministing*[8] the field of STS in Latin America, as it brings new—feminist—perspectives into the study of science and technology.

Acknowledgments

We would like to offer our thanks to the six women geneticists who agreed to share their stories with us; to Elisa Arond, Natalia Niño, and Mauricio Caviedes for their insightful reading of earlier versions of this chapter; and to Zoe Goldstein for her impeccable editing.

Notes

1. The order of the authors does not represent any hierarchy. This paper is a collaborative work.

2. On average, 9 out of 10 bacteriologists and 7 out of 10 microbiologists are women (Daza-Caicedo 2010).

3. In Colombia only 37% of the scientists registered in the database of Colciencias (National Council of Science and Technology) are women.

4. We used pseudonyms to protect the privacy of the interviewed women geneticists.

5. The early years of the 2000s marked the beginning of an unprecedented process in Colombia, in which the country began a "transitional justice without transition" (Uprimmy Yepes et al. 2006). That is, while still in the midst of war, the country embarked on a process that uses mechanisms designed primarily for post-conflict situations. This process materialized with the enactment of Law 975 in 2005, one of whose pillars is victim reparation, which includes acts

such as "effective collaboration to locate abducted or missing persons and the location of the bodies of victims" and "the search for the missing and the remains of people killed, and assistance in the identification and reburial by family according to their community traditions" (Law 975/2005, our translation). Since the law was enacted, it has proven highly controversial. National and international actors such as human rights organizations and academics have problematized it with regard to—among other issues—who is considered a victim, who is regarded as perpetrator, who is responsible for reparation, and what kind of "truth" and history can be reconstructed from the process.

6. According to the Royal Spanish Academy, Spanish nouns and adjectives can be classified according to their grammatical gender as feminine or masculine. Some nouns are ambiguous (any gender is accepted), common (a single form is used for both feminine and masculine individuals), or *epicenos* (the grammatical gender is independent of the sex of the individual). Also according to the academy, the masculine form is used in the plural to refer to all individuals regardless of their sex. See http://buscon.rae.es/dpdI/SrvltGUIBusDPD?lema=g%E9nero2. Some critical discussions of this use of the language can be found in Castellanos (2010).

For examples of these news articles, see http://www.eltiempo.com/archivo/documento/CMS-7113333; http://www.eltiempo.com/archivo/documento/CMS-7294405; http://www.eltiempo.com/archivo/documento/MAM-3458954; http://www.eltiempo.com/archivo/documento/CMS-6040707; http://www.eltiempo.com/archivo/documento/CMS-5183832; http://www.eltiempo.com/archivo/documento/CMS-7170808.

7. We use "modesty" following Haraway's (2004) description of the scientist as a modest, invisible, neutral witness who reveals reality through standardized experiments.

8. We borrow the term *feministing* from "feministing.com"; the politically committed blog that provides a platform for feminist writing (http://feministing.com). Here we use the term as a verb for bringing a feminist perspective to varied themes and discussions regarding the field of STS.

References

Arango Gaviria, Luz. 2011. El trabajo de cuidado: ¿Servidumbre, profesión o ingeniería emocional? In Arango Gaviria and Molinier 2011, 91–109.

Arango Gaviria, Luz, and Pascale Molinier. 2011. *El trabajo y la ética del cuidado*. Medellín/Bogotá: La Carreta Editores—Universidad Nacional de Colombia.

Bartra, Eli. 1997. Estudios de la mujer. ¿Un paso adelante, dos pasos atrás? *Política y Cultura* 9:201–214.

Blinder, Daniel. 2011. Tecnología misilística y sus usos duales: Aproximaciones políticas entre la ciencia y las relaciones internacionales en el caso del V2 alemán y el Cóndor II argentine. *Revista CTS* 6 (18):9–33.

Castellanos, Gabriela. 2010. ¿"Lenguajes incluyentes", o lenguajes "políticamente correctos"? Cómo construir equidad en el discurso. *Labrys, Études Féministes/ Estudos Feministas*,

July-December 2010. Available at http://www.tanianavarroswain.com.br/labrys/labrys%2018/libre/gabriela.htm (accessed January 30, 2013).

Cutcliffe, Stephen. 2003. *Ideas máquinas y valores. Los estudios de ciencia, tecnología y sociedad.* Mexico City: Anthropos.

Dagnino, Renato. 1988. Cuando negocios no son negocios. Los aviones de guerra del Brasil. *Nueva Sociedad* 97:178–187.

Daza-Caicedo, Sandra. 2010. Las mujeres en el SNCTI. Balance de una década en condiciones diferentes. In *Indicadores de ciencia y tecnología*, ed. Mónica Salazar, 279–318. Bogotá: Observatorio Colombiano de Ciencia y Tecnología.

Díaz del Castillo Hernández, Adriana, María Fernanda Olarte Sierra, and Tania Pérez-Bustos. 2012. Testigos modestos y poblaciones invisibles en la cobertura de la genética humana en los medios de comunicación colombianos. *Interface: Comunicação, Saude, Educação* 16 (41):451–467.

Fernández Rius, Lourdes. 2006. Género y mujeres académicas: ¿Hasta dónde la equidad? In *Ciencia, tecnología y género en iberoamérica*, ed. Eulalia Pérez Sedeño et al., 55–66. Monografías 29. Madrid: CSIC.

Gómez, Yuri. 2011. La ingeniería del delito: Una propuesta alternativa para la comprensión de la innovación en contextos periféricos. Paper presented at the Coloquio Nacional Esamblado a Colombia I: Naturalezas, Culturas, Tecnologías, Universidad Nacional de Colombia, Bogotá, August 10–13.

Haraway, Donna. 1988. Situated knowledges: The science question in feminism and the privilege of partial perspective. *Feminist Studies* 14 (3): 575–599.

Haraway, Donna. 2004. *Testigo_Modesto@Segundo_Milenio. HombreHembra©_Conoce_Oncoratón.* Barcelona: Editorial UOC.

Haraway, Donna. 2007. *When Species Meet.* Minneapolis: University of Minnesota Press.

Kruse, Corina. 2010. Producing absolute truth: CSI science as wishful thinking. *American Anthropologist* 112 (1):79–91.

Lalouf, Alberto. 2004. Desarrollo tecnológico en países periféricos a partir de la cooptación de recursos humanos calificados. Aviones de caza a reacción en la Argentina. *Convergencia. Revista de Ciencias Sociales* 11 (35):221–248.

Latour, Bruno. 1993. *We Have Never Been Modern.* Cambridge: Harvard University Press.

Mol, Annemarie. 2008. *The Logic of Care: Health and the Problem of Patient Choice.* New York: Routledge.

Olarte Sierra, María Fernanda, and Adriana Díaz del Castillo H. 2013. "We are all the same, we all are mestizos": Imagined populations and nations in genetics research in Colombia. *Science as Culture* (October 18): 1–27. doi:10.1080/09505431.2013.838214. http://dx.doi.org/10.1080/09505431.2013.838214.

Olarte Sierra, María Fernanda, and Adriana Díaz del Castillo. Forthcoming. From the lab to the literature: The travels of humans, categories and other genetic products. A case study of the practice of population genetics in Colombia. Unpublished manuscript, personal archive of the authors, last modified November 15, 2013.

Pérez Sedeño, Eulalia. 2011. Mujeres y pioneras en las ciencias: Una Mirada a la realidad iberoamericana. In *Ciência, tecnología e gênero*, 213–232. Abordagens iberoameicanos. Curitiba: UTPFR.

Puig de la Bellacasa, Maria. 2009. Touching technologies, touching visions: The reclaiming of sensorial experience and the politics of speculative thinking. *Subjectivity* 28:297–315.

Puig de la Bellacasa, Maria. 2011. Matters of care in technoscience: Assembling neglected things. *Social Studies of Science* 41 (1):85–106.

Puig de la Bellacasa, Maria. 2012. "Nothing comes without its world": Thinking with care. *Sociological Review* 60 (2):197–216.

Rose, Hilary. 1983. Hand, brain, and heart: A feminist epistemology for the natural sciences. *Signs* (Chicago) 9 (1):73–90.

Singleton, Vicky. 2011. When contexts meet: Feminism and accountability in UK cattle. *Science, Technology and Human Values* 37 (4):404–433.

Uprimmy Yepes, Rodrigo, Catalina Botero Marino, Esteban Restrepo, and María Paula Saffon. 2006. *Justicia transicional sin transición? Reflexiones sobre verdad, justicia y reparación en Colombia*. Bogotá: Centro de Estudios de Derecho, Justicia y Sociedad. Dejusticia.

Usaquén, William. 2012. Validación y consistencia de información en estudios de diversidad genética humana a partir de marcadores microsatélites. PhD diss., Universidad Nacional de Colombia.

Vessuri, Hebe. 2007. *O inventamos o erramos. La ciencia como idea-fuerza en América Latina*. Buenos Aires: UNQ Editorial.

Wajcman, J. 2006. *El tecnofeminismo*. Madrid: Cátedra.

Wotipka, Christine Min, Francisco Ramírez, and Capitolina Díaz Martínez. 2007. A transnational analysis of the rise and institutionalization of women's studies. *Revista Española de Sociología* 117:35–59.

5 Ontological Politics and Latin American Local Knowledges

Ivan da Costa Marques

Ontological politics! So what? What are the implications of ontological politics for issues involving distributions of privileges and disadvantages in engagements with the world? What does it mean to have an ontological political perspective, and what is gained with it? How does one come to recognize ontological political options after they have been made? How does one open ontological political controversies?

One way to make ontological politics visible is to look closely at how realities of the world are configured and stabilized. Everyone's world or reality is populated by entities (people, things, subjects, objects, theories, and practices) that are ontologically configured. A Yanomami, an Amazonian "caboclo" in South America, or a Sora in India, for instance, lives in a world that results from ontological processes in which options are selected that are quite different from those made in the ontological processes constituting the world of a Western European. Pragmatically, ontological processes involve choices and decisions that result in the establishment and stabilization or obduracy of frames of reference, which are the tools people use to situate their engagements with the world (Callon 1998). Frames of reference are historically contingent, not natural, universal, or neutral, and ontological politics plays a crucial role in their formation.

In this chapter, we look closely at two cases to discuss how, through the exercise of ontological politics, Western realities become "the reality" and other people's realities are considered merely different interpretations of reality. It is a widely held Western belief that the West has knowledge, while other people have mere beliefs about reality. This kind of Western common sense is established as a result of a specific ontological political perspective adopted in the production of modern scientific knowledge. This ontological perspective approaches reality through a process of discoveries of forms or objects (entities) that are assumed to already exist in Nature (e.g., oxygen, microbes, atoms) or, alternatively, in Society (e.g., homo economicus, law of supply and demand, "the iron laws of society"), although Society is presumed to have a Nature that is the subject of study of the social sciences. These entities are seen as part of the world of things-in-themselves (Nature) and as being there independently of the world of humans-among-themselves (Society) (Latour 1993). As things-in-themselves, seen as having an independent isolatable existence, these presumably preexistent entities become universal and neutral.

Examining the practice of ontological political perspectives highlights that other approaches to reality may produce other choices of different valuable knowledge, other objects, other subjects, other worlds. Other ontological political perspectives do not endow modern Western science with the attributes of universality and neutrality—rather, they render the facts and laws of Western science (e.g., oxygen, microbes, atoms, homo economicus, "the iron laws of society") historically contingent. From this perspective, modern sciences do not discover objects and laws that were "already there." If the results of Western modern sciences lose the privilege of universality and neutrality and are instead seen as situated and local knowledges of the West, then the relation between Western sciences and technologies and other practices of engagement with the world can change substantially. In such a case, it becomes no longer a matter of being right or wrong depending on whether the proposed engagement with the world agrees or does not agree with the scientific knowledge of the West. In these terms, realizing that other ontological political perspectives can also be enacted means recognizing that Western modern sciences have from the start provided a cage that confines the space in which Latin Americans can search for solutions of their practical problems.

What is to be gained by acknowledging and understanding the often invisible operations of ontological politics? The gain may be immense for non-Western and Western peoples alike. Different ontological political perspectives may enact a more dialogical world, a world that strongly makes viable and legitimates relations with Western science that are relations neither of complete subjection nor of complete rejection. Different ontological political perspectives enact different ways for Latin Americans and other people to relate to the West, ways that may greatly increase dialogues and symmetries—and consequently, though certainly with more than a grain of optimism, increase choices, selection, respect for differences, negotiation, and peace. So far, Western science and technology has been able to maintain a dialogue (to discuss "facts" or "reality") only in its own terms. Western modern sciences have dialogued only with people who can set up equally powerful laboratories to propose and stabilize new or different scientific facts by the production of new inscriptions and/or new selections and combinations of inscriptions. Latin Americans and other non-Western people usually lack the resources to build such "counterlaboratories" (Latour 1987). Only different ontological political perspectives may accomplish the very delicate operation of bringing Western science down from its pedestal of universality and neutrality to place it more simply among other valued and valuable knowledges. Of course, this will eventually change the ontological political perspective of Western sciences and technologies.

In Latin America, despite criticisms since or even prior to the 1990s, dominant approaches to technoscientific issues are still based on reductionist development and diffusion models. The first configures development as a process formed by discoveries or inventions of facts and artifacts in successive stages, and the second configures those facts and artifacts of modern (i.e., Western, Euro-American) technosciences as stable entities that propagate themselves throughout societies by causing "impacts" (Basalla 1988; Rostow 1962; introduction to this volume). These models are radically committed to the dominant ontological political perspective of the West and leave no space for enacting other ontological political perspectives. Ontological politics focuses on the work of marking a division in a previously undifferentiated reality, that is, on those activities that demarcate what is natural rather than social, technical rather than political.

During the 1970s and the 1980s, social scientists in the United States and Europe began to go into laboratories to study the production processes of scientific and technological knowledges (Latour and Woolgar 1979; Knorr-Cetina 1981; Lynch 1985; Traweek 1988). This work problematized the "God's-eye" trick that places scientists in the advantageous position of being able to observe/study everything without having their own methods and results observed/studied by other disciplines (Haraway 1989, 189–195). In 1987, Wiebe Bijker, Thomas Hughes, and Trevor Pinch published a book with the bold subtitle "New Directions in the Sociology and History of Technology." After a couple of decades, I argue, these "new directions" have led science and technology studies (STS) to a place from which there is a possibility of attributing respectability and value to different kinds of knowledges. Although these "new directions" originated in the United States and Europe, where modern sciences had historically been conceded the attributes of universality and neutrality, they have led research to matters very important for Latin Americans. Universality and neutrality have been and still are powerful sources of value and respectability when conflicts arise in disputes involving the judgment of Western and other kinds of knowledges. The new directions taken by STS have questioned this presumed universality and neutrality and lessened their epistemic power. At the same time they have also stressed the value and increased the respectability of local knowledges. Indeed, the previously presumably universal facts or truths have become the Western local facts and truths that have spread all over the planet, which sets a level of equivalence (not equality) with different types of knowledge and makes them easier to assess and compare.

Ontological political perspectives based on more symmetrical and dialogical approaches will attempt to juxtapose, although not necessarily oppose, different kinds of knowledge. In Latin America itself, as one might say, non-Western knowledges are typically seen, at least by the most educated part of the population, as opposed to, rather than just different from, supposedly universal and neutral Western knowledges and therefore as of little value. In general, educated Latin Americans are very wary of the proposition that what is accepted as universal or global knowledge is the result of large-scale social and intellectual construction that does not transcend the systems and individuals that collectively produce it.

This chapter takes dialogical and symmetrical approaches seriously, conferring respectability on the multiplicity of the real (Latour 1999, 1–23; Mol 1999; Mol 2002). This is certainly a risky quest, beginning with the fact that it is written in English. Indeed, the use of English as an assumed lingua franca already reflects the existing relations of power, for the knowledges and practices of "the other" (Latin American) will be translated into the terms of "the one" (Western). As Helen Verran reminds us in her compelling study of an African logic, "the differential power and privilege associated with the two [knowledges] at issue, matters" (2001, 6). In attempting a more dialogical and symmetrical approach, using one people's discursive strategies to analyze other people's knowledges risks merely recreating and reconstituting the interpretative frames of reference from which these same strategies come and derive their strength (ibid.). In our case, this is the risk of failing to distance ourselves from dominant Western forms of knowledge. Moreover, one might argue, perfect symmetry or dialogue does not exist. But the quest for symmetry and dialogue does, and it is this quest that makes investigations more or less symmetrical and dialogical. The more symmetrical and dialogical a study is, the more potential lines of flight it will offer from hegemonic notions of what constitutes knowledge, leading to understandings that grant more respect to local or situated knowledges when they are compared with so-called global or universal knowledges. The more dialogical and symmetrical a study is, the more lines of flights it opens up for different ontological political perspectives and their associated affects. A line of flight from Western frames of reference is not a way to simply escape to another space in the same territory, but rather a way to get to and enact spaces or worlds that extend beyond Western ontological political perspectives, or at least beyond those hegemonic and official scientific entities (objects, subjects, laws) or frameworks that form the basis of what is considered reliable knowledge in industrialized Western nations.

To explore the importance of symmetrical and dialogical studies, I next present the two cases mentioned above, which, although of very different kinds, both throw light on issues that may arise when local and situated Brazilian knowledges and practices are configured in Euro-American frames of reference and compared to modern, global, and universal knowledges. The first and shorter case involves Jeca Tatu, a very popular fictional figure in Brazil. Jeca Tatu is the main character of *Urupês*, a collection of tales and chronicles, first published in 1918, that is the masterpiece of the Brazilian writer and farmer José Bento Renato Monteiro Lobato. Brazilian intellectuals have transformed the figure of Jeca Tatu from a literary object into an object of sociological research. I will argue that how it has been deployed exemplifies how both a configuration of the local and its contraposition can reinforce dominant Western ontological political perspectives. In contrast, the second and more detailed case shows three different worlds or realities (not just three different interpretations of reality) based on varying frames of reference applied to a child nutritional proposal involving a food additive, multimistura, made from native ingredients in response to the serious problem of malnutrition in Brazilian children. I then draw on an anthropologically informed ontological political perspective to set a scale of increasing dialogical capacity. By examining a local practice dating from the 1970s that has resisted the hegemonic ontological political perspective of Western science, I demonstrate a line of flight from the Western global frame of reference leading toward a more dialogical space that can provide Brazilian local practices and organizations with greater respectability in their agreements and disputes with science.

Jeca Tatu

In Monteiro Lobato's first depiction of Jeca Tatu, his ontological political perspective built frames of reference that regarded mestizos with utter contempt. In this version, Monteiro Lobato depicts Jeca Tatu as the indolent, heedless, and parasitical character who typifies the hillbilly from the interior of São Paulo, Brazil.[1] Lobato's first image of Jeca Tatu can be summed up by the expression "it just isn't worthwhile":

> Confronted with the problems on the farm where he was an agregado [unproductive farmhand] or the great changes in national political life, such as the abolition of slavery or the proclamation of the Republic, the caboclo continued to "squat, dozing, on his haunches" ... all the caboclo's ignorant philosophizing prattles in this world transfixed by fatalism and apathy. Nothing is worthwhile. Neither culture nor conveniences. Life goes on anyway. (Cited in Lima and Hochman 2004, 510–511)

As Jeca Tatu is transformed from a literary object into an object of sociological research, his image develops into that of the caboclo, a Brazilian mestizo of European and Native South American blood. In his second version, the frame of reference produced by Monteiro Lobato's ontological political perspective is one of integrating the people of the interior into national life, which results in another configuration of Jeca Tatu. The means of his escape from the first configuration appears to have been Monteiro Lobato's encounter with intellectuals and proposals to support the expansion of basic sanitation in Brazil from 1916 to 1920. Alongside the sanitation campaigns and medical scientists' reports of their trips to the interior, Monteiro Lobato revives Jeca Tatu, now attributing the caboclo's laziness to diseases. Jeca Tatu's second image can be summed up thus: "Jeca is not really like that; it's just the way he is now" (in Portuguese, *Jeca não é assim, está assim*):

> When he began to believe in medical science and its prescriptions, Jeca Tatu was transformed. Free from hookworm disease, and the state of permanent despondency it caused, he became

productive and, in no time at all, a prosperous farmer, competing with his Italian neighbor and rapidly overtaking him. More than that: he modernized his farm, introduced new crops and technologies, and learned to speak English. (Lima and Hochman 2004, 511)

Jeca Tatu's symbolic importance rests in the change of the frame of reference. As Lima and Hochman point out, "Jeca Tatu's strength lies precisely in the loss of his specific reference to the agregado or unproductive rural worker when he achieves the dimension of a national symbol" (2004, 511–512).

Seeking to open up a line of flight within either of these frames of reference from the Western hegemonic ontological political perspective, however, my starting question is: with whom does Jeca Tatu's strength ally? A different ontological political perspective would focus on and explore the choices and decisions involved in the construction of the frames of reference in which the complex unity of the two Jeca Tatus is configured. Who do these frames of reference benefit? Which options and decisions are naturalized and thus do not appear as options or decisions, silencing discussion of other alternatives? Which options and decisions lead to the distribution of preferences, benefits, and drawbacks, and are thus political but represented and discussed as if they were "technical" problems that only specialists can handle? These questions lead us to issues of ontological politics or "empirical philosophy" (Mol 1999). Thus, rather than attributing a meaning (a truth) to those frames of reference, the main concern becomes to reveal the mechanisms through which those frames of reference are stabilized. After the ontological political choices and decisions are made, the entities involved either remain naturalized as universal and neutral elements belonging to Nature or are erased in the diffusion of the resulting frames of reference.

Where we might look for entrances to lines of flight that could lead us to territories in which the hidden materiality of the controversies and conflicts among different ontological political perspectives could be found is suggested by controversies over the figure of Jeca Tatu, including Ruy Barbosa's condemnation "of those who wanted to eliminate Jeca Tatu":

But gentlemen, if this [colonized Jeca Tatu] is what they [the colonizers] see, is this really what we [the Brazilians] are? Aren't the Brazilian people more than just a specimen of the most ignorant caboclo, who can't even stand on his feet, or even sit, a manifestation of all the stigmas of indolence and stupidity, whose vote can be bought with a roll of tobacco, a suit of coarse cloth, and a swig of rum? (Barbosa 1981, 174)

Barbosa's questions raise others that can help reveal the ontological politics at play. What entities operate but are not represented in the universe inhabited by Jeca Tatu? Moreover, what hypotheses were naturalized when choosing the entities these two versions used to describe the two Jeca Tatus? Why choose fatalism, the instinctive distrust of civilization, a habitual indolence, the idea of the predatory exploitation of the land, productivity? In terms of materiality, what is and what is not included in economic accounting and comparisons? In the advantageous modernization of agriculture? In the overarching view of progress as a universal blessing? What translations were performed to establish the equivalence between health and modern development? In adopting an American farmer as a model or idealization of the rural "entrepreneur"? In positioning English as Esperanto?

In all this, a work of division is performed surreptitiously. Frontiers and hierarchies are established covertly between the good and the evil. Options, decisions, and questions that enact the resulting frames of reference rarely appear, which makes them seem natural, neutral, and universal. No evaluation is ever made of the distribution effects of the options and decisions strongly conditioned by entities naturalized by universal and neutral knowledge originated in the West. Left unasked is whether these options and decisions are better or worse for whom, where, and when.

Although these two figures of Jeca Tatu appear to be opposed, they enact the same Western ontological political perspective and privilege the same values: the dominant values of European colonization, the first denying them to the despised Jeca Tatu and the second attributing them to the praised Jeca Tatu. Thus, instead of adopting a different ontological political perspective that would enable and enact a line of flight from the frame of reference in which the caboclo is depicted as worthless, Jeca Tatu's story does not perform a line of flight. In fact, it reinforces that frame of reference, world, or reality.

Multimistura

The multimistura case deals with conflicts between scientific knowledge and a local nutritional practice that in its current form started in the 1970s in Brazil.[2] As noted earlier, different ontological political perspectives construct different realities or worlds (Mol 1999), and in this chapter we shall visit three worlds of multimistura. These three worlds or stories do not represent different ways to interpret or tell the same story or suppose the existence of a single story that is the "real thing." Rather, we can view the three stories as devices that build upon and take us to spaces that result from different ontological options and decisions and that are thus inhabited by different entities; that is, the three stories enact three different worlds. If the three stories were different interpretations of the same "real" world, then the ontological political perspective that constructs this "real" world would enjoy universality and neutrality, precisely the God's-eye trick of the West that posits its reality as "the reality" and other knowledges as just beliefs about reality. In contrast with Jeca Tatu's story, it is the construction of realities rather than interpretations of reality that is at stake in the multimistura case.[3]

To better understand the process through which this construction takes place, I first present a series of statements from the body of literature about multimistura that I am calling historiographical inscriptions. My intention in doing so is to argue that historiographical inscriptions play, for those who practice history, the same role that the inscriptions made on laboratory instruments play for natural scientists.[4] For physicists, for example, the photographed trace that a physical particle makes in a bubble chamber is an inscription. It is through the juxtaposition, selection, and combination of inscriptions that physicists create the reality of a particle. Similarly to the way that inscriptions in physics are the result of the encounter between laboratory instruments and possible future physical entities (particles) that at the time of the encounter do not yet have a form, historiographical inscriptions are the result of the encounter between an instrument—that is, the author of the historiographical inscription—and possible histories that do not yet have a form. It is through the juxtaposition, selection, and combination of historiographical inscriptions that historians and social scientists construct historical facts, or, one may say, a historical reality.

The statements about multimistura below have been extracted from scientific reports and articles, interviews, and manifestos from leading newspapers and the Internet aimed at the general public. They are, in sum, disseminated or published text fragments or building blocks of narratives or texts about multimistura. The equivalence between these historiographical inscriptions and the inscriptions obtained in laboratories is important because it allows us to speak of all facts in the same terms, be they physical, chemical, nutritional, economic, social, or historical facts. To be able to understand and analyze facts of different modern disciplines in the same terms is not only a very important result of the science studies of the last few decades, but also a crucial tool in enacting different ontological political perspectives.

So, just as inscriptions obtained in labs are selected and combined to produce the reality of a physical reality or world (for example, a particle), historiographical inscriptions are selected and combined to produce a historical reality or world (for example, the collapse of the Soviet Union or multimistura). The historiographical inscriptions below are not properly a story of multimistura but rather a set of elements, like preliminary collected material or building blocks, to be selected, combined, and possibly added to other historiographical inscriptions to develop the three stories of multimistura that will follow.

Historiographical Inscriptions

In the mid-1970s, pediatrician Clara Brandão observed a drastic reduction in diarrhea in malnourished children in thirteen daycare centers in Santarém, Pará, after experimenting over a three-day period with a nutritional supplement obtained from brans and dark leaves as well as other ingredients such as seeds and eggshell powder.[5]

Interviewing the local population to gain more knowledge of their eating habits, she had begun searching the local produce for "alternative foods" that had high nutritional value and that weren't being regularly consumed by the population.[6]

From there, Clara Brandão became influential via the spread of the use of multimistura, as the product obtained from these alternative foods came to be known.[7]

The movement became prominent. In 1984, a consultant appointed by UNICEF to evaluate the nutritional value of multimistura presented a favorable report.[8]

In its work in the area of basic assistance, the Children's Pastoral Service of the Brazilian National Bishops' Conference (CNBB) began to distribute multimistura and gave national prominence to the practice of using it.[9]

In 1989, awareness of what multimistura had to offer increased even more with the transfer of Clara Brandão and her husband, Dr. Rubens Brandão, to the Health Ministry in Brasília.[10]

In 1990, the Federal Board of Nutritionists (CFN) pointed out several technical and conceptual errors in a leaflet authored by Clara Brandão, saying that the leaflet's information should be proven by scientific research and soliciting the Health Minister to "halt the release of the leaflet 'Alimentação alternativa' [Alternative diet] to make necessary corrections."[11]

In 1994, doctors Jaime Amaya-Farfán and Hilda Torin wrote a technical report that set off a campaign to discredit multimistura, calling attention to the dangers of using a diet composed of elements whose efficiency was questioned by the results of their research.[12]

According to *Veja* magazine in October 1996, "the Children's Pastoral Service alone had more than 80 thousand community leaders providing assistance to 2 million poor children all over the country."[13]

In December 2002, multimistura was "designated by the State Secretary for Social Action and Municipal Councils from the state of Tocantins (in direct acquisition) for the program of donations to the underprivileged" and made exempt from payment of the ICMS tax.[14]

In 2006, Zilda Arns Neuman, coordinator of the Children's Pastoral Service, received "the Opus prize (from the Opus Prize Foundation and Catholic University of Notre Dame, Indiana, US) that awards US $1 million to people or organizations that unite entrepreneurship and faith in their work and who are committed to the profound transformation of social problems such as injustice, poverty, hunger, illiteracy and disease."[15]

The Children's Pastoral Service stopped letting its business identification number (CNPJ) be used to open multimistura factories, giving as reasons the doubt cast on multimistura's efficacy by research results, the need to ensure public health standards, and its preference for having its leaders working in the communities rather than in factories. "The multimistura flour, as well as any other food, when produced to be consumed on a large scale, outside of the home or the community, must follow the standards set by the sanitary agencies."[16]

In October 2006, Clara Brandão declared that "they have already advised me that I am now clandestine in the government."[17]

Nonetheless, in June 2008 the AABB Community Program of the Banco do Brasil Foundation inaugurated a multimistura factory in Bom Conselho, Pernambuco. Three hundred and ninety-two municipalities participate in the AABB Community Program, which includes more than 50,000 children and youths aged between seven and eighteen years of age and almost 4,000 educators.[18]

Clara Brandão declared that the government had begun to exclude multimistura from school meals to make room for Mucilon, produced by Nestlé, and powdered milk, whose market is divided between Nestlé and Proctor & Gamble. "It is political genocide to replace multimistura with industrialized food," said the pediatrician.[19]

The national coordinator of the Children's Pastoral Service, Zilda Arns, recognized that multimistura has played an important role in reducing the rate of infant malnutrition: "Multimistura really helped, but by itself, it is not capable of decimating anemia; importance must also be given to breastfeeding."[20]

In 2007, the Health Ministry declared that "multimistura, a composition of bran and other ingredients, was never adopted as a national strategy for the treatment of infant malnutrition. Moreover, the Health Ministry neither buys nor distributes foods to the population. Therefore, there is no foundation to the reports that the portfolio would have substituted multimistura with industrialized foods."[21]

First Story, First World

The first story regarding multimistura results from an ontological political perspective that enacts a frame of reference that is radically epistemologically divided between science and belief, wisdom and ignorance, and correspondingly selects, treats, and combines historiographical inscriptions. In this story, in its initial phase in Pará in 1974 until the end of the decade, Clara Brandão's project spread in daycare centers and poor schools in an ever-increasing number of Brazilian municipalities with no scientific evaluation. It is hard to establish precise quantitative information about the adoption of multimistura, especially in its first years, for it was carried out in a quite decentralized way and mostly by very poor people with little schooling.[22] In all likelihood it had reached the scale of millions of children by the late 1980s.[23]

Brandão counted on the volunteers of the Brazilian Legion of Assistance to spread the word about multimistura. Later, in 1983, the Children's Pastoral Service of the Catholic Church adopted multimistura almost "spontaneously"—that is, from the point of view of scientists, with no scientific explanation or reliable evidence of efficacy.[24] This adoption took multimistura to thousands of communities. From there, after also being adopted "spontaneously" by the Banco do Brasil Foundation, in 1990 multimistura was considered for inclusion in a government program looking to resolve the problem of infant malnutrition in Brazil, and also spread to other countries.

But eventually, academic groups analyzed the composition of various local preparations of multimistura and concluded scientifically, through the use of inscriptions obtained in their laboratories, that multimistura could not have the effects that it was claiming to have. This conclusion, thus far not having been substantially contested by other scientists in laboratories, is a scientific fact. Multimistura did not pass the "reality test" of the first story's ontological political perspective, options, and decisions, and multimistura's nutritional capacity was pronounced fiction and not fact. As a result, multimistura's capacity to garner support declined drastically. Multimistura's practices and associated knowledges were transformed into belief. Multimistura is no longer viable as part of a government program in a modern state.

In this first story, there is no relativism as to the truths established in the laboratories about the capacity of multimistura. The initial "spontaneity" of its adoption can be easily explained as a result of intervention by Clara Brandão and "interested" activists. The academic groups in opposition to them are aligned and taken as "disinterested" actors or "interested only in discovering the truth."

In the first story's frame of reference, a universal and neutral Nature ("things-in-themselves") decides what is true and what is false, and the sole legitimate spokespersons of Nature are scientists. In the first story, the "interested" activists work in the field, are ignorant or lack scientific knowledge about nutrition, and increase their level of influence by mixing in "subjective" (rather than "objective") and emotional elements that are present in Society (economic, political, and social factors absent from scientific truths, according to the first story's ontological political perspective). To show evidence of this expedient combination by interested activists, is it not enough to point out their resorting to sentiments of solidarity in their use of photographs, as in figure 5.1?

Second Story, Second World
A line of flight may escape from the restricted frame of reference of the first story and lead to an amplified ontological space by attending to demands of symmetry and dialogue. The establishment of scientific truths about multimistura by the academic groups in relation to dietary and nutritional questions is not an untainted, disinterested, and isolated process. The possibility of adopting multimistura for school meals in a national program, involving several hundred millions of reais annually, put multimistura into an arena where other actors were already well established in the Brazilian food market.[25] These actors had important relations with academic groups. The Children's Pastoral Service established an accord with large companies in the food sector at the same time that they abandoned the use of multimistura, and its coordinator, Zilda Arns, was thanked by well-connected institutions in the business world

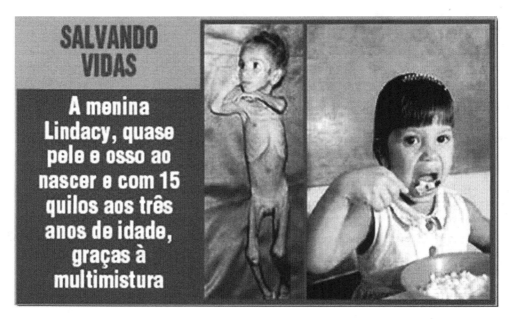

Figure 5.1
Cover story of *Veja* (a Brazilian equivalent of *Time* magazine, which sells over one million issues a week), October 30, 1996. Translation: "SAVING LIVES—The girl Lindacy, almost skin and bones at birth and weighing 15 kg [about 27 pounds] at three years old, thanks to multimistura."

with an award of one million dollars. The academic groups of professional nutritionists involved came in large part from the southern and southeastern regions of Brazil, regions that are more Westernized and where the paradigm that equates scientific knowledge and reliable knowledge is more solid and amply installed in opposition to so-called popular beliefs. Nutritionists as a professional group perhaps had a special interest in marking their territory in possible conflict with medical doctors—a much more powerful professional group.

This second story is a typical STS story. It is more relativist and symmetric and engages the natural sciences in dialogue with the social sciences. The causal intensity and the classification of each actor must be established empirically for the case in question. When its historicity is noted, a scientific fact becomes relative. Activists in favor of multimistura try to show its effectiveness by weighing children at regular intervals and recording narratives, thus creating their own inscriptions as evidence that multimistura works, as shown in figure 5.2. However, their procedures do not satisfy the requirements of laboratory experiment protocols. Their inscriptions, of the kind shown in box 5.1 and box 5.2, are not taken seriously by scientists, who consider

Figure 5.2
Monthly weighing of children on April 25, 2013, by the volunteers of the Children's Pastoral Service (Comunidade Santa Luzia) at Igreja de Santa Luzia in Volta Redonda, RJ, Brazil. Photo by Lucimeri Ricas Dias.

their methods precarious and their measurements imprecise. Their results are not scientifically convincing, although they insist that the symptoms of malnourishment disappear and the children become healthy.

Yet a limit to such relativism appears here for the analyst, anthropologist, sociologist, or historian of science who adopts another ontological political perspective but is under the aegis of what John Law calls "Euro-American metaphysics" (Law 2004, passim). This limit is fixed at the closure of scientific controversies. If there is no controversy among scientists specialized in the nutritional field regarding the nutritional characteristics of multimistura, then these characteristics become a stabilized (obdurated) scientific fact, and this is as far as he or she can go in seeking a line of flight leading to a more dialogical space. Period. In the words of Bruno Latour, "We cannot be more relativist than scientists about these [facts] and keep on denying evidence where no one else does. Why? Because the cost of dispute is too high for an average citizen, even if he or she is a historian and sociologist of science" (Latour 1987, 100).[26]

The requirements of symmetry and dialogue with other sciences may help in gaining respectability for multimistura, but they alone are not strong enough to establish facts that are hard enough to penetrate certain domains such as those of the state. To convince scientists and the state to change their minds, the activists who support multimistura would have to mobilize concentrated resources, such as "counterlaboratories" capable of problematizing and provoking fissures in the scientific knowledge established in the laboratories regarding the nutritional qualities of multimistura, thus creating strong "properly scientific" controversies. To continue their struggle in this ontological territory, the activists face a task that may be well beyond what they can do, at least in the short term.

But how, then, can multimistura survive? How does one find lines of flight from the heuristic limits of the modern epistemology derived from Euro-American metaphysics? How can one hold various options of reality instead of merely taking it for granted that all the cards are laid on the table with the inscriptions of Western sciences?

Third Story, Third World
To survive, multimistura counts on the strength and resilience of allies whose practices or knowledges are consistently invalidated by Western ontological political perspectives and downplayed or even relegated to worthlessness by Western sciences and technologies. Multimistura's activists may endure and resist the opposition of scientists and compel them to dialogue because other ontological political perspectives come into play and open lines of flight to other territories where the weapons of the nutritional scientists, their powerful enemies, do not have the same effect. The nutritionists assert that multimistura is not proper food because it does not contain nutrients, but, in spite of this, lay people say that sick children are cured by it. The activists collect declarations from mothers, relatives, friends, and volunteers that contain "anecdotal evidence," evidence that scientists disregard, at least temporarily, as lacking in reliability or explanatory power.[27] Anecdotal evidence reveals a tension between the scientific knowledge of specialists and the knowledge of lay people. It usually refers to specific, local conditions, such as knowledge gained from intimate connections involving a mother and her child that is not—indeed cannot be—taken into account in general evaluations based on "typical" circumstances. According to Moore and Stilgoe, "The key point here is that knowledge of these particular social conditions

[of the anecdotal evidence] needs to come from the people that are more intimately involved" (2009, 657).

The people who adhere to the practice of multimistura choose to live a reality in which the effects of laboratory truths do not have the ontological (mobilizing or demobilizing) force they usually have in collectives whose worlds are configured and stabilized in scientific frames of references. Accounts like the following from the "declaration" (*depoimentos*) section at www.multimistura.org.br are common among multimistura practitioners:

A simple and miraculous mixture that continues to save lives. The multimistura saved Tiago da Silva, who is now 21 years old. Malnourished, he weighed only 900 grams at birth due to lack of care during his mother's (the housewife Maria Aparecida da Silva) pregnancy. The baby spent 28 days in a hospital in Porecatu (North). "Sometimes, owing to his desperate condition, the nuns would not even let me see him. He was at death's door. Nobody said he would recover," recalls his mother. The nuns Maria Aparecida refers to are Ana Maria and Mother Eugênia. When he was six months old, they added multimistura to his milk and food and he gradually began to recover his health. "If it wasn't for the help of God and the Children's Pastoral Service, my son would not have survived," says the mother, proud to have a man in the house today. "Looking at Tiago today you couldn't imagine he almost died." Tiago finished high school and works in the boiler sector of the Porecatu Sugar Mill. Married and the father of a fifteen-month-old daughter, he insists on giving multimistura to his daughter as a food supplement. "I know how important it was for me." After her son had recovered, Maria Aparecida began to help the Pastoral Service and is currently the leader of a sector. "I help children just as I was helped when I needed it."

Edileusa Martins de Oliveira, twenty-four years old, is another young person who recovered. She was also hospitalized with malnutrition at birth, and as her condition was more serious, she stayed for three years in the Nutritional Center. Today, there is always multimistura in Edileusa's house. She has three children—João Lucas, four, Bruna Stefane, two, and Bruno Ariel, who is six months old. It's a twenty-five-year-old story, and together with Marcia Michelim, ... who has been working as a project supporter for twelve years, she goes from house to house, orienting mothers and monitoring their children's development. "It is blessed work that we perform lovingly."

"When they go there, they feel fulfilled, they live a dream," said Edilene Oliveira Ferro referring to the presence of girls and boys in the program.

By Way of Conclusion

As we have seen, Jeca Tatu's story nurtures very little dialogical capacity, for in fact it promotes a monologue in the imperative mode. The two apparently opposing configurations or frames of reference in which Monteiro Lobato depicts the two versions of Jeca Tatu, indolence or industriousness, result from the Western hegemonic ontological political perspective. Only one voice can be heard there, that of the "modern

Euro-American citizen" who can have different interpretations of his supposedly only real world.

The multimistura case, in contrast, nurtures dialogical capacities, for it makes a point of the importance of understanding knowledge within the particular conditions of its use. This point is dear to anthropologists who study and discuss the ways in which local knowledge is discarded by specialists (in this case, the nutritionists) who often have no idea that there are alternative, well-thought-out ways of dealing with the problems of the world (i.e., alternative ontological political perspectives), alternative ways of acting and knowing, or "temporal relations of cosmopolitanism" (Fischer 2009). The anthropology of development, according to Hobart, tells us that "both knowledge and ignorance, rather than describing well-defined states or situations of meaning, constitute peculiarly ideal and atemporal notions that some people attribute to others in particular situations, often with moral connotations. In other words, the proponents of a 'system' try to eliminate other knowledge, portraying those that use it as being not only wrong but also dark and evil" (Hobart 1993, 21).

Anthropologist Pier Vitebsky offers a way of assessing and comparing dialogical capacities of different worlds that I draw upon to claim that in the multimistura case, each story and the world it enacts become more dialogical by incorporating more heterogeneous elements—the body in the first; body and soul in the second; body, soul, and spirit in the third. Respecting the right to be different and accepting the knowledge that emerges from the third story can enable one to break the shackles of Euro-American metaphysics by forgoing a centered reality, instituting multiple realities and laying the ground for a more inclusive dialogue among different realities.

In his own work, Vitebsky contrasts Freudian psychoanalysis with shamanism among the Sora in India, showing how each deals with the unknown through dialogue.[28] According to Vitebsky, Freud's article "Mourning and Melancholy" presents "the most coherent and influential secular theory of the mental processes of loss by death in the industrialized West" (Vitebsky 1993, 102). The ontological political perspective of Freud's therapeutic model and practice is the certainty that the person who died has ceased to exist in a "profound ontological sense." Freud's "reality test" should soon convince anyone that the dead person is no longer a subjective being and that any continuing attempt to interact with the deceased person is thus based on an illusion. If this "verdict of reality" is not accepted, a pathological state of melancholy installs itself, constituting a withdrawal into a hallucinatory psychosis dominated by desire in which the existence of the dead person is psychically (and erroneously) prolonged in excess. The dialogue with the psychoanalyst must, after a certain period of time, help the patient to recognize this error (ibid.).

But the Sora have another ontological political perspective. According to Vitebsky, for them the dead continue to exist fully although they have been qualitatively transformed. Where Freud contrasts normal and pathological states of mind in the grieving person, the Sora contrast benign or aggressive states of mind in the dead person and locate them in various distinct parts of the landscape. The dead reside in these places according to their state of mind at the moment and that of the living person who meets them. The dead person involves himself with the living as they move around the landscape. The shaman provides a channel through which the dead and the living establish a dialogue. These dialogues take place in divinations and cures and funeral rituals. Here the living and the dead explore each other's animi to modify them. To be cured, the living person invites the dead to engage in a dialogue in order to discover how the dead feel about him and why they have attacked him. The living person tries to persuade the dead to adopt a less aggressive state of mind, while the dead, in turn, can persuade the living person to change something in himself (Vitebsky 1993, 103).

Vitebsky asserts that the difference here is not one of observed fact or empirical evidence: "in both cases, one could argue from outside these beliefs that they are mistaken. Freud's 'reality testing' does not actually test reality, but rather, tests propositions against a preconceived notion of reality. Similarly, the entire practice of dialogues with the dead could plausibly (though I think thinly) be interpreted on the assumption that these dialogues are no more than a theatrical stunt" (Vitebsky 1993, 103–104).

The first, second, and third stories of multimistura, respectively, correspond to the three types of articulations between techniques and truths within the "domains of sickness and health, body, mind and soul" configured by Vitebsky (1993, 112). The first story's reality limits the body to the ontological space inhabited by the entities created and stabilized by the ontological political perspectives of Brazilian nutritionists in the laboratories and unsurprisingly corresponds to the materialist techniques (biophysical or biochemical) of psychiatry. The second story's reality understands and promotes the success or failure of multimistura in terms of scientific facts that do or do not stabilize themselves, adopting ontological political perspectives that include the laboratory and society, and corresponds to Freudian psychoanalysis, which expands being by giving the body a mind (affects or soul without spirits). Going yet further, the third story, which enacts the "magic of multimistura," corresponds to the sorcery of the Sora, a ritual that creates an ontological dialogical space inhabited by bodies and affects or souls with spirits, opening a line of flight from the limits imposed on the dialogue in the second story when it encounters the closure of scientific controversies (or faces the black boxes of science).

Thus a hierarchy in the dialogic capacity of each world can be established by recognizing that only from outside can one can argue the incompleteness or relativity (as opposed to the absoluteness) of the realities of such different worlds. By incorporating increasingly heterogeneous elements, the three stories are also increasingly inclusive and therefore increasingly dialogical. For example, as many practitioners in the STS field will acknowledge, someone living in the world of the second story is outside the world of the first story but can entertain a dialogue with it. In the reality enacted by the third story, multimistura's allies are not isolated a priori and despised as equivocating practitioners of a false recipe to cure malnutrition. Rather, multimistura coexists and dialogues with science and may even, by modifying itself and science in a complex dynamics, tame it. In the words of the Banco do Brasil Foundation manager for multimistura, "The popular education movement introduced practices that provide an alternative to the dominant market-oriented and biologistic model, especially from the 1970s. The participation of professionals introduced a culture of establishing relationships with the popular classes, which helped overcome the authoritarian and normalizing tradition of educational practice" (Frota, Albuquerque, and Linard 2007, 248; Dias 2010, 22).[29]

In this formulation, I do not make use of the entities "soul" or "spirit" to compensate for a lack of cultural wealth or power, hegemony, or reason, as is often done, but to configure a growing dialogism between the stories. Whereas Monteiro Lobato's Jeca Tatu's reality fell prey to Western ontological political options and decisions, the multimistura case, more than half a century later, shows a different, more complex relation with the modern Western sciences and technologies. Jeca Tatu's case does not reject questions such as "Which of the two Jeca Tatus represent progress?" In contrast, the multimistura case is conducive to actions that derive from a much broader question, that is, "Which narrative shall we follow?" The multimistura case is able to further reveal the consequences of different replies to Barbosa's questions facing the first Jeca Tatu in 1919: "Is THIS who we are? Is THIS how we are perceived?" (Barbosa 1981, 174). Jeca Tatu's case indicates limited possibilities of change, while multimistura draws on an open universe of possibilities. It is in this sense that Jeca Tatu's case is about different interpretations of the same world, while the multimistura case is about the construction of different worlds.

Several chapters in this book address the relations between worlds constructed by scientific knowledge and other worlds. Ultimately, they render problematic the notions of fact and fiction. In effect, they bring to the fore examples of multiple and heterogeneous relations between *entities* that enjoy scientific credibility and other *entities*. They show that Latin American strategies for taking advantage of scientific and technological knowledges (Western *knowledge entities*) involve all kinds of different uses and heterogeneous relations with them, and sometimes workarounds to enable one to do without them. Here it is worth highlighting the different relations that appear in Hagood's chapter about the bottling of atomic energy and Chan's and Ames's chapters about the OLPC projects in Peru and Paraguay, respectively. These three chapters show a symmetrically inverted relation when compared to the multimistura case: interested actors draw upon possible scientific evidence to stabilize the bottling of atomic energy project and the OLPC project, whereas scientific evidence is used to destabilize the project of multimistura. Hagood's study claims that, given what was then known about atomic energy and the world availability and distribution of scientists, the local development and application of atomic energy might have appeared to make sense in the 1950s and to be an achievable goal. Hagood claims that the heterogeneous relations between knowledges about atomic energy and Argentina in the 1950s allowed for the provisional stabilization of the "bottling of atomic energy," an entity that was configured in a meeting that Perón held with a German scientist. Chan and Ames show that the impetus for continuing the OLPC projects in the lifeworld does not come from their results but rather from, among other translations, the promises envisioned by scientific hypothesis about the learning capacities of children formulated at MIT. It is easy to grasp an inverted relation when compared to the multimistura case: the OLPC and the bottle of atomic energy came to play similar roles in the lifeworld by means of support from scientific hypotheses.

People in the third story of multimistura adopt different ontological political perspectives that go beyond those of the second story to test practices that, to use Callon's term, overflow the frame of reference of a nutritional apparatus separated from the rest of the world. By resisting surrendering to the black boxes of science, they allot respectability and value to popular knowledge. When the black boxes of science are in place, the political ontological options and decisions that have led to them are hard to see. By partially rejecting a preconceived absolute commitment to the normative and prescriptive values of scientific truths enacted by the hard-to-see ontological political options of the nutritionists, people in the third story problematize the closure of controversies and seek to open (without guaranteed success) lines of flight from the more limited and purified scientific paradigms of nutritional processes configured in laboratories, where the relations with the body are seen in terms of formal models, toward broader conceptions of the inclusion of life and the body in their practices and ways of mobilizing providence.

Box 5.1
Examples of the declarations recorded by multimistura activists

> Meeting on November 26, 2006
> Grupo Fraterno Estrela do Oriente—support for Federal District leprosy sufferers
> Declarations of people attending who consumed multimistura for a month (a soupspoon per day)
> (1) Ana Rita Leitão. Previous symptoms: pains in the joints, dizziness and constipation. Results: her condition improved and she lost 3.5 kg (out of a total of 5 kg, including the loss of weight in October).
> ...
> (15) Severino Fernandes da Silva. Previous symptoms: loss of appetite, insomnia, and dormancy in the legs. Results: is eating better and the dormancy has diminished.

Box 5.2
Example of monitoring carried out and evidence collected by multimistura activists

> Monitoring the use of multimistura by people who take part in community therapy at Paranoá Health Center no. 1 (Thursdays from 14:45 to 17:00)
> August 23, 2007—First distribution of 21 packets to 13 women and 1 man, involving 73 people (10 copies of recipes using multimistura were handed out).
> August 30, 2007—Declarations of 6 people (6 were absent and 2 gave no feedback):
>
> - 5-year-old child who was underweight and depressed began to eat the mixture and weight increased from 17.4 kg to 18.2 kg. Mother reduced her consumption of sweets. Grandmother began to eat it at meals: her constipation improved.
> - 2 people haven't observed any change yet.
> - Better appetite, has more energy.
> - Better appetite, runny nose and cough stopped.
> - Boy is putting on weight.
>
> Distribution of 27 packets of multimistura to 19 people, involving 101 people.
> With these two distributions a total of 129 people were involved in the consumption of multimistura.
> [Similar weekly monitoring declarations follow until:]
> November 8, 2007—Declarations of 4 people:
>
> - Intestine before 7/7d now 4/4d. Feels less starved.
> - Is stronger, putting on weight, better appetite. Girls have a better appetite, putting on weight, clear skin.
> - Bellyache, weak legs and diarrhea—everything improved.
> - Swollen belly diminished and daughter's facial blemishes improved.
>
> On this final day, a mother came with her 2-year-and-3-month-old son, who weighed 6.3 kg and was 71 cm high, with a catheter due to malnutrition. (According to Dr. De Lamare's book—boy 2 years 6 months old—minimum weight 11.4 kg and height 86.5 cm.) The same mother has another son who is 4 years old and weighs 10 kg (De Lamare—minimum weight 13.65 kg). She took the 9 remaining multimistura packets for herself and her 2 sons for 3 months (7 people live in her home).

Notes

1. Besides being a very popular folk image and literary character, Jeca Tatu immortalized and was immortalized in films by the actor Amácio Mazzaropi. Even breaking all domestic box office records, Jeca Tatu was hated by the critics. http://www.caiman.de/jecapo.html.

2. I thank Lucimeri Ricas Dias for sharing the research material of her master's thesis about multimistura.

3. My theoretical standpoint here relies on actor-network theory (ANT) approaches developed by Michel Callon (1999), Bruno Latour (1987; 2005; 2013), John Law (1986; 1987; 2004), and Annemarie Mol (1999). ANT grants ontological power to stories.

4. An inscription is a raw image, a hybrid produced in the interface between the narrated world of texts and the world of laboratory instruments. See Latour (1989, 157).

5. Shrimpton (1984); Velho and Velho (2002).

6. Brandão and Brandão (1996); Velho and Velho (2002).

7. Beausset (1992); Velho and Velho (2002).

8. Shrimpton (1984); Velho and Velho (2002).

9. Pastoral da Criança (the Children's Pastoral Service) in 1985 began work on what it called the alternative diet. In the introduction of the book *Alimentação alternativa*, published in 1988, Dr. Clara Takaki Brandão draws attention to the local peculiarities that, according to her, are constituent parts of multimistura, claiming that "only through a combination of the most diverse kind possible—multimistura—can one utilize all the potential nutrition of foods" (Brandão 1988).

10. Vizeu, Feijó, and de Campos (2005, 254).

11. Conselho Federal de Nutricionistas, "Posicionamento do Conselho Federal de nutricionistas quanto à multimistura" (Brasília, February 1996, 5p); available at http://www.cfn.org.br/novosite/conteudo.aspx?IDMenu=61 (accessed September 2008).

12. Torin, H.R., Domene, S.M.A, Fárfan, J.A. P. (1994). Informe técnico: programas emergenciais de combate à fome e o uso de sub-produtos de alimentos. Rev Ciências Médicas—PUCCAMP, Campinas.

13. *Veja*, October 30, 1996, 58.

14. First clause of the National Council of Farm Policy (CONFAZ) in its 108th ordinary meeting, held in Natal, RN, December 13, 2002.

15. Portal Agência Brasil, April 23, 2007, available at http://www.agenciabrasil.gov.br/noticias/2007/04/23/materia.2007-04-23.9418914205/view (accessed September 2008).

16. "Multimistura—posicionamento oficial da Pastoral da Criança" (2007), available at http://compastoraldacrianca.blogspot.com/2007/10/multimistura-posicionamento-pastoral-da.html (accessed August 2011).

17. *Isto É*, September 19, 2007, available at http://www.agenciabrasil.gov.br/noticias/2007/04/23/materia.2007-04-23.9418914205/view (accessed September 2008). By "clandestine" (*clandestina*), Brandão seems to mean that, in spite of continuing as a public employee, no one should know about this continuation. In other words, she was not dismissed but she could not act.

18. Fundação Banco do Brasil, available at http://www.fbb.org.br/portal/pages/publico/expandir.fbb?codConteudoLog=5938 (accessed September 2008).

19. *Isto É*, September 19, 2007, available at http://www.agenciabrasil.gov.br/noticias/2007/04/23/materia.2007-04-23.9418914205/view (accessed September 2008).

20. *Isto É*, September 19, 2007, available at http://www.istoe.com.br/reportagens/2931_A+VITORIA+DOS+ENLATADOS (accessed January 2013).

21. Portal of Health Ministry, October 26, 2007, available at http://www.saudedafamilia.rs.gov.br/v1/clipping/fullnews.php?id=390#topo (accessed September 2008).

22. To complicate matters, twenty years ago, much more than now, many children died of malnutrition and were buried illegally by their parents for not paying the costs of a public notary to register them.

23. *Veja*, October 30, 1996, 58.

24. Today the Children's Pastoral Service alone provides assistance to 1.4 million families in all Brazilian states. Deutche Welle; see http://www.dw.de/sobras-de-alimentos-viram-aliadas-no-combate-%C3%A0-obesidade-infantil-no-brasil/a-16843733 (accessed June 15, 2013).

25. As of March 2011, US $1.00 was equivalent to approximately R $1.70.

26. It is possible to consider this a timid position for a social scientist. In such cases of systematically underfunded science, it is presumably possible for a social scientist to engage more in the politics of funding—even if it is not explicitly worded as ontological politics or politics of construction of scientific facts. Where there is great opposition to "stabilized" scientific facts, there is presumably the possibility of leveraging that into scientific funding. I am grateful to Christina Holmes and David Hess for bringing this point to my attention at the workshop "Studying Science and Technology in Latin America: Beyond Imported Magic" (Indiana University, Bloomington, August 24, 2012).

27. We can relate the idea of anecdotal evidence to Thomas Kuhn's notion of anomaly. Anecdotal evidence has managed to change scientists' attitudes regarding the effects of electromagnetic waves on people in the case of cell phones, but not in the case of triple viral vaccine (against chicken pox, mumps, and German measles). For a detailed account of these cases, see Moore and Stilgoe (2009).

28. "The Sora are a 'tribal' people living historically on the margins between shifting political centers in Central India. They think of themselves as *adivasi* (tribal), but also as 'Hindu,' in conscious opposition to the small enclaves of Christian Soras. Culturally, Soras in the plains are similar to surrounding castes but in the hills they retain a distinctive character." See http://www.everyculture.com/South-Asia/Sora-Orientation.html (accessed January 9, 2012).

29. More than a grain of optimism is also supported by other examples that show that "overcoming the resistance of professional scientists and formal research institutions is a slow and difficult process. Soil conservation techniques known as minimum cultivation, developed and disseminated by farmers in the south of Brazil from 1972, had already been adopted in 10 million hectares of farmland in 1997, when the Embrapa [the government agricultural research institution] validated them as an agricultural procedure" (Fioravanti 2010, 26).

References

Barbosa, R. 1981. [1919]. A questão social e política do Brasil. *Ciência e Trópico* 9 (2):171–178.

Basalla, G. 1988. *The Evolution of Technology*. New York: Cambridge University Press.

Beausset, I. 1992. *Estudio de las bases cientificas para el uso de alimentos alternativos en la nutricion humana. Mimeo*. Brasília: Inan/Unicef.

Bijker, W. E., T. P. Hughes, and T. Pinch. 1987. *The Social Construction of Technological Systems: New Directions in the Sociology and History of Technology*. Cambridge, MA: MIT Press.

Brandão, C. T. 1988. *Alimentação alternativa*. Brasília: Pastoral da Criança.

Brandão, C. T., and R. F. Brandão. 1996. *Alimentação alternativa*. Brasília: INAN/Ministério da Saúde.

Callon, M. 1998. *The Laws of the Markets*. Oxford: Blackwell.

Callon, M. [1986] 1999. Some elements of a sociology of translation: Domestication of the scallops and the fishermen of St. Brieuc Bay. In *The Science Studies Reader*, ed. M. Biagioli, 67–83. New York: Routledge.

Dias, L. R. 2010. A "multimistura" entre conhecimento científico e conhecimento leigo. Master's thesis, Programa de Pós-Graduação em Informática, Universidade Federal do Rio de Janeiro.

Fioravanti, C. H. 2010. Fungos, instituições, máquinas e pessoas em negociação: O percurso do fármaco P-MAPA. PhD diss., Instituto de Geociências, UNICAMP, Campinas.

Fischer, M. M. J. 2009. *Anthropological Futures*. Durham: Duke University Press.

Frota, M., C. Albuquerque, and A. Linard. 2007. Educação popular em saúde no cuidado à criança desnutrida. *Texto Contexto Enferm* 16 (2):246–253.

Haraway, D. J. 1989. *Primate Visions: Gender, Race, and Nature in the World of Modern Science*. New York: Routledge.

Hobart, M., ed. 1993. *An Anthropological Critique of Development: The Growth of Ignorance.* London: Routledge.

Knorr-Cetina, K. 1981. *The Manufacture of Knowledge: An Essay on the Constructivist and Contextual Nature of Science.* New York: Pergamon Press.

Latour, B. 1987. *Science in Action: How to Follow Scientists and Engineers through Society.* Cambridge, MA: Harvard University Press.

Latour, B. 1989. *La science en action.* Paris: Gallimard.

Latour, B. 1993. *We Have Never Been Modern.* Cambridge, MA: Harvard University Press.

Latour, B. 1999. *Pandora's Hope: Essays on the Reality of Science Studies.* Cambridge, MA: Harvard University Press.

Latour, B. 2005. *Reassembling the Social: An Introduction to Actor-Network-Theory.* Oxford: Oxford University Press.

Latour, B. 2013. *An Inquiry into Modes of Existence: An Anthropology of the Moderns.* Cambridge, MA: Harvard University Press.

Latour, B., and S. Woolgar. 1979. *Laboratory Life: The Social Construction of Scientific Facts.* Beverly Hills, CA: Sage Publications.

Law, J. 1986. *Power, Action, and Belief: A New Sociology of Knowledge?* London: Routledge and Kegan Paul.

Law, J. 1987. Technology and heterogeneous engineering: The case of Portuguese expansion. In *The Social Construction of Technological Systems: New Directions in the Sociology and History of Technology*, ed. W. E. Bijker, T. P. Hughes, and T. Pinch, 111–134. Cambridge, MA: MIT Press.

Law, J. 2004. *After Method: Mess in Social Science Research.* London: Routledge.

Lima, N. T., and G. Hochman. 2004. "Pouca saúde e muita saúva": Sanitarismo, interpretações do país e ciências sociais. In *Cuidar, controlar, curar—Ensaios históricos sobre saúde e doença na América Latina e Caribe*, ed. G. Hochman and D. Armus, 493–533. Rio de Janeiro: Editora Fiocruz.

Lynch, M. 1985. *Art and Artifact in Laboratory Science: A Study of Shop Work and Shop Talk in a Research Laboratory.* Boston: Routledge and Kegan Paul.

Mol, A. 1999. Ontological politics: A word and some questions. In *Actor Network Theory and After*, ed. J. Law and J. Hassard, 75–89. Oxford: Wiley-Blackwell.

Mol, A. 2002. *The Body Multiple: Ontology in Medical Practice.* Durham: Duke University Press.

Moore, A., and J. Stilgoe. 2009. Experts and anecdotes: The role of "anecdotal evidence" in public scientific controversies. *Science, Technology & Human Values* 34 (5):654–677.

Rostow, W. W. 1962. *The Stages of Economic Growth.* London: Cambridge University Press.

Shrimpton, R. 1984. *Uma avaliação ex-post do programa de atenção primária em saúde sendo realizada em Santarém*. Pará: Sociedade de Estudios e Aproveitamentos dos Recursos da Amazônia (Seara).

Traweek, S. 1988. *Beamtimes and Lifetimes: The World of High Energy Physicists*. Cambridge, MA: Harvard University Press.

Velho, L., and P. Velho. 2002. A controvérsia sobre o uso de alimentação "alternativa" no combate à subnutrição no Brasil. *Historia, Ciencias, Saude--Manguinhos* 9 (1):125–157.

Verran, H. 2001. *Science and African Logic*. Chicago: University of Chicago Press.

Vitebsky, P. 1993. Is death the same everywhere? Contexts of knowing and doubting. In *An Anthropological Critique of Development: The Growth of Ignorance*, ed. M. Hobart, 100–115. London: Routledge

Vizeu, V. E., M. B. S. Feijó, and R. C. de Campos. 2005. Determinação da composição mineral de diferentes formulações de multimistura. *Ciência e Tecnologia de Alimentos* 25 (2):254–258.

6 Technology in an Expanded Field: A Review of History of Technology Scholarship on Latin America in Selected English-Language Journals

Michael Lemon and Eden Medina

In *One Hundred Years of Solitude,* Gabriel García Márquez uses technology to tell the history of a fictional Latin American town named Macondo.¹ The novel begins with gypsies bringing "great inventions" such as ice, magnets, a telescope, and false teeth to the tiny town. Technological innovations connect the isolated town and the outside world throughout the story, most notably through "the innocent yellow train" that brings a foreign-owned banana company, rational forms of production, armed troops, and later a massacre. García Márquez uses technology as a trope to explore solitude and connection, changing social and economic relationships, ideas of modernity, the legacy of colonialism, the notion of identity, different ways of knowing (as seen by the regular coexistence of science and magic), and the history of inequality and violence in Latin America.²

Magical realism, the literary genre exemplified by García Márquez's work, has been criticized for propagating stereotypes about Latin American life that are distinct from the urban, high-tech, pop-culture and consumer-oriented reality of Latin America today (Fuguet 1997). Yet works of magical realism, such as *One Hundred Years of Solitude,* can also draw our attention to the many ways technology has shaped, and been shaped by, Latin American history. Indeed, García Márquez's focus on and employment of technology in what is perhaps his best-known work points to a rich and fruitful way of studying Latin America. However, technology is rarely positioned as the central object of study in historical research on the region. The converse is also true in the history of technology, which as a field has not given significant attention to Latin America. For example, of the 272 articles published in the first twenty-one years of the premier history of technology journal, *Technology and Culture,* only five—2 percent of all articles published—pay central attention to a Latin American country (Staudenmaier 1985).³

The reasons for this lack of engagement between Latin American studies and the history of technology are outside the scope of this essay.⁴ We wish to focus instead on the benefits of bringing these two literatures together and on how this kind of cross-fertilization can contribute to conversations in the larger field of history. We argue that historians of technology and scholars of Latin America stand to gain important new perspectives and avenues of inquiry by drawing from the methods, questions, and analytical framings that are central to each field. This review points to various ways in which researchers in these fields can find common ground for greater collaboration and engagement with one another.

This essay reflects, in part, a shift within the history of technology, which to date has largely focused on the inventions, innovations, and technological experiences of the United States and Europe. Today, however, the field is moving toward a more geographically inclusive understanding of technological development and change, reflecting a greater use of transnational, international, and global frames of study in the field of history more generally (Turchetti, Herran, and Boudia 2012; van der Vleuten 2008; Roberts 2009; Saraiva 2013; Hecht 2011). As Palmarola and Alonso (this volume) and Mateos and Suárez-Díaz (this volume) demonstrate, national views on science and technology reveal only part of a larger story. Artifacts and ideas regularly cross borders, are modified in their travels, and take on multiple meanings in different contexts.

We further argue that the political, economic, and cultural specificities of Latin America make it an important site for the study of technology. The growing importance of Latin American economies (most notably those of Brazil, Mexico, Peru, and Chile) on the global stage has moved Latin America back onto the radar screen of policy makers and scholars to a degree not seen since the consolidation of the "Washington Consensus" neoliberal economic model in the early 1990s.[5] Moreover, Latin American scholarship is centrally concerned with themes that are frequently absent from existing scholarship in the history of technology, including social justice, indigenous identities and indigenous rights, the causes and effects of economic dependency, neoliberalism, and human rights—themes central to a full understanding of technology and its role in society. In addition, studying technology in a Latin American setting foregrounds international relations, political economy, colonialism, imperialism, hegemony, and the materiality of these relationships as elements of analysis. Although these considerations can (and should) also form part of the history of technology in the United States, the peripheral economic and political status of Latin America provides a different vantage point from which to connect local experiences with technology to global political and economic structures and for developing international and transnational frameworks for studying such history. Engaging the specificities of Latin American experiences with technology can thus add new nuances to our understanding of terms such as *global South*, *postcolonial*, and *developing* and define more precisely the relationships among technology, politics, and power.

Conversely, studying technology can also increase our knowledge of Latin American historical processes more generally. As previous scholarship has shown, examining technology can help us understand the possibilities for human action; the value systems associated with production and consumption practices; the construction of gender, race, ethnicity, and class; the organization of labor and institutions;

and how ideologies and belief systems become part of material life. Technologies also make visible the oft-competing visions of progress and modernity articulated by various sectors of society and different areas of the world. In a broad sense, studying technology allows us to see material aspects of historical change and the social, economic, and political negotiations that have been embedded in the making of the past.

Because Latin America has often been framed as technologically backward or lagging behind more industrialized nations, some may find studying Latin American technology counterintuitive. Yet such beliefs are not grounded in history. In fact, the region has been central to processes of economic globalization since the sixteenth century. The extraction, processing, and transportation of such goods as hides, wool, rubber, sugar, silver, and other minerals spurred the development of technological capabilities in areas such as mining and agriculture. These activities, moreover, helped fuel the industrial revolution on the other side of the Atlantic as well as within the Latin American region (Moya 2011). Cuba, for example, built the seventh railroad in the world in 1837 to help move agricultural products, such as sugar, to export markets (Zanetti and García 1998).

This assumption of Latin America's technological backwardness is precisely what needs to be interrogated, as it marginalizes Latin Americans' experiences with technology and implicitly justifies giving greater attention to more industrialized areas of the world. Giving greater attention to Latin America and other areas of the global South, however, forces investigators to examine how Western ideologies affect the way we study technology and confront how and why we frame certain experiences as the local, the other, or backward. Indeed, technology has occupied a central place in the development discourse that has set Latin American nations on trajectories of progress modeled after those of more industrialized nations, shaped how Latin American peoples viewed their place in the world, and opened spaces for resistance (Escobar 2011).

We have three purposes in this essay. The first of these is to argue that the history of technology in Latin America enriches our understanding of Latin American history and technological change. The second is to review a subset of the published literature on the history of technology in Latin America in English-language journals. Although it is tempting to think that not much scholarship has been produced on this topic, we assert that a literature does exist but can be difficult to identify because it spans disciplines and is often written by people in disciplines other than history. The last is to use this analysis of published literature to identify and discuss new areas for future scholarship.

We conducted this research by reviewing articles published on Latin American technology history in five leading academic journals. This focus on academic journal articles excluded relevant books and monographs not simply to narrow the scope of our study but because we view these journals as key sites of knowledge production in the field.[6] Indeed, these journals both illustrate the state of the field and guide its future direction. Moreover, publishing their research in a given journal forces authors to vest themselves in a particular discipline or topic and thus commit to addressing a specific audience in a way that monographs do not. By focusing on journals, we therefore analyze not only the content of the literature being produced but also what editors and practitioners in the field consider relevant and interesting to others in the community.

Our period of analysis spans 1970 to 2011.[7] All five journals are published in the United States or England, and all are in English (which Cueto, in the foreword to this volume, refers to as the "Latin of our day").[8] Three of these journals—*Hispanic American Historical Review*, *Journal of Latin American Studies*, and *Latin American Research Review*—focus on Latin America and are widely regarded as premier venues for scholarship on Latin American history and Latin American studies. The remaining two journals—*Technology and Culture* and *History and Technology*—are considered top venues for work in the history of technology. We recognize that other journals may also publish research relevant to this study and that limiting our study to journals published in English overlooks a crucial corpus of literature published in Spanish and Portuguese, which, although smaller than the literature in related areas such as the history of science or the history of medicine, is nonetheless important. Thus, we view this essay as a preliminary assessment of the field and hope that this review of the English-language literature will identify both strengths and absences and foster greater conversation among scholars working in different fields and in multiple languages.

Three recurring findings that emerged from this analysis illuminate how scholars have approached technology history in the past and may also guide future research. The first is that the literature on technology in Latin America frequently uses technology to connect the experiences of historical actors to international relations and economic policies, often through discussions of modernization, industrialization, and dependency. The second is that the literature emphasizes the role of labor and focuses attention on workers as agents of historical and technological change. The third and final finding is that the literature is less concerned with moments of technological invention and innovation than with processes of technological adaptation and use.

We conclude by highlighting specific examples of scholarship that bring Latin American history and the history of technology into fruitful conversation with one another and may push both fields in productive directions. These articles address topics of interest shared by both fields, such as materiality, culture, and political and economic structures, and demonstrate how human actions, ideologies, and discourses condition and are conditioned by economic policies, industrialization programs, and the development of technological systems. As *One Hundred Years of Solitude* illustrates, technology can provide a meaningful way to analyze and understand historical change in Latin America. Latin American experiences, in turn, can expand our understanding of technology history and open new areas for study.

Methodology

Although the history of technology and the history of Latin America are both mature fields that have coalesced around identifiable questions, concerns, and debates, such is not the case for historical studies of technology in Latin America. Research related to this topic therefore is spread across multiple disciplines, which makes it challenging to identify. Latin Americanists from such disciplines as history, political science, anthropology, and economics may study topics that relate to technological artifacts and systems, yet most do not consider the history of technology their primary field of inquiry.[9] Many articles cited here do not reference literature from the history of technology, nor are these articles typically identified as histories of technology.

Because this is a nascent field that has yet to produce a clearly defined body of literature, we found it necessary to adopt an expansive definition of both history and technology for this review. We define historical analysis as the use of methods from the humanities and social sciences to study change over time. We did not include articles that described something that happened in the past without employing such methods. To qualify as a history of technology, the article had to use change over time as a central explanatory force in its argument, and thus we omitted from our review several studies that focused heavily on technology but not on change over time (see, for example, Boas 2005; Martz 1994; and Newell 2009). Many studies of technology that did not qualify as historical investigations addressed such themes as policy analysis and contemporary technology development.

Technology is more difficult to define because it encompasses such a broad range of topics and concepts. These may include technological artifacts, bodies of technical knowledge, technological support networks, technical traditions, technological systems, and the technical aspects of scientific work. For the purposes of this study, we adopt a necessarily broad view of technology that includes studies of artifacts, economic modernization and industrialization programs, technocratic management philosophies, instances of technology transfer, and technical education programs.[10]

This broad view of technology forces us to reconsider the relationship between the history of technology and other historical subfields such as labor history. As Hess (2011) has argued, historical scholarship that focuses on class, labor, social movements, and democratic participation has been marginalized within the history of science and science and technology studies. Scranton has similarly observed that "historians of technology have often allowed their fascination with men, ideas, and machines to yield studies of 'objects' shorn of their human and production dimensions" (1988, 731), a point recently echoed by such scholars as Ensmenger (2004). Yet the opposite is true in Latin American studies, where class, labor, and social movements have been central concerns. A significant portion of the literature reviewed here examines the relationship between technologies and workers and how this relationship contributes to class identity, forms of organization, and resistance. Studies of technology in Latin America thus not only broaden the geographic scope of the history of technology, but also may encourage scholars of technology history to give greater weight to people and their labors and diversify the kinds of work they study.

The degree to which the articles included in our review investigate technology differs. Some feature technology as a central explanatory force in their argument, while others bring in technology and technological developments as evidence in the service of a different argument. We therefore developed a taxonomy that distinguishes between studies in which technology is used as a driving explanatory force and studies in which technology is more peripheral. For example, we classified Colloredo-Mansfeld and Antrosio's (2009) study of the economic relations of textile production in two mountain towns in the Ecuadorian Andes as a peripheral study of technology because, although it describes the mechanization of production and the relationship of industrial production to artisanal production, it does so to contextualize a broader argument about why these economies should be studied as a cultural commons rather than an industrial cluster.

Quantitative Analysis

A quantitative analysis of the number of articles produced during the period under study provides a general sense of research activity in this area over time. Figure 6.1 illustrates the total number of articles on technology in Latin America published in each decade by the Latin America journals and by the history of technology journals selected for this study. Figure 6.2 shows the total number of articles on technology in Latin America that each journal published from 1970 to 2011. Since 1970, the five journals under review here published one hundred articles that we consider historical studies of technology in Latin America, seventy-two of which have focused centrally on technology. *Latin American Research Review* has published the greatest number of studies on technology in Latin America, arguably because its interdisciplinary scope and commitment to publishing studies of public policy outcomes has welcomed different perspectives on such themes as industrialization and economic modernization that overlap with studies of technology.

As figure 6.1 shows, the 1980s saw the largest number of articles meeting the criteria for this study published by the five journals during the period, which was twenty-eight. We attribute this large number in part to the decline in popularity of import substitution industrialization (ISI) policies from the late 1960s to the early 1980s and to a desire within the academic community to explain the shortcomings of this model. For example, during the 1980s *Latin American Research Review* published fourteen articles on the history of technology, seven of which focused centrally on technological support networks such as policy initiatives, technocratic management, and science and technology education (Langer 1989; Adler 1988; Frieden 1987; Camp 1985; Bunker

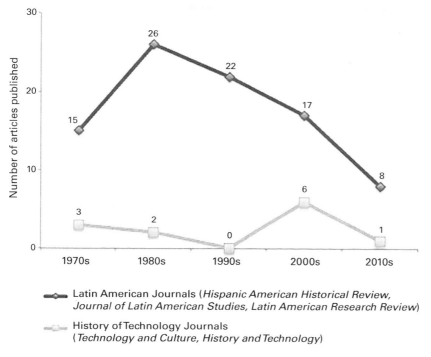

Figure 6.1
Number of Latin America technology articles published by decade (1970–2011) in Latin America journals and history of technology journals. Data for the decade of the 2010s covers only two years (2010 and 2011), so the downward trend at the end of this graph is apparent rather than actual.

1983; Erber 1981; Wionczek 1981). Another six focused on development, economic modernization, and industrialization (Bergad 1989; Caballero 1984; da Rosa 1983; Joseph and Wells 1982; Muñoz 1981; Wallerstein 1980), and one addressed labor and worker movements (Saint 1981). Political and economic transformations shape the kinds of questions scholars ask of the past, and in this case the decline of an industrialization model that had held sway in Latin America for nearly forty years prompted scholars to critically evaluate its role in promoting (or undermining) national technological development.

Figures 6.1 and 6.2 reveal that the two leading journals in the history of technology have produced significantly fewer articles on the history of technology in Latin America than the leading journals on Latin America. Since 1970, *Technology and Culture* has published seven articles on this topic, and *History and Technology* has published five. Yet interest in Latin America has been increasing in these journals. Although *History and Technology* published only one article on Latin America before 2000, it has

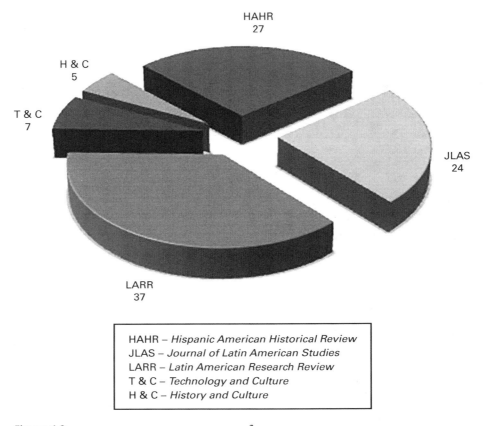

Figure 6.2
Number of technology history articles by journal, 1970–2011.

published four in the past twelve years. Similarly, *Technology and Culture* published four articles between 1970 and 2000 but has published three since then. These are small gains that may stem from a realization by those in the field that the history of technology has disproportionately focused on technology in the United States and Western Europe. These numbers suggest that interest in the history of technology in Latin America is on the rise but that much work on this important topic remains to be done.

Development, Dependency, and Modernization

The articles in the history of technology journals and Latin America journals examined for this analysis diverged sharply with regard to how they treated the relationship of technology to processes of development and economic modernization. The Latin

America journal articles focused predominantly on questions of broad economic structure and dependency and often framed technology as a form of Western economic hegemony in the region. The majority of these studies presented technology as an abstraction seemingly divorced from the complex political, social, and cultural universes in which it circulates, thereby leaving unexamined the social construction of its inner workings and the ways these inner workings shape society. This was especially true of earlier articles in the Latin America journals, which were not centrally concerned with how politics and culture influenced the spread of technology, how technology shaped and was shaped by labor, or how Latin Americans created or adapted technologies to suit local needs and limitations.

In contrast, a substantial portion of the scholarship that appeared in the history of technology journals was dedicated to revealing the inner social and technical workings of an artifact or technical system and how they illustrate the coproduction of technology and society. This type of analysis did not often engage with international political and economic relationships and the inequalities and forms of power they produce, significant themes within scholarship on Latin America more generally. Moreover, the history of technology journals gave greater emphasis to the elites within economic modernization (e.g., inventors, champions of industry, and high-ranking members of the political, economic, and engineering worlds), specific technological artifacts (especially successful ones), and changes in cultural values associated with economic and technological progress. The history of technology journals thus offered a nuanced view of the social production of technology, but one that tended to provide a view from above.

Articles published early in the history of *Technology and Culture* often documented processes of technology transfer from the industrial north to the less industrial south, and some identified Latin American cultures and institutions as resistant to progress and technological transformation (e.g., Gade 1971). Claxton (1973), for instance, claimed that conservative and reactionary governments had slowed technological development in the region during the nineteenth century, despite the work of innovative Latin American scientists and engineers. In contrast, more recent scholarship in *Technology and Culture* and *History and Technology* has been more critical of equating access to technology with development (e.g., Smith 2010; Arnold 2005). For example, Valderrama and colleagues (2009) demonstrated that metrics of societal progress have been rooted in a development discourse largely based on the kinds and quantities of technology a society possesses.

Articles by historians of technology in this set of journals also paid attention to the role of scientific and technical education in promoting national development. Lucena (2007), for instance, examined how national politics in Mexico, including a desire to rid the nation of foreign influence, can promote engineering education programs and consolidate a national professional identity for engineers. In another example, Valderrama and colleagues (2009) illustrated how the creation of new engineering schools in Colombia in the twentieth century fueled the development of a positivist national identity rooted in technological progress. This new corps of engineers viewed itself as literally building the nation by extending roads, bridges, and communications technology to the countryside and were lauded by policy makers and the public for bringing Colombia into the modern age.

The articles in the Latin America journals, in contrast, tended to address technology in its relation to broader processes of economic modernization and industrialization. Of the eighty-eight articles we identified as historical studies of technology in these three journals, sixty-three concentrated on industrialization and economic modernization. These studies analyzed the industrialization of a national economy (e.g., Dosal 1988; Gómez-Galvarriato and Williamson 2009; Beatty 2000; Weiss 1999; Haber 1992; Libby 1991; Caban 1989; Korol and Sabato 1990), agricultural modernization (e.g., Baud and Koonings 1999; Binford 1992; Robles-Ortiz 2009; Chinea 2010; Soluri 2011; Lacoste 2007; Marquardt 2002; Ayala 1995; Caballero 1984; Joseph and Wells 1982; Price 1971), and the development of a national mining industry (e.g., Ortega 1982; Hillman 1984; Valenzuela 1992). The focus on economic industrialization and modernization found in the Latin America journals often includes discussions of authoritarianism, exploitative wage relationships, and the inequalities of capitalism. This differs from the emphasis historians of technology have placed on specific technological artifacts and education programs.[11]

As noted earlier, from the 1960s to the 1980s, studies on the rise and fall of import substitution industrialization occupied a significant space in the scholarly literature on Latin America. A substantial body of work explored the history of the ISI model, its ability to promote the early stages of industrialization and technological development, and its inability to help nations achieve technological independence through an expanded industrial base in sectors such as steel and machinery (e.g., Baer 1972; Dosal 1988; Tafunell 2007; Weiss 1999; Wallerstein 1980; Korol and Sabato 1990). Other articles use the framework of dependency theory to examine how the importation of foreign machinery contributed to the inability of Latin American nations to sustain industrialization processes (O'Brien 1980; Haber 1992; Muñoz 1981; Beatty 2000; Eakin 1986; Véliz 1975). Scholars in the Latin America journals also used education and the national promotion of technical expertise to explain dependent economic relationships, but this line of inquiry was not as proportionally significant as within the history of technology journals. Articles in the Latin America journals also examined the development of national technical education programs (e.g., Murray 1994), the rise of nationalist technocrats (e.g., Langer 1989; Ortiz Rocha de Aragão and Vanneman 1990; Camp 1985; Adler 1988), and collaborations between Western experts and Latin American professionals (e.g., Medina 2006).

Both historians of technology and scholars of Latin America have used technological artifacts as a way to explain historical processes. Generally speaking, this approach was not as pronounced in the Latin America journals as in the history of technology journals, although multiple articles in the Latin America journals employed the spread of Western innovation to study the history of economic modernization and industrialization. Within this group, the railroad attracted by far the most attention as a technological innovation that brought modernity to Latin America's untouched backwaters, which is understandable given that more than half of British and American investment in Latin America was dedicated to the railroad until well after the turn of the twentieth century (Coatsworth 1974, 48). Moreover, railroad history provided researchers with a way to study modernization in Latin America and the encounter between "civilization and barbarism" at the region's frontiers.[12] The role of the railroad in the "civilizing mission" of colonialism has also been a recurring topic among historians of technology who have studied other areas of the global South (Adas 1997).

Scholars have sometimes adopted contrary positions on the role of the railroad in Latin America, such as Mattoon (1977), who argued that the railroad promoted only export growth, and Coatsworth (1974) and Wells (1992), who contended that the railroad also modified land use patterns even in areas not directly linked to export agriculture and drastically remade the countryside. Others have used the railroad to investigate the relationship between technological innovation and economic growth, such as Goodwin (1977), who posited that technological innovation does not necessarily lead to economic growth but that economic growth and technological innovation fuel one another. Other articles analyzed the role of institutions and ideologies in the success or failure of railroad development projects (Ficker 2000; Parlee 1984; Regalsky 1989; Horna 1982).

Agricultural modernization also attracted significant attention in the Latin America journals. This included studies on the bureaucratic rationalization of land management (Ervin 2007; Klubock 2006), the experiences of modernization in specific agricultural enterprises (Henderson 1993; Baud and Koonings 1999), the effect of modernization on peasant life (Joseph and Wells 1982; Caballero 1984), reasons for the success or failure of agricultural modernization programs (Ridings 1978), and the relationship of modern agriculture to the development of pre-Hispanic civilization (Price 1971).

Histories of Labor and Technology

In his review of the first twenty-one years of *Technology and Culture*, John Staudenmaier observed that the journal had largely overlooked the perspective of workers. As noted earlier, history of technology journals have published a wide range of studies on inventors, engineers, managers, and political elites but have not given the same level of attention to other forms of labor, including blue-collar labor, and to such topics as shop floor political organization and worker-management relations. Staudenmaier cited this absence of attention to labor as one of the major reasons for his conclusion that *Technology and Culture* has embodied a Whig perspective that positions technology as both a central agent in historical change and a force for progress in the world. He argued that "if technological changes simply embody autonomous progress it would be consistent to treat workers as marginal and historically insignificant figures. If, on the other hand, technical designs embody cultural, economic, and political values, then the experience of workers who are intimately and sometimes violently affected by production system designs would seem to be an important perspective for interpreting these value-laden design changes" (Staudenmaier 1985, 177). Our review of the literature shows that labor was still largely absent from the articles on Latin America in the two history of technology journals. Of the twelve articles published on technology in Latin America during the period studied, we identified only one that focused on labor (Portuondo 2003), an examination of slavery as an institution on Cuban sugar plantations and the relationship of technology to worker experiences, class identities, organized resistance, and discourses of collective empowerment.

In contrast, critical studies of technology and labor appeared repeatedly in the Latin America journals. For instance, articles on labor (particularly slave labor) in Latin American sugar economies appeared in all three journals. These articles investigated innovations in agricultural processing technology and its relationship to the viability of slave economies (Bergad 1989), the impact of abolition on agricultural modernization programs (Chinea 2010; Ayala 1995), and the Mexican revolution's effect on promoting small-scale sugar production by local artisans (Binford 1992). In one example, a compelling study of labor organization and resistance in the Puerto Rican sugar industry, Martínez Vergne (1988) argued that although postabolition modernization programs in the late nineteenth century displaced former slaves and increased the proletarianization of the rural population, sugar workers were often effective in resisting the new regimes of wage labor by suing landowners for infringements of property rights, moving to new areas beyond the reach of sugar factories, and organizing to purchase land collectively. Examining the advent of wage labor along with changes in sugar production, economic modernization policies, and technology, Martínez Vergne concluded that technological advancement and shifts in patterns of capital investment had a strong impact not only on production processes but also on Puerto Rican social and cultural relations.

Studies of labor and technology in the Latin America journals also presented a range of other topics, exploring how new technologies and rationalized systems of management shaped worker identity, worker organization, and the proletarianization of the workforce (e.g., Klubock 2006; Robles-Ortiz 2009; Soluri 2000; Saint 1981). Other studies documented specific conflicts between workers and management, such as Schmink's (2011) examination of alternative development programs used by Brazilian rubber tappers to form a new collective identity and body of citizen rights that resisted bureaucratic management and deforestation. Rosenthal (1995) studied conflicts between streetcar workers and management to investigate contested meanings of progress in early twentieth-century Uruguay, while Winn (1979) investigated the implementation of the Taylor system in a Chilean textile mill and how antipathy toward the system galvanized worker resistance and generated support for socialist revolution. Marquardt (2002) documented the role of union activities among Costa Rica banana workers in forcing fundamental changes to pesticide application regulations.

These studies of technology and labor expand our understanding of technology in Latin America by broadening our view from a narrow focus on the elite sectors of society to include the struggles among workers, managers, and society at large, the manners in which these struggles shaped the social meaning of technology, and the ways that technology shaped the experience of work. Rosenthal, for example, provides a nuanced analysis of the broader social importance of the development of the streetcar in Montevideo, Uruguay, by focusing less on the design of the cars and the extent of the lines and instead emphasizing how different sectors of society viewed the streetcar and used it as a potent symbol for larger battles about urban geography and social stratification.

Adaptation

Studies of technological adaptation constitute the third topic we identified in the literature. We use the term *adaptation* to refer to the modification of a technology to suit local conditions (which may or may not involve the invention of new technologies) or to the changing meaning of a technology as it travels. Put another way, what we are calling scholarship on technological adaptation studies the ways that local political, economic, and social realities condition and change technological artifacts, discourses, and meanings. This focus on technological adaptation in developing world contexts forces a reconsideration of which aspects of technology are worthy of historical study. Scholars in the history of technology have dedicated much attention to the study of invention (the creation of something new) and innovation (the first use of something new) and less attention to the physical and discursive adaptations that occur as technological artifacts travel between metropole and periphery and among supposedly peripheral nations, regions, and communities. Inquiries into processes of technological adaptation, which appeared in both the history of technology and Latin America journals, challenge these invention- and innovation-centric approaches and reveal critical new perspectives on how technologies, and the meanings that surround them, move and change over time.

Historians of technology have produced a wealth of literature on innovation and invention. Staudenmaier observed, for example, that the study of emerging technology and technological innovation were the two most prominent themes of articles published in *Technology and Culture* between 1959 and 1980 (1985, 35). Yet these have not been prevalent themes in the history of technology scholarship on Latin America. Indeed, of the one hundred articles under review here, only three focused specifically on innovation (Brading and Cross 1972; Platt 2000; Lechtman 1984), all of which examined mining and metallurgy before the twentieth century and challenged common assumptions about European prowess in metallurgy technology and that metallurgy practices traveled unchanged from north to south.

In the two history of technology journals selected for this study, adaptation (or the lack thereof) has been used to explain the success or failure of technology transfer from the developed to the developing world (Gade 1971). Other articles explored how technologies have been adapted for political, cultural, and agricultural projects that are markedly different from their originally intended uses (Smith 2010; Arnold 2005; DeWalt 1978). DeWalt, for example, traced the development of a revised tilling and planting tool in Mexico and argued that this technology is more appropriate for local needs than fuel-intensive methods of cultivation, such as the use of tractors. Smith studied how indigenous activists use video technology to define and disseminate an alternative vision of indigeneity in Mexico for cultural and political projects that run counter to prevalent colonial representations of otherness and the exotic.

Articles in the Latin America journals similarly addressed instances of technological adaptation. Sesto (2002), for instance, described how Argentine cattle ranchers adapted new technologies such as wire fencing, water pumps, waterwheels, plows, and harvesters to local agricultural and livestock needs. Medina (2006) investigated how a group of Chilean and British engineers used mainframe computers and telex machines in innovative ways to create a new system of economic management within the confines of Chile's limited technological resources. Boone (1995) traced contract negotiations between the Canadian Light and Power Company and the municipal council of Rio de Janeiro to examine how the city successfully adapted Canadian technology to meet local development goals of urban renewal and beautification. The focus on adaptation in these articles highlighted an important line of inquiry into how technologies change in physical form as they travel and how the social meanings surrounding a technology shift as these objects cross borders. These studies point to rich new avenues of scholarship that can make significant contributions to our understanding of transnational flows of technology and the discourses that surround them.

Although the history of invention and innovation in Latin America is barely mentioned in the articles we surveyed, a closer examination of technology in Latin America is likely to reveal that Latin American nations are sites not only of adaptation but also of invention and innovation (although such activities are sometimes problematically referred to as "indigenous innovation"). Moreover, studies of Latin America may broaden what constitutes innovation to include instances of adaptation, as well as activities such as recycling, renovation, and maintenance. This, in turn, may allow us to interrogate why certain forms of innovation have greater legitimacy than others, and the historical significance of these value judgments.

Conclusion

This discussion has broadly addressed how scholarship on Latin America can enrich investigations of technology and how the history of technology can broaden studies of Latin American history. Moreover, we have argued that this form of cross-fertilization promises to benefit both fields. The history of technology offers Latin American studies a powerful way to study materiality and connect studies of resistance, discourse, affect, ideology, and social movements to studies of business, policy, economics, culture, and international relations. At the same time, a greater engagement with the history of Latin America will push historians of technology to broaden the geography of the field, make technology history more human (through studies of labor and marginalized groups), and engage with international political and economic configurations and inequalities. This may require giving more attention to international and transnational frames of analysis, or concentrating on how artifacts and practices move within and among Latin American nations and regions. Such actions would not only propel the history of technology and Latin American studies in productive ways but also broaden the potential audience for history of technology scholarship on the Latin American region. A literature on the history of technology in Latin America does exist, but the opportunities and avenues for future investigation are many.

The divergent approaches of historians of technology and historians of Latin America, moreover, offer meaningful ways to challenge, expand, and connect these fields. Giving greater attention to marginalized, impoverished, or oppressed peoples, as did the journals on Latin America, would allow historians of technology to enrich the existing literature with a much-needed view from below. Shifting the emphasis of analysis from invention to adaptation and use may provide ways for historians of technology working in Latin America to broaden the kinds of historical actors they study. Similarly, closer scrutiny on the part of Latin American historians of how technological artifacts are produced, stabilized, and used, as exemplified by historians of technology, may help us better understand the formation of empire, while challenging the notion that technology in Latin America is from elsewhere and broadening our understandings of invention and innovation.

Although our analysis has focused on the different approaches to technology history used by historians of technology and scholars of Latin America, they nonetheless share important commonalities. Both recognize that technological systems cannot be divorced from the cultural, social, and political environments they inhabit and that we shape and are shaped by our technologies and technological systems. These commonalities also suggest avenues for increasing engagement between these historical subfields.

Recent scholarship offers insights into how we might proceed in addressing issues that are of relevance to both the history of technology and the history of Latin America. For example, Sesto's (2002) local study of technological adaptation among Argentine cattle ranchers also connected these specific actions and adaptions to a larger story about Latin American industrialization, showing how these cattle ranching technologies improved the quality of Argentine herds, fueled the growth of the cattle industry in Argentina, and helped create one of the primary engines of the Argentine economy. Valderrama and colleagues' (2009) study of engineering education as a way to understand the development of a national economy and the formation of national identity also challenged the power relationships and structural inequities embodied by the resulting discourse of progress. Medina (2006) used the development of a cybernetic computer system as a way to read the limitations and contradictions of Chile's socialist revolution, calling attention to the manner in which Latin American political projects shape the design of technological systems and how the study of technological systems can help us understand important moments in Latin American history. In a final example, Smith's (2010) analysis of indigenous video production in Oaxaca, Mexico, explored how indigenous activists adapted a technology to fit their specific cultural and political projects, thereby offering historians of technology a model for moving beyond the study of technological elites and early adapters to the study of subaltern and indigenous subjects, and offering Latin Americanists a possible framework for tying isolated moments of discourse analysis, identity politics, and social movements to global transfers of material goods and technical expertise.

Now is a timely moment to expand the field of technology history. Economic growth in Latin America in the twenty-first century has made the region increasingly relevant to international public policy, economic agreements, and scholarly debates. Technology has also become increasingly central to how we understand twenty-first-century life, as seen in the increasing prevalence of cell phones and social media, the looming threat of an energy crisis, and the increasing dependence on global supply chains to produce an abundance of consumer goods. This in turn raises questions about how technology in different contexts contributes to or undermines social justice, political freedom, environmental sustainability, and economic growth. Across the academic community, topics such as globalization, transnationalism, and postcolonialism are now receiving greater attention from humanists and social scientists in a number of fields, including the history of science and technology. Within these frameworks, Latin America is an essential site for historical analysis. The heightened interest in technology and global connectedness further makes the historical study of technology in Latin America an activity of increasing importance.

Acknowledgments

We wish to thank Lessie Jo Frazier, Ilana Gershon, Emily Maguire, Nathan Ensmenger, Ivan da Costa Marques, Hugo Palmarola, Jonathan Hagood, Anita Chan, Morgan Ames, and Eve Buckley for commenting on a draft version of this essay. A special thanks to Will Odom for his work compiling the initial version of this bibliography.

Notes

1. It is widely believed that García Márquez based Macondo on his childhood hometown of Aracataca, Colombia, although he does not give the specific location of Macondo, allowing the fictional town to represent Latin American life anywhere.

2. We cite *One Hundred Years of Solitude* as an example of how Latin American literature has used technology to address themes of interest to historians, not to argue that all Latin American experiences are defined by magic realism or that Macondo provides a useful stereotype of Latin American life. Other contemporary examples of technology in Latin American literature are discussed in Brown (2007). For a critique of technology and magic realism, see the discussion of the McOndo movement in O'Bryen (2011).

3. The Society for the History of Technology (SHOT) was formed in 1958, and its journal, *Technology and Culture*, began publication in 1959. John Staudenmaier writes that around 1963, "both society and journal began to show evidence of emerging from their early frailty into a durable identity" (1985, 2).

4. The historian Carolyn de la Peña provides one possible answer, arguing that technology does not often appear in American studies literature because of the "antipathy of the new left to technology and the propensity, particularly among early technology scholars, to construct uncritically white narratives of nation and progress" (de la Peña 2006, 920). A similar explanation is offered by Philip Scranton in his discussion of why labor historians and historians of technology have not been more engaged with one another (1988). Although we cannot judge the collective mindset of those in Latin American studies, the literature in this field frequently critiques technology as a tool of economic, social, and cultural imperialism, as the term *antipathy* might suggest. This connection between technology and ideas of progress that come from elsewhere may have lessened technology's appeal as a central focus of historical analysis for some, just as it has provided a locus of analysis for others.

5. The term *Washington Consensus* refers to the neoliberal economic model that achieved hegemonic status throughout Latin America with the fall of military dictatorships and the spread of democratization beginning in the late 1980s and 1990s. This economic structure, promoted by the United States and the International Monetary Fund, championed the privatization of government services and state-owned enterprises, trade liberalization, the deregulation of the financial sector, and strict fiscal discipline.

6. Judging from the book reviews that appeared in the journals we studied, relevant books on the history of technology in Latin America were published during the period under study. For example, *Technology and Culture* reviewed books on such topics as technology transfer and diffusion in Latin America, the Bolivian and Mexican mining industry, attempts by scientists to modernize Guatemala in the eighteenth and nineteenth centuries, the evolution of sugar production technology in the former Spanish empire, the philosophy of technology in Spanish-speaking countries, Aztec cosmology, Incan architecture, and the history of Mexican radio.

7. The decision to start our analysis in 1970 was based on the noticeable absence of earlier articles on technology history in Latin America. For example, *Technology and Culture* published only one article on Latin America before 1970; see Drucker (1963). (This article accounts for the difference in Staudenamier's count of five articles on Latin America before 1980 and our count of four.) In the other journals, only one article on technology history in Latin America (Pearse 1966) appeared before 1970. The journal *History and Technology* was founded in 1984, fourteen years after our analysis begins.

8. One article (Lacoste 2007) was published in Spanish by *Latin American Research Review*.

9. As a result, some authors mentioned here might be surprised to see their work described as belonging to the history of technology.

10. Although we might have broadened our definition of technology to make it synonymous with "things," a move Edgerton persuasively argues will produce more inclusive histories (2007), this is not how the term is currently understood or employed in either academic community and deemphasizes aspects important to our analysis, such as practices and forms of expertise. We do, however, agree that broadening what is considered technology to include the things used in everyday life, and activities such as use and maintenance, will lead to analyses that give greater attention to women, members of poor and marginalized groups, and those living in less industrialized nations.

11. Of the twelve articles identified in *Technology and Culture* and *History and Technology*, three addressed the relationship between national identity and engineering education (Valderrama et al. 2009; Lucena 2007; Claxton 1973) and six focused on a particular technological artifact such as the telegraph (Britton 2007), the horizontal waterwheel (Gade 1971), or a tube-fed planting device (DeWalt 1978).

12. "Civilization and barbarism" is a phrase made famous by nineteenth-century Argentine scholar-turned-president Domingo Faustino Sarmiento in describing the clash between Sarmiento's ideal of liberal, European-inspired progress and what he saw as his conservative, provincial, reactionary enemies.

References

Adas, Michael. 1997. A field matures: Technology, science, and western colonialism. *Technology and Culture* 48 (2):478–487.

Adler, Emanuel. 1988. State institutions, ideology, and autonomous technological development: Computers and nuclear energy in Argentina and Brazil. *Latin American Research Review* 23 (2):59–90.

Arnold, David. 2005. Europe, technology, and colonialism in the twentieth century. *History and Technology* 21 (1):85–106.

Ayala, César J. 1995. Social and economic aspects of sugar production in Cuba, 1880–1930. *Latin American Research Review* 30 (1):95–124.

Baer, Werner. 1972. Import substitution and industrialization in Latin America: Experiences and interpretations. *Latin American Research Review* 7 (1):95–122.

Baud, Michiel, and Kees Koonings. 1999. A lavoura dos pobres: Tobacco farming and the development of commercial agriculture in Bahia, 1870–1930. *Journal of Latin American Studies* 31 (2):287–329.

Beatty, Edward. 2000. The impact of foreign trade on the Mexican economy: Terms of trade and the rise of industry, 1880–1923. *Journal of Latin American Studies* 32 (2):399–433.

Bergad, Laird W. 1989. The economic viability of sugar production based on slave labor in Cuba, 1859–1878. *Latin American Research Review* 24 (1):95–113.

Binford, Leigh. 1992. Peasants and petty capitalists in southern Oaxacan sugar cane production and processing, 1930–1980. *Journal of Latin American Studies* 24 (1):33–55.

Boas, Taylor C. 2005. Television and neopopulism in Latin America: Media effects in Brazil and Peru. *Latin American Research Review* 40 (2):27–49.

Boone, Christopher G. 1995. Streetcars and politics in Rio de Janeiro: Private enterprise versus municipal government in the provision of mass transit, 1903–1920. *Journal of Latin American Studies* 27 (2):343–365.

Brading, D. A., and Harry E. Cross. 1972. Colonial silver mining: Mexico and Peru. *Hispanic American Historical Review* 52 (4):545–579.

Britton, John A. 2007. "The confusion provoked by instantaneous discussion": The new international communications network and the Chilean crisis of 1891–1892 in the United States. *Technology and Culture* 48 (4):729–757.

Brown, J. Andrew. 2007. Tecno-escritura: literatura y tecnología en América Latina. *Revista Iberoamericana* 73 (221):735–741.

Bunker, Stephen G. 1983. Policy implementation in an authoritarian state: A case from Brazil. *Latin American Research Review* 18 (1):33–58.

Caballero, José María. 1984. Agriculture and the peasantry under industrialization pressures: Lessons from the Peruvian experience. *Latin American Research Review* 19 (2):3–41.

Caban, Pedro. 1989. Industrial transformation and labour relations in Puerto Rico: From "Operation Bootstrap" to the 1970s. *Journal of Latin American Studies* 21 (3):559–591.

Camp, Roderic A. 1985. The political technocrat in Mexico and the survival of the political system. *Latin American Research Review* 20 (1):97–118.

Chinea, Jorge L. 2010. Confronting the crisis of the slave-based plantation system in Puerto Rico: Bureaucratic proposals for agricultural modernisation, diversification and free labour, c. 1846–1852. *Journal of Latin American Studies* 42 (1):121–154.

Claxton, Robert H. 1973. Miguel Rivera Maestre: Guatemalan scientist-engineer. *Technology and Culture* 14 (3):384–403.

Coatsworth, John. 1974. Railroads, landholding, and agrarian protest in the early Porfiriato. *Hispanic American Historical Review* 54 (1):48–71.

Colloredo-Mansfeld, Rudi, and Jason Antrosio. 2009. Economic clusters or cultural commons? The limits of competition-driven development in the Ecuadorian Andes. *Latin American Research Review* 44 (1):132–157.

da Rosa, J. Eliseo. 1983. Economics, politics, and hydroelectric power: The Paraná River basin. *Latin American Research Review* 18 (3):77–107.

de la Peña, Carolyn. 2006. "Slow and low progress," or why American studies should do technology. *American Quarterly* 58 (3):915–941.

DeWalt, Billie R. 1978. Appropriate technology in rural Mexico: Antecedents and consequences of an indigenous peasant innovation. *Technology and Culture* 19 (1):32–52.

Dosal, Paul J. 1988. The political economy of Guatemalan industrialization, 1871–1948: The career of Carlos F. Novella. *Hispanic American Historical Review* 68 (2):321–358.

Drucker, Peter F. 1963. Modern technology and ancient jobs. *Technology and Culture* 4 (3):277–281.

Eakin, Marshall G. 1986. Business imperialism and British enterprise in Brazil: The St. John d'el Rey Mining Company, Limited, 1830–1960. *Hispanic American Historical Review* 66 (4):697–741.

Edgerton, David. 2007. *The Shock of the Old: Technology and Global History since 1900*. Oxford: Oxford University Press.

Ensmenger, Nathan. 2004. Power to the people: Toward a social history of Computing. *IEEE Annals of the History of Computing* 26 (1):94–96.

Erber, Fabio S. 1981. Science and technology policy in Brazil: A review of the literature. *Latin American Research Review* 16 (1):3–56.

Ervin, Michael A. 2007. The 1930 agrarian census in Mexico: Agronomists, middle politics, and the negotiation of data collection. *Hispanic American Historical Review* 87 (3):537–570.

Escobar, Arturo. 2011. *Encountering Development: The Making and Unmaking of the Third World*. Princeton: Princeton University Press.

Ficker, Sandra Kuntz. 2000. Economic backwardness and firm strategy: An American railroad corporation in nineteenth-century Mexico. *Hispanic American Historical Review* 80 (2):267–298.

Frieden, Jeffry A. 1987. The Brazilian borrowing experience: From miracle to debacle and back. *Latin American Research Review* 22 (1):95–131.

Fuguet, Alberto. 1997. I am not a magic realist. *Salon*, June 11. Available at http://www.salon.com/1997/06/11/magicalintro/ (accessed March 18, 2014).

Gade, Daniel W. 1971. Grist milling with the horizontal waterwheel in the central Andes. *Technology and Culture* 12 (1):43–51.

Gómez-Galvarriato, Aurora, and Jeffrey G. Williamson. 2009. Was it prices, productivity or policy? Latin American industrialisation after 1870. *Journal of Latin American Studies* 41 (4):663–694.

Goodwin, Paul B., Jr. 1977. The central Argentine railway and the economic development of Argentina, 1854–1881. *Hispanic American Historical Review* 57 (4):613–632.

Haber, Stephen. 1992. Assessing the obstacles to industrialization: The Mexican economy, 1830–1940. *Journal of Latin American Studies* 24 (1):1–32.

Hecht, Gabrielle. 2011. *Entangled Geographies: Empire and Technopolitics in the Global Cold War.* Cambridge, MA: MIT Press.

Henderson, Peter V. N. 1993. Modernization and change in Mexico: La Zacualpa rubber plantation, 1890–1920. *Hispanic American Historical Review* 73 (2):235–260.

Hess, David J. 2011. Bourdieu and science and technology studies: Toward a reflexive sociology. *Minerva* 49:333–348.

Hillman, John. 1984. The emergence of the tin industry in Bolivia. *Journal of Latin American Studies* 16 (2):403–437.

Horna, Hernan. 1982. Transportation modernization and entrepreneurship in nineteenth-century Colombia. *Journal of Latin American Studies* 14 (1):33–53.

Joseph, Gilbert M., and Allen Wells. 1982. Corporate control of a monocrop economy: International Harvester and Yucatán's henequen industry during the Porfiriato. *Latin American Research Review* 17 (1):69–99.

Klubock, Thomas Miller. 2006. The politics of forests and forestry on Chile's southern frontier, 1880s–1940s. *Hispanic American Historical Review* 86 (3):535–570.

Korol, Juan Carlos, and Hilda Sabato. 1990. Incomplete industrialization: An Argentine obsession. *Latin American Research Review* 25 (1):7–30.

Lacoste, Pablo. 2007. Complejidad de la industria vitivinícola colonial: Crianza biológica de vino (Reino de Chile siglo XVIII). *Latin American Research Review* 42 (2):154–168.

Langer, Erick D. 1989. Generations of scientists and engineers: Origins of the computer industry in Brazil. *Latin American Research Review* 24 (2):95–111.

Lechtman, Heather. 1984. Andean value systems and the development of prehistoric metallurgy. *Technology and Culture* 25 (1):1–36.

Libby, Douglas C. 1991. Proto-industrialisation in a slave society: The case of Minas Gerais. *Journal of Latin American Studies* 23 (1):1–35.

Lucena, Juan C. 2007. De criollos a Mexicanos: Engineers' identity and the construction of Mexico. *History and Technology* 23 (3):275–288.

Marquardt, Steve. 2002. Pesticides, parakeets, and unions in the Costa Rican banana industry, 1938–1962. *Latin American Research Review* 37 (2):3–36.

Martínez Vergne, Teresita. 1988. New patterns for Puerto Rico's sugar workers: Abolition and centralization at San Vicente, 1873–92. *Hispanic American Historical Review* 68 (1):45–74.

Martz, John D. 1994. Technological elites and political parties: The Venezuelan professional community. *Latin American Research Review* 29 (1):7–27.

Mattoon, Robert H., Jr. 1977. Railroads, coffee, and the growth of big business in São Paulo, Brazil. *Hispanic American Historical Review* 57 (2):273–295.

Medina, Eden. 2006. Designing freedom, regulating a nation: Socialist cybernetics in Allende's Chile. *Journal of Latin American Studies* 38 (3):571–606.

Moya, José C. 2011. Introduction: Latin America—the limitations and meaning of a historical category. In *The Oxford Handbook of Latin American History*, ed. José C. Moya 1–24. Oxford: Oxford University Press.

Muñoz, Heraldo. 1981. The strategic dependency of the centers and the economic importance of the Latin American periphery. *Latin American Research Review* 16 (3):3–29.

Murray, Pamela. 1994. Engineering development: Colombia's National School of Mines, 1887–1930. *Hispanic American Historical Review* 74 (1):63–82.

Newell, Peter. 2009. Bio-hegemony: The political economy of agricultural biotechnology in Argentina. *Journal of Latin American Studies* 41 (1):27–57.

O'Brien, Thomas F. 1980. The Antofagasta Company: A case study of peripheral capitalism. *Hispanic American Historical Review* 60 (1):1–31.

O'Bryen, Rory. 2011. McOndo, magic neoliberalism and Latin American identity. *Bulletin of Latin American Research* 30 (1):158–174.

Ortega, Luis. 1982. The first four decades of the Chilean coal mining industry, 1840–1879. *Journal of Latin American Studies* 14 (1):1–32.

Ortiz Rocha de Aragão, Paulo, and Reeve Vanneman. 1990. Technology transfers and managerial-professional employment: Brazilian manufacturing, 1960–1975. *Latin American Research Review* 25 (1):87–101.

Parlee, Lorena M. 1984. The impact of United States railroad unions on organized labor and government policy in Mexico (1880–1911). *Hispanic American Historical Review* 64 (3):443–475.

Pearse, A. 1966. Agrarian change trends in Latin America. *Latin American Research Review* 1 (3):45–69.

Platt, Tristan. 2000. The alchemy of modernity: Alonso Barba's copper cauldrons and the independence of Bolivian metallurgy (1790–1890). *Journal of Latin American Studies* 32 (1):1–54.

Portuondo, María M. 2003. Plantation factories: Science and technology in late-eighteenth-century Cuba. *Technology and Culture* 44 (2):231–257.

Price, Barbara J. 1971. Prehispanic irrigation agriculture in nuclear America. *Latin American Research Review* 6 (3):3–60.

Regalsky, Andrés M. 1989. Foreign capital, local interests and railway development in Argentina: French investments in railways, 1900–1914. *Journal of Latin American Studies* 21 (3):425–452.

Ridings, Eugene W. 1978. Class sector unity in an export economy: The case of nineteenth-century Brazil. *Hispanic American Historical Review* 58 (3):432–450.

Roberts, Lissa. 2009. Situating science in a global history: Local exchanges and networks of circulation. *Itinerario* 33 (1):9–30.

Robles-Ortiz, Claudio. 2009. Agrarian capitalism and rural labour: The hacienda system in central Chile, 1870–1920. *Journal of Latin American Studies* 41 (3):493–526.

Rosenthal, Anton. 1995. The arrival of the electric streetcar and the conflict over progress in early twentieth-century Montevideo. *Journal of Latin American Studies* 27 (2):319–341.

Saint, William S. 1981. The wages of modernization: A review of the literature on temporary labor arrangements in Brazilian agriculture. *Latin American Research Review* 16 (3):91–110.

Saraiva, Tiago. 2013. The history of cybernetics in McOndo. *History and Technology* 28 (4):423–430.

Schmink, Marianne. 2011. Forest citizens: Changing life conditions and social identities in the land of the rubber tappers. *Latin American Research Review* 46 (3):141–158.

Scranton, Philip. 1988. None-too-porous boundaries: Labor history and the history of technology. *Technology and Culture* 29 (4):722–743.

Sesto, Carmen. 2002. The vanguard landowners of Buenos Aires: A new production model, 1856–1900. *Hispanic American Historical Review* 82 (4):719–754.

Smith, Laurel C. 2010. Locating post-colonial technoscience: Through the lens of indigenous video. *History and Technology* 26 (3):251–280.

Soluri, John. 2000. People, plants, and pathogens: The eco-social dynamics of export banana production in Honduras, 1875–1950. *Hispanic American Historical Review* 80 (3):463–501.

Soluri, John. 2011. Something fishy: Chile's blue revolution, commodity diseases, and the problem of sustainability. *Latin American Research Review* 46 (3):55–81.

Staudenmaier, John M. 1985. *Technology's Storytellers: Reweaving the Human Fabric*. Cambridge, MA: MIT Press.

Tafunell, Xavier. 2007. On the origins of ISI: The Latin American cement industry, 1900–30. *Journal of Latin American Studies* 39 (2):299–328.

Turchetti, Simone, Néstor Herran, and Soraya Boudia. 2012. Introduction: Have we ever been "transnational"? Towards a history of science across and beyond borders. *British Journal for the History of Science*, first view article (August): 1–18.

Valderrama, Andrés, Juan Camargo, Idelman Mejía, Antonio Mejía, Ernesto Lleras, and Antonio García. 2009. Engineering education and the identities of engineers in Colombia, 1887–1972. *Technology and Culture* 50 (4):811–838.

Valenzuela, Luis. 1992. The Chilean copper smelting industry in the mid-nineteenth century: Phases of expansion and stagnation, 1834–58. *Journal of Latin American Studies* 24 (3):507–550.

van der Vleuten, Erik. 2008. Toward a transnational history of technology: Meanings, promises, pitfalls. *Technology and Culture* 49 (4):974–994.

Véliz, Claudio. 1975. Egaña, Lambert, and the Chilean Mining Associations of 1825. *Hispanic American Historical Review* 55 (4):637–663.

Wallerstein, Michael. 1980. The collapse of democracy in Brazil: Its economic determinants. *Latin American Research Review* 15 (3):3–40.

Weiss, John. 1999. Trade reform and manufacturing performance in Mexico: From import substitution to dramatic export growth. *Journal of Latin American Studies* 31 (1):151–166.

Wells, Allen. 1992. All in the family: Railroads and henequen monoculture in Porfirian Yucatán. *Hispanic American Historical Review* 72 (2):159–209.

Winn, Peter. 1979. Oral history and the factory study: New approaches to labor history. *Latin American Research Review* 14 (2):130–140.

Wionczek, Miguel S. 1981. On the viability of a policy for science and technology in Mexico. *Latin American Research Review* 16 (1):57–78.

Zanetti, Oscar, and Alejandro García. 1998. *Sugar and Railroads: A Cuban History, 1837–1959*. Chapel Hill: University of North Carolina Press.

Further Reading

To assist readers, this list contains articles from the five journals that we classified as histories of technology in Latin America but that were not cited in this review article. Please note that technology may occupy a central or a peripheral role in these articles. Many are written by nonhistorians, and historical analysis may constitute only part of the text. However, all these articles use change over time as a central explanatory force in their arguments.

Alcañiz, Isabella. 2010. Bureaucratic networks and government spending. *Latin American Research Review* 45 (1):148–172.

Bauer, Arnold J. 1990. Industry and the missing bourgeoisie: Consumption and development in Chile, 1850–1950. *Hispanic American Historical Review* 70 (2):227–253.

Beyersdorff, Margot. 2007. Covering the earth: Mapping the walkabout in Andean Pueblos de Indios. *Latin American Research Review* 42 (3):129–160.

Deustua, José. 1994. Routes, roads, and silver trade in Cerro de Pasco, 1820–1860: The internal market in nineteenth-century Peru. *Hispanic American Historical Review* 74 (1):1–31.

Downes, Richard. 1992. Autos over rails: How U.S. business supplanted the British in Brazil, 1910–28. *Journal of Latin American Studies* 24 (3):551–583.

Furtado, Andre. 1986. Long-term scenarios for Latin America and the new technological revolution. *History and Technology* 3 (1):1–23.

García Heras, Raúl. 1987. Hostage private companies under restraint: British railways and transport coordination in Argentina during the 1930s. *Journal of Latin American Studies* 19 (1):41–67.

García Heras, Raúl. 1994. State intervention in urban passenger transportation: The Transport Corporation of Buenos Aires, 1939–1962. *Hispanic American Historical Review* 74 (1):83–110.

Gouvea Neto, Raúl de. 1991. How Brazil competes in the global defense industry. *Latin American Research Review* 26 (3):83–107.

Green, Cecilia. 1998. The Asian connection: The U.S.-Caribbean apparel circuit and a new model of industrial relations. *Latin American Research Review* 33 (3):7–47.

Hilton, Stanley E. 1982. The armed forces and industrialists in modern Brazil: The drive for military autonomy (1889–1954). *Hispanic American Historical Review* 62 (4):629–673.

Kropf, Simone Petraglia, and Gilberto Hochman. 2011. From the beginnings: Debates on the history of science in Brazil. *Hispanic American Historical Review* 91 (3):391–408.

Libby, Douglas. 1997. Reconsidering textile production in late colonial Brazil: New evidence from Minas Gerais. *Latin American Research Review* 32 (1):88–108.

Lindo-Fuentes, Héctor. 2009. Educational television in El Salvador and modernisation theory. *Journal of Latin American Studies* 41 (4):757–792.

Mendonça de Barros, José Roberto, and Douglas H. Graham. 1978. The Brazilian economic miracle revisited: Private and public sector initiative in a market economy. *Latin American Research Review* 13 (2):5–38.

Morse, Richard M. 1972. A prolegomenon to Latin American urban history. *Hispanic American Historical Review* 52 (3):359–394.

O'Donnell, Guillermo. 1978. Reflections on the patterns of change in the bureaucratic-authoritarian state. *Latin American Research Review* 13 (1):3–38.

Pineda, Yovanna. 2006. Sources of finance and reputation: Merchant finance groups in Argentine industrialization, 1890–1930. *Latin American Research Review* 41 (2):3–30.

Radcliffe, Sarah A. 2010. Re-mapping the nation: Cartography, geographical knowledge and Ecuadorean multiculturalism. *Journal of Latin American Studies* 42 (2):293–323.

Ringrose, David R. 1970. Carting in the Hispanic world: An example of divergent development. *Hispanic American Historical Review* 50 (1):30–51.

Sanders, James E. 2011. The vanguard of the Atlantic world: Contesting modernity in nineteenth-century Latin America. *Latin American Research Review* 46 (2):104–127.

Santiago, Myrna. 2011. The Huasteca rain forest: An environmental history. *Latin American Research Review* 46 (3):32–54.

Stevis, Dimitris, and Stephen P. Mumme. 1991. Nuclear power, technological autonomy, and the state in Mexico. *Latin American Research Review* 26 (3):55–82.

Street, James H. 1977. The internal frontier and technological progress in Latin America. *Latin American Research Review* 12 (3):25–56.

Super, John C. 1976. Querétaro *obrajes*: Industry and society in provincial Mexico, 1600–1810. *Hispanic American Historical Review* 56 (2):197–216.

Tone, John Lawrence. 2002. How the mosquito (man) liberated Cuba. *History and Technology* 18 (4):277–308.

Part II Local and Global Networks of Innovation

7 South Atlantic Crossings: Fingerprints, Science, and the State in Turn-of-the-Twentieth-Century Argentina

Julia Rodriguez

In the course of one of the most infamous murder cases in late nineteenth-century Argentina, prosecutors obtained in 1892 the world's first criminal conviction based on fingerprint evidence. Immersed in the ghoulish facts of the case, in which two small children were stabbed to death in their beds, the coastal villagers of Necochea in Buenos Aires province hardly noted this high-water mark of transatlantic science. But it was from here that the first practical applications of fingerprinting burst forth, a vital eddy in the thick currents of people, ideas, and technologies surging across the Atlantic at the turn of the nineteenth century. The case had been initially vexing: no one had seen the crime and interrogations had yielded contradictory evidence. Amid the gore, however, was a single bloody fingerprint left on a doorjamb. How, short of finding blood on the suspect, could a match be proved? Several weeks into the trial, the investigating officer, Eduardo M. Alvarez, shocked observers with a novel brand of evidence, a method of linking finger marks to police records of known or suspected criminals. He demonstrated a match between the bloody mark and the prints of the children's mother, Francesca Rojas, who promptly confessed to the crime.

Just months before, a local police scientist named Juan Vucetich, who had recently emigrated from Croatia, had developed a new classification and filing system for fingerprint records that made the match possible. Confronted in his daily work with rising numbers of crimes as the Argentine population grew increasingly heterogeneous, urban, and mobile, Vucetich effectively focused his attention on the means of comparing prints from a crime scene with ever expanding individual identification records. The genius of his system of *dactyloscopy*, as he called it, was not its accuracy (detailed measurements of finger marks were by then common), but the efficiency with which his classification system could be married to an emerging bureaucratic archive of individual fingerprints. The ability to classify and retrieve large, if not unlimited, numbers of record cards had many potential applications beyond crime solving. Civil uses, he argued, ranged from general population records to prostitutes' registers to immigrant tracking systems. As news spread through legal and political circles of the successful use of fingerprint evidence to solve the Necochea child murders—a full ten years before such evidence was used in criminal trials in London and Paris—Vucetich was hailed first locally and then abroad. Argentina's scientific community, self-consciously peripheral, basked in its newfound prestige as a cradle of advanced criminology.

No one knew better than the Argentines that the development, implementation, and dissemination of fingerprint technology was a distinctively transatlantic scientific event. Historians of the United States and modern Europe have recently documented the significance of the linkages and cross-fertilizations of ideas across the northern Atlantic that shaped social policy, legislation, and state formation in the nineteenth and twentieth centuries. They revealed how ideas and their applications evolved not in sterile theoretical laboratories, but by virtue of their vigorous empiricism and through their interaction with diverse, fertile, local contexts (Rodgers 1998; Kloppenberg 1986; Gläser and Wellenreuther 2002; see also chapters in this volume by Cukierman, Palmarola and Alonso, Hagood, and Mateos and Suárez-Díaz). In the process, these ideas were transformed more often than not as they were carried back and forth between the Americas and Europe. We know much less, however, about south Atlantic crossings and the science of Latin America (Gerbi 1973; Arciniegas 1986; Arciniegas 2001; Cañizares-Esguerra 2001). Shrouded in assumptions about their remoteness, poverty, and Catholic traditionalism, the mere existence of world-class science in Latin American nations had to be "discovered" (Cañizares-Esguerra 2001; Cueto 1989; Cueto 1995).

Among the many untold stories of exchange between Europe and South America, I offer here one episode that not only belies common myths of a "peripheral science" that is purely imitative and underdeveloped, but also extends southward the important story of collaboration between scientists and modernizing statebuilders. Argentina's contribution to modern fingerprint science does not fit our traditional models of scientific progress, which sees a "diffusion" of ideas largely from center to periphery. Under that model, the precocious development of fingerprinting in Argentina might be explained as the result of chance immigration from Europe of a clever and creative mind. But as this article makes clear, Vucetich's dactyloscopy was at once part of a new "scientific" and intercontinental approach to social problems and a very particular response to the local problems of crime in his adopted country. The early application of dactyloscopy reveals a chapter in the history of the free flowing exchange of novel ideas and state practices, not only from east to west, but also west to east and south to north. The story of the Argentine contribution to the now ubiquitous technique of fingerprinting was staged in a Latin American nation intimately connected to the northern scientific world, poised to exploit one of the world's richest economies, and led by a forward-looking class of state builders bent on a thoroughgoing modernization. The central importance of the state in promoting, supporting, and selectively applying science cannot be overlooked in this case; and echoes of state science at the turn of the nineteenth century reverberate in other episodes described in this volume, such as Cukierman's chapter on expeditions and Brazilian nation building (for twentieth-century examples, see Alonso and Palmarola on Cuban-Soviet technological collaboration, Mateos and Suárez-Díaz on cold war physics in Mexico, and Hagood's analysis of atomic science in Peronist Argentina).

Moreover, though Vucetich's work was part of a transatlantic scientific conversation, its success was rooted in the empirical opportunities of the Argentine environment. Historians of natural science, and above all medicine, have often remarked that world sites such as Argentina offered an open field to experimental scientists (Cañizares-Esguerra 2001, 7; Cañizares-Esguerra 1999, 35). This was no less true for the emerging "science" of crime and other social problems such as prostitution, alcoholism, and political violence. Far from being a barren or primitive cultural and intellectual landscape, Argentina was a veritable hothouse for scientific innovations essential to the era's statebuilding projects. With its uncharted terrain, "exotic" populations, and novel social and political formations, it contained observational data unknown to the old world. Argentina's educated and trained scientists who came of age after 1880 were intimately shaped by the dramatic political events of their times and they in turn directed some of that nascent state's most consequential projects. Chief among them was this postcolonial state's effort to harness the surging forces of mass immigration and rapid economic expansion. The social "pathologies" plaguing France, Germany, the United States, and other nations—increasing crime, urban crowding, ethnic factionalism, and social and class conflict—were to be found in extreme form in Argentina. Moreover, its oligarchical political culture and less seasoned constitutional structures gave elites with scientific credentials far greater influence and power. It was these two elements—the Argentine state's strong urge to engineer order at a time of tumultuous social transformation wrought by its accelerated modernity, and its patronizing of science—that created a fertile field for the development and application of fingerprinting.

Ivan Vucetić was born on the Dalmatian island of Lesina, off the Adriatic coast of Croatia, in 1858. He immigrated to Argentina, adopting the name Juan Vucetich, as he would be internationally known, at the age of twenty-six. It would be interesting to know more about Vucetich's background, but there is a dearth of information on his life before his appointment in the Argentine police. It is likely that that he came from an educated background in his native Croatia, for within four years of his arrival, he was hired into the scientific wing of the Buenos Aires provincial police. His first post there was to direct the *Monthly Statistical Bulletin* that collected arrest and crime statistics for the province; a few years later, he was hired to run the Office of Identification. Vucetich followed in the long tradition of European explorers and adventurers drawn to the new world with its rich and abundant, if not exotic, natural and social formations. Scientists emigrated from Europe to the Americas and back again,

Figure 7.1
Frontispiece with photo, signature, and thumbprint of Juan Vucetich, in *Homenaje a Juan Vucetich* (Buenos Aires, 1938).

often in search of new ground for scientific observation. These scientists came equipped with the most recent theories, approaches, and objectives defined in the northern metropolitan centers, but were highly aware of the fresh data that the new world offered them. They thus arrived to measure, study, and cure the local populations, but also expected to develop and generate new theories based on that data.

"Criminal anthropology," as the new positivist discipline was known, coalesced out of a trans-European intellectual movement investigating the causes of criminal, antisocial, and antinational behavior (Davis 1988; Gibson 2002). It distinguished itself from the so-called "classical school" of criminology, elaborated by Cesare Beccaria and others in the late eighteenth century, by emphasizing the supposedly "scientific" (as opposed to metaphysical) basis of its theory; by focusing on the individual offender and belief in the lack of free will in criminal behavior; by emphasizing "humane" approaches to criminals, replacing punishment with "treatment"; and finally by being concerned with "social defense," or a privileging of social stability over individual rights. The inventors of criminology had attained both scientific and political status in Europe by 1880. In Italy, the leading crime theorist was Cesare Lombroso, who was succeeded by his student Enrico Ferri. The 1876 publication of his *Criminal Man* (*L'uomo delinquente*) was understood as a turning point in approaches to crime and was widely read in Latin America (Johnson 1990; Piccato 2001; Salvatore and Aguirre 1996; Salvatore, Aguirre, and Joseph 2001). Lombroso's main—and most controversial—contribution to criminological theory was his idea of the "born criminal," a type of human being believed to represent a distinct species, *homo delinquens*. Regardless of their stance on the biological origins of criminality, Lombroso and his critics alike considered crime to be a "natural" and inevitable problem, like illness, but also preventable if approached scientifically. Europe forwarded experts who famously developed aspects of the criminological enterprise, among them the French police scientist Alphonse Bertillon, the Austrian criminalist Hans Gross, and the British polymath Sir Francis Galton.

The expertise of these men, and others, was appealing to states that sought to contain what many perceived as a rising tide of crime and social danger riding the heels of modernity (Cole 2001; Caplan and Torpey 2001). They felt that scientists and governments could cooperate to eliminate, or at least mitigate, the effects of criminality. In turn, the state became the paying client of the self-designated social pathologists. Legislatures supported the development of new methods of healing an ailing social body while vesting the practitioners with extraordinary power to control their patients. A new "science" based on immutable features of the body and devoted to protecting social norms from encroaching waves of crime appealed to legislators and bureaucrats in a wide range of Atlantic states. Burgeoning cities and the arrival of large numbers of immigrants in many countries raised the specter of criminals who could escape detection by hiding in sprawling slums behind a wall of foreignness. One of the most imperative needs was for tools to combat the anonymous or hidden criminal. Governments began building up their criminal justice apparatus, including police, prisons, and courts procedures, as part of a broader attempt in the mid-nineteenth century to rein in those threatening forces.

Against this backdrop of transatlantic tension and scientific engagement, Juan Vucetich arrived in Argentina and took his position in the provincial police of Buenos Aires. La Plata, the provincial capital, was a boomtown in the nation's most prosperous region. It benefited from the wealth flowing through it from the pampas to the port. Yet La Plata was an inauspicious site for scientific innovation. It was a cultural and intellectual satellite of the cosmopolitan city Buenos Aires, a mere 100 km to its north. In Argentina, as in many former colonies, all roads lead to the central city. A bureaucrat such as Vucetich, as he rose through the ranks of the police, was fluent in the theoretical research issues at stake and focused on the activities and needs of his colleagues and collaborators in the capital city. Part of the overlapping circles of the national intelligentsia and the political elite, he rubbed shoulders with the top research scientists in Buenos Aires, sharing the podium at conferences and providing crime data and fingerprints for experts in the city. The scientific community—criminologists, medicolegists, pathologists, and statisticians—he met with in Buenos Aires was not a sleepy one. The university enjoyed rapid expansion, research institutes were forming, and scientific publication was booming. Argentine scientists in all fields knew that they had an abundance of valuable data. By producing world-class research, they hoped to show the rest of the world's scientists that empiricism knew no nationality.

While today this South American country is often plotted on the distant shore of the scientific world, it was at the turn of the nineteenth century a much larger and promising station in the ocean of transatlantic ideas. Latin American and North Atlantic scientists alike recognized researchers in the up-and-coming nation, particularly in the fields of public health, physiology, and anthropology, as well as criminology and criminalistics. The Argentine state, in turn, took pains to promote its investment in science by sending ambassadors and reams of published research abroad. Coming largely from the privileged class, Argentine scientists had the linguistic skills, the resources for travel, and not least, the motivation to contribute to international scientific knowledge (Babini 1986; Biagini 1985; Montserrat 2000; Terán 2000). Among the most educated members of their society, they had much in common with their North Atlantic counterparts; they hailed largely from the liberal and reformist wing of the elite and upper middle class intelligentsia and sought to advance democratic, progressive reforms in public health, education, and the law. Thanks to their close cultural and social connections to the seats of power, they had a disproportionate influence on government policies forged in the first decades after national consolidation in 1880. At this time too, Argentine scientists consciously built a scientific infrastructure—universities, research institutes, and government agencies—that they thought essential for the development of the nation at the awesome pace of their northern neighbor, the United States. Forward-looking state officials in Argentina were equally determined to exploit "science"—broadly defined as modern engineering techniques, public health campaigns, and new educational and penological regimes—to achieve monumental national goals, commonly understood as democratic progress and peaceful modernization (Rodriguez 2006).

Chief among the state's new responsibilities was the policing of its borders. Turn-of-the-century Argentine society was marked by class and ethnic tension as wealthy elites sought to tame the torrential waves of immigration after 1870. Between 1871 and 1914, nearly six million immigrants arrived, about half of whom remained in the country. Buenos Aires was especially transformed by the population explosion. Government policies aimed at increasing population hoped to shuttle new arrivals to the countryside to fuel the nation's agricultural boom, but they did not make land available, and so most immigrants chose to remain in the big cities, especially in Buenos Aires (Rock 1987, 141; Moya 1998, 149; Solberg 1970, 33–38; Kleiner 1983, 30). The percentage of foreign-born residents of the city, already a notable 36 percent in 1855, increased dramatically to 52 percent by 1895. In 1914, nearly 30 percent of Buenos Aires's population was foreign born, a higher proportion than in any other immigrant city at that time, including New York. The majority of these immigrants were working-class and from Italy or Spain; in 1914, native Italians represented 39 percent of foreign-born Argentine residents and Spaniards over 35 percent.

This large-scale immigration shaped anxieties about modern mass anonymity. Ironically, the immigrant was at once the nation's food and its toxin, driving Argentina's new economy and its bid to become a powerful nation, but at the same time threatening to destroy all it achieved. Statebuilders hoped to find the right alchemy of immigration to keep the country in equilibrium. To do so, they relied increasingly on experts to determine the balance. Elite discourse about immigration was rife with theories then fashionable on the European continent. Scientists were excited by the challenge of studying the features of this unprecedented social phenomenon. In their view, criminal immigrants to Argentina were especially dangerous—they may have been from the same genus as other delinquents, but were a more advanced species.

In 1890, Vucetich advanced to the head position in the Office of Identification at the provincial police. His move from statistics to identification signaled a new phase in hands-on policing as well as a shift in emphasis from "diagnosis" to "cure" in his work. He viewed his labors in the police department as scientific in the positivist sense—objective, measurable, and verifiable. In his view, the scientific approach carried forth in his laboratory could directly address what he and many other state officials saw as Argentina's symptoms of social pathology, crime foremost among them. This stage had several dimensions to it, including Vucetich's work as an individual scientist; the institutional and bureaucratic reception of dactyloscopy; and finally, state patronage on the national level, or the nationalization of his system. The remainder of this article will describe these phases. The newly opened office followed Bertillon's meticulous system of body measurement, notation, and classification (Rhodes 1968, 71–101). As the official in charge of the filing and day-to-day application of this Byzantine record system, Vucetich became convinced that modern police forces needed a more effective method (Vucetich 1904, 15). In 1891, after reading an article about Galton's work on fingerprints in the French journal *Revue Scientifique*, Vucetich began advocating the use of fingerprints as a clearly superior method to Bertillonage (Cole 2001, 133; Chapman 1992, 288, 294; Wilton 1938, 78). His recommendation to the provincial police of Buenos Aires to adopt fingerprinting made him the first public official in the world to replace the dominant anthropometric methods with fingerprint identification. His influence was felt immediately. In 1891, the Identification Service was installed in the provincial prisons of Buenos Aires; its officials strove to record and archive the fingerprints of all potential recidivists, including each prisoner and all arrestees in police stations (Vucetich 1931, 29; Vucetich to Doyenhard, May 1, 1903, correspondence, AV).[1] One year later, the governor of Buenos Aires province, Julio Costa, mentioned in his official address to the provincial legislature that fingerprinting had been added to the anthropometric system (Vucetich 1931, 30; Registro Oficial 1893, 184).

At the same time, however, Vucetich also found fault in Galton's system with its three-part organization of finger pads because it was too general to be of any use. Vucetich, who thought of Galton as the "father of fingerprinting" and considered himself but an improver and proselytizer of the system, had within a few months done what Galton could not—he created a universal classification scheme based on a manageable number of subdivisions of finger pad patterns. Initially he called his revised system *icnofalangométrica* (derived from the Greek term for "finger track measurement"), but in 1896 he renamed it with the more accessible *dactiloscopía* (Latin for "finger description"). The name change reflected not only a streamlining in classification of types of fingertip patterns, but also a shift from scientific measurement of fingerprints to classification—the difference between the observation and study of information and its ordering in a useful way (Cole 2001, 128). Vucetich's innovation was to subclassify among loops, for example, identifying an "internal" inclination, or tracks leaning to the left with the delta to the right of the observer, as well as the reverse, "external" inclining loops. Vucetich's simple classification consisted of the assignment of four single letters (AIEV) to thumbprints and numbers (1234) to the types of marks on the remaining four fingers. An individual's fingerprints could thus be classified, filed, and easily located by each finger's code. For example, the designation V2443 meant (from thumb to pinky) whorl, inner loop, whorl, whorl, outer loop. Finally, Vucetich used a secondary classification to further subdivide sets of fingerprints, in which he assigned five subtypes to each of the four primary types. Then, he

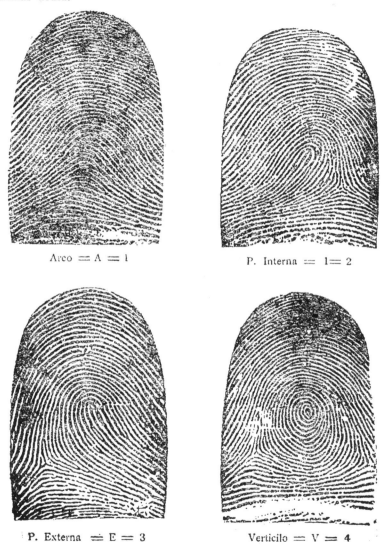

Figure 7.2
Juan Vucetich's fingerprint classification (from his *Dactiloscopía comparada*, 1904).

added an additional distinction to each by counting the ridges between the core (the point at the center of the pattern) and the delta (the outer ridge of the core pattern) on two of the fingers and adding them in parentheses to the fingerprint code (Reyna Almandos 1909, 34–37; Cole 2001, 128–129; Boolsen 1935). It was the limited number of these subdivisions that rendered his method of fingerprinting viable for mass identification and classification. The subclasses of patterns were few enough to be easily defined, yet were diverse enough to be filed and selected, or easily matched. After fingerprinting, police officials would analyze and type each print, recording the five-digit code on an identification card that held a photograph or other personal information. An officer would then file the cards by their print type in multidrawer cabinets, each drawer clearly labeled. Record cards were copied and sent to police stations in other cities, or even other countries, to aid in prosecuting criminal cases and in identifying at-large recidivists.

In 1896, Vucetich announced his new system in a pamphlet entitled "General Instructions for the 'Province of Buenos Aires' System of Filiation" (Vucetich 1896). Mindful of his intellectual debt to Galton, Vucetich sent the British scientist his article and wrote in his accompanying letter:

In my capacity of Chief and Founder of the office of Statistics and Anthropometric Identification of the Province of Buenos Aires, I have the satisfaction to share with you that the Provincial Police are in the only state in South America that has implemented fingerprinting. (Vucetich to Galton, December 11, 1896, AR)[1]

In his reply, Galton acknowledged the Argentine's breakthrough improvements to his efforts, writing, "it is a great satisfaction to learn that you have found the Finger Print Method to be of real, practical service." His next words revealed a grudging acceptance of the South American's findings and also foreshadowed the difficulties that Latin Americans—no matter how well funded by their local institutions—would have establishing a place for themselves in the international community of scientists. Galton wrote,

My knowledge of Spanish is unfortunately too limited to enable me to properly read your volume, but I have understood its main features and fully recognized the great pains you must have taken to compose it. It affords another instance of that statistical zeal for which Buenos Ayres is so justly renowned. (Galton to Vucetich, January 11, 1897, AR)

No doubt stung by Galton's lukewarm response, Vucetich was nonetheless confident that once scientists and police officials around the world heard of his system they would adopt it, and he persisted in promoting dactyloscopy as a universally comprehensible system of identification. Publishing in Spanish, he argued that his method was compatible with the needs of large-scale, urban police forces, whose case loads might number in the thousands and whose suspects might have crossed the Atlantic many times. He predicted in 1901 that fingerprinting would soon become a "universal police language." Three years later, Vucetich published *Dactiloscopía comparada*, a book he hoped would convince not just local officials, but the international community as well, of what he considered dactyloscopy's unique ability to provide that universal language. Dedicated to Galton but subtitled "The New Argentine System," the book described as its goal "to set a new course in modern personal identification, an urgent and arduous problem of universal necessity" (Vucetich 1904, 14). It also represented Vucetich's hope that dactyloscopy could enhance his adopted country's international reputation. Argentina's role in this process, he believed, was confirmed by the fact that ever more North Atlantic scientists, such as Galton, were beginning to recognize the applicability and advantages of fingerprinting. In an appendix to *Dactiloscopia comparada* he reprinted letters of praise for his method from three preeminent figures—Galton, Lombroso, and Gross—all attesting to his international reputation (Vucetich 1931, 36, 44). By 1904, Vucetich could proudly claim that "today, all the well-organized police departments on earth use [fingerprints]. ... The Province of Buenos Aires was the first of them. ... Our police have obtained a complete success" (Vucetich 1904, 41).

Officials relied on new technology combined with a racialized definition of criminal types that spoke to the new needs of the nonintimate, nonlocal society that Buenos Aires had become (Vucetich 1904, 55). Just as the perceived need to distinguish individuals from the amorphous mass reached its peak, fingerprinting promised to isolate them and see through their aliases. Guillermo J. Nunes, the chief of police of the province of Buenos Aires, wrote in his 1891 annual report:

The vast territorial extension of the province, with its borders open to the desert and neighboring countries, and with its dispersed population, presents great obstacles to good policing with relatively scarce personnel. It also offers opportunity for criminals to escape the reach of authority.

Moreover, he cited the difficulties of criminal investigation "given the constitutive elements of our population, essentially heterogeneous and cosmopolitan." He explained that the province of Buenos Aires was

composed of men from all parts of the globe, who belong to various nationalities and distinct races, with ignorant life habits, native tendencies, peculiarities, characteristics ... whose forms are consequently unknown to the majority of agents. This difficulty, in most cases, can cheat success in important investigations. (Nunes 1892, 5)

Police reports and correspondence reveal that virtually all officials in the capital and province of Buenos Aires embraced dactyloscopy by the mid-1890s. Early skepticism about dactyloscopy from within the ranks of the police was swept away within a few years (Ruggiero 2001, 185–188). Police officials praised dactyloscopy for its organization and accessibility. Nunes, for example, called in 1892 for both a universal system of identification and the sharing of data on criminals throughout the country

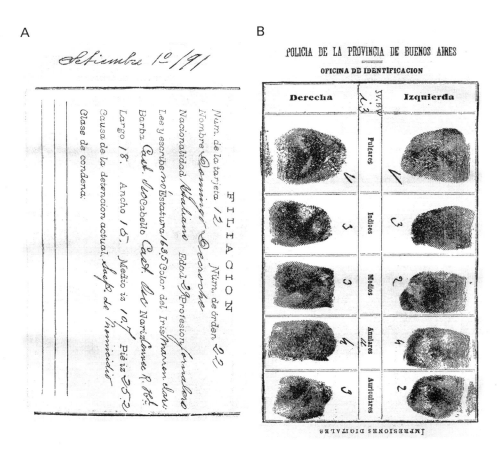

Figure 7.3
One of the first criminal identification cards from the provincial police of Buenos Aires, 1891. One side exhibited the convict's fingerprints, the other recorded anthropometric data. (From Reyna Almandos 1909.)

(Nunes 1892, 5). In 1894, Bibiano S. Torres published a pamphlet aimed at convincing police chiefs around the country to use the Vucetich system of identification. He wrote of dactyloscopy: "If the police ... have as their mission the maintenance of the public order, liberty, individual property and security ... we have proposed to bring to the government's awareness an improvement that we consider a very necessary compliment to police service" (Torres 1894, 5).

In 1895 police official Ernesto Weigel Muñoz recommended dactyloscopy over Bertillonage, pointing out that the old anthropometric cards caused great confusion among different officers due to their vague and imprecise information (Weigel Muñoz in Vucetich 1896, 3). In September 1901, Buenos Aires City Police Commissioner of Investigation José Rossi wrote his chief, Francisco Beazley, that he had attended a lecture by Vucetich and then reported in detail the differences between dactyloscopy and Bertillonage, pointing out in his letter the relative simplicity and ease of the Argentine system. He recommended its general adoption as "an acquisition of a new universal language, simple and efficient" (Rossi to Beazley, September 9, 1901, AV). Two years later, Rossi oversaw the introduction of the Vucetich system in the Buenos Aires City Police. He wrote to Vucetich requesting books and instructions, after which he mandated the adoption of dactyloscopy (Rossi to Vucetich, correspondence 1901–1907, AV; *Orden del día*, November 9, 1903, 1066). Police officials in the capital began soliciting Vucetich regularly for advice, instruction, and the loan of fingerprint records (Belasario Otamendi to Vucetich, May 11, 1901, correspondence, AV). In celebrated cases, they exchanged prisoner fingerprint files by telegram to snare fugitives. For example, in 1910 the records of Simon Radowisky, an anarchist who had assassinated the Buenos Aires City police chief one year earlier, flew between the capital police and local stations (Rossi to Ricardo Rojas, January 19, 1910, correspondence folder, AV).

These new possibilities of identifying and tracking criminals fed the enthusiasm of officials such as Rossi, who as chief of police had called Vucetich's work "one of the greatest discoveries in applied policing in recent years" (Rossi to Vucetich, December 31, 1907, AV). The capital police's Investigative Division hoped to utilize dactyloscopy to carry out its mission to maintain public order "by means of knowledge, preventative observation, and repression of socially disruptive elements"—in other words, to expand the practice to the noncriminal population (Memoria del Departamento de la Policía 1909, 298). Officials in the capital city police department expected to increase the number of records; indeed, in 1916 alone, they took 141,691 fingerprints, bringing the total to 619,553 records (of a total city population of about 1,500,000 in 1916) (Memoria de la División Técnica 1917, 18). They had demonstrated the ease with which they could take prints from large numbers of people in a short period of time. Buoyed by the triumphs of dactyloscopy as used by the police, state officials recognized that its use in court trials was only one, relatively modest application of the new technique. As officials began to think more broadly, they began to realize fingerprinting's promise to help efficiently exercise state discipline in commissaries, jails, public spaces, and at the nation's borders.

Armed with their new identification technique, police joined forces with their partners in law enforcement, the nation's legislators. Thanks to journalistic accounts trumpeting the novelty of dactyloscopy and its use in high-profile cases, and important meetings between officials in the province and the capital, political leaders joined forces with police to apply it on the federal level to specific national problems. Some recommended the broadest possible civic applications of fingerprinting. In 1909, Luis Reyna Almandos, one of Vucetich's most influential followers, suggested that the government apply dactyloscopy in nearly every realm of human experience—births, deaths, and marriages; public testimonies, legislation, and contracts; public health, welfare, residency; and in the control of prostitution and vagrancy (Reyna Almandos 1909, 109). Beginning in 1916, bolstered by their national successes in police work, Vucetich and Reyna Almandos spent years lobbying first the provincial and then the national government to create a national register of fingerprints (Vucetich 1916; Rodriguez 2006; Ruggiero 2001, 192–196). Constitutionalist objections delayed the creation of a national identification register until 1935, but a program of targeted fingerprinting of select groups proceeded at full force. This moment revealed both the ambitions that state-building modernizers had for fingerprinting as well as its dangers. Exciting as the new practice was, it presented the risk, as its critics pointed out, of violating civil liberties and discriminating against "honest" immigrants.

Police and other government officials were also increasingly convinced that large-scale, mandatory fingerprinting of immigrants was a necessary complement to the exclusionary residence law. In promoting that view, they targeted not only foreign criminals but all newcomers. In 1908, the criminologist Nicéforo Castellano proposed a new National Identity Office, including in his proposal an article establishing the routine recording of foreigners' personal information as well as fingerprint exchanges with foreign offices. He called for the identification of "those immigrants who desire a citizenship card," since "our laws demand morality and good habits in conferring the privileges of citizenship … the manner of searching for that background is to identify them." He recommended the routine fingerprinting of immigrants in third class passage, warning that recent criminal and anarchist events, such as the bombing of the elegant Colon Theater, were reminders that the state needed to be vigilant. Argentina tempted the criminal population with what he cited as a set of "unique conditions, already recognized before disembarking," including unexploited land, available jobs, and a tolerant state and liberal constitution that promised to protect them (Castellano 1908, 21–22).

Four years later, in 1912, the national immigration service incorporated fingerprinting into its procedures. At the same time, the government established the "Dactyloscopic Register of Immigrants" (*Registro Dactiloscópico del Inmigrado*) as a repository for immigrants' fingerprint records. Upon entering the country, most newcomers were given a numbered "immigrant's book" (Dirección General de Inmigración, "Resolución No. 292," May 21, 1912, AV). This passport-like document, which included information about their rights and responsibilities, held a place for personal description and a fingerprint. The immigration service ordered officers at the Office of Work and Dispatch and the Immigrants' Hotel to compile extensive personal identification data for each immigrant. The form they used, titled "Description of the Immigrant," included anthropometric data and fingerprints. Each form was linked by number with the booklet issued to the immigrant. Newcomers were required by law to carry their identification book with them at all times; without it, they could claim no protection under the law. While officials hoped to use dactyloscopy to turn back criminal "bad seeds" at the borders and permit entrance only to assimilable, productive immigrants, in practice it served primarily as a warning to immigrants of their subordination to the state (Nunes to the President of the Supreme Court of the Province of Buenos Aires, November 24, 1891, AV). The goal, in the words of the "immigrant's book," was the "assimilation and permanent rootedness" of the foreign population (Dirección General de Inmigración 1911, 3). The register was a tangible reminder of the requirements of participation in Argentine society—a broad disciplinary device applied at entry to all newcomers regardless of their past.

Perched in the provincial police and immersed in Argentina's accelerated conditions of modernity, Juan Vucetich understood well that the Atlantic was a system of migratory movement. He repeatedly insisted that in order for dactyloscopy to truly succeed, nations beyond Argentina's borders would have to universally and uniformly apply it. He knew the interconnected nature of Argentine crime and structures abroad. Proposing to help nations track people across borders, he anticipated the advent of international policing bodies such as Interpol and its international fingerprint database (Anderson 1989; Bresler 1992). As he lobbied for the application of universal fingerprinting in Argentina, Vucetich simultaneously proselytized abroad. From La Plata, he corresponded regularly with police officials, scientists, and government officials in Europe, the Americas, and Asia. He was especially close with other Latin American police scientists, including Felix Pacheco, the chief of identification in Rio de Janeiro, one of the most enthusiastic promoters of dactyloscopy. At regional police congresses, they explained the Vucetich system and often did public demonstrations together (Anonymous 1906, 354).

By the first decade of the twentieth century, when most North Atlantic countries started using fingerprints more consistently, Vucetich was a household name in international police circles. Most scientific commentators in Europe viewed Vucetichism favorably when compared with Bertillonage, and on equal footing with the new Henry system (Cole 2001, 81–83; Sengoopta 2003, 138–145). One year after Vucetich had published his 1896 article, Edward Henry, a British colonial administrator in India, introduced a system that was similarly based on Galton's ideas. With three basic categories of print types, as opposed to four in the Vucetich system, the Henry system was considered by many experts to be less convenient. In a typical assessment, French rapporteur M. Dastre wrote in 1907 that compared with Henry's achievement, "the definitive progress appears to have been realized by M. Vucetich." Some specialists argued strongly for conversion to the Vucetich system, which they believed was technically superior (Dastre 1907, 45; Daae 1906, 26–44; Spirlet 1910, 7–34; Fosdick 1915).

Vucetich traveled widely to North America, Europe, and Asia, hoping to spread his method throughout the "civilized world" by visiting police departments and presenting his system at international conferences. His main goal at such meetings, beyond advocating universal fingerprinting, was to convince representatives of other governments to establish "Intercontinental Identification Offices" to facilitate the exchange of criminal records. His plan had resonated strongly in Argentina, where a growing number of officials imagined that their country, with its multiple points of entry, was permeable and vulnerable. But fingerprinting had obvious applications in other countries as well, such as the United States, with its own migratory waves, and the colonial outposts of the various European empires. In 1913 Vucetich suggested that Argentina could provide a model for others:

I have been able to observe in this great country [the United States] that the lack of connection of the [identification] service with the national needs is notorious, owing to the independence of each state which establishes its own service without any possibility of coordination. The result of this is that a fugitive criminal, by merely removing from one state to another, is sometimes more secure than if he had emigrated to another hemisphere where a uniform system of identification is in vogue. (cited in Dilworth 1977, 94)

The Vucetich system eventually made inroads to regions far beyond its borders as dozens of police departments around the world imported it wholesale or used elements of it for classification purposes. It dominated South America, was adopted by the Peking (Beijing) City Police after a personal visit from Vucetich, and was applied in modified form in at least twenty-one other countries. Observers in the early twentieth century noted that the dispersion of the two competing systems broke down along language lines: Vucetich dominated the "Latin" world (including Italian- and Portuguese-speaking countries), Henry ruled the English-speaking world (including police departments in a number of US cities), and all other countries were split between the two (Cole 2001, 134).

While Argentina's turn-of-the-century political culture offered rich incentives for the rapid development and adoption of dactyloscopy, officials in other countries did not embrace it with the same enthusiasm until some years later. Imperatives of the nascent federal state in Argentina, such as the rapid search for new technological solutions to racialized social problems, mass immigration, and political unrest, had encouraged the funding of state offices that would deploy those new disciplinary tools. A combination of prosperity, immigration, and an infatuation with scientific solutions created a hothouse environment in which the development and implementation of modern fingerprinting flourished. It was in this setting that the formation of modern techniques of surveillance and social control were worked out, anticipating broader international concerns about immigrants and the mobility of criminals (Cole 2001; Sengoopta 2003).

Argentines took pride in the knowledge that they had pioneered the creation of "Vucetichism," a practice that marked a turning point in global police methods and expanded in due course any state's ability to track, control, and confine its citizens and visitors alike. The "success" of Argentine dactyloscopy, trumpeted as it was by nationalists, had its dark side as well. The formation of powerful paradigms like the universal reliance on fingerprint identity not only highlights the role of social and political exigencies in scientific "discoveries." It also reminds us, across the board, of the dangers of vesting in our scientists omniscience about social problems. As this story of science crossing borders shows, the intertwined relationship of scientific knowledge and political imperatives, while representing a genuine step forward for forensic science, had both intentional and unforeseen consequences on the ability of immigrants and citizens alike to move freely in society.

Notes

An extended version of this chapter appeared as Julia Rodriguez, "South Atlantic Crossings: Fingerprints, Science, and the State in Turn-of-the-Century Argentina," *American Historical Review* 109 (2) (2004): 387–416, © Oxford University Press. It has been republished in edited form with permission from Oxford University Press.

1. "AV" refers to the Archivo Vucetich, Museo de la Policía de la Provincia de Buenos Aires (La Plata).

2. "AR" refers to the Archivo Romay, Policía Federal, Centro de Estudios Policiales, Buenos Aires.

References

Anderson, Malcolm. 1989. *Policing the World*. Oxford: Oxford University Press.

Anonymous. 1906. La dactiloscopía: Sus resultados en el Tercer Congreso Científico Latino-Americano de Rio-Janeiro y en el Convenio Policia Sud-Americano de Buenos Aires. *Archivos de Psiquiatría* 5: 354.

Arciniegas, Germán. 1986. *America in Europe: A History of the New World in Reverse*. San Diego: Harcourt Brace Jovanovich.

Arciniegas, Germán. 2001. *Cuando América completó la tierra*. Bogotá: Villegas Editores.

Babini, José. 1986. *Historia de la ciencia en Argentina*. Buenos Aires: Solar.

Biagini, Hugo. 1985. *El movimiento positivista argentino*. Buenos Aires: Belgrano.

Boolsen, Frank M. 1935. Fifty-one fingerprint systems: An outline of the various methods in different countries for classifying and filing fingerprints under the ten-finger and single-print systems. Unpublished typescript, ed. Burtis C. Bridges, Berkeley.

Bresler, Fenton. 1992. *Interpol*. London: Sinclair-Stevenson.

Cañizares-Esguerra, Jorge. 1999. New world, new stars: Patriotic astrology and the invention of Indian and creole bodies in colonial Spanish America, 1600–1650. *American Historical Review* 104 (1):33–68.

Cañizares-Esguerra, Jorge. 2001. *How to Write the History of the New World: Histories, Epistemologies, and Identities in the Eighteenth-Century Atlantic World.* Stanford: Stanford University Press.

Caplan, Jane, and John Torpey, eds. 2001. *Documenting Individual Identity: The Development of State Practices in the Modern World.* Princeton: Princeton University Press.

Castellano, Nicéforo. 1908. *Oficina Nacional de Identidad. Proyecto presentado al Ministerio del Interior.* Buenos Aires: Escuela Tipográfica del Colegio Pío IX.

Chapman, Carey. 1992. Dr. Juan Vucetich: His contribution to the science of fingerprints. *Journal of Forensic Identification* 42 (4):286–294.

Cole, Simon. 2001. *Suspect Identities: A History of Fingerprinting and Criminal Identification.* Cambridge, MA: Harvard University Press.

Cueto, Marcos. 1989. *Excelencia scientifica en la periferia.* Lima: Grupo de Analisis para el Desarrollo.

Cueto, Marcos. 1995. Laboratory styles in Argentine physiology. *Isis* 85 (2):228–246.

Daae, A. 1906. Die daktyloskopische Registratur. *Archiv für Kriminal Antropologie und Kriminalistik* 24:26–44.

Dastre, M. 1907. Des empreintes digitales comme procédé d'identification. *Comptes Rendus Hebdomadaires des Séances de l'Académie des Sciences* 45.

Davis, John A. 1988. *Conflict and Control: Law and Order in Nineteenth-Century Italy.* Atlantic Highlands, NJ: Humanities Press International.

Dilworth, Donald C., ed. 1977. *Identification Wanted: Development of the American Criminal Identification System, 1893–1943.* Gaithersburg, MD: International Association of Chiefs of Police.

Dirección General de Inmigración. 1911. *Libreta del inmigrado.* Buenos Aires: Dirección General de Inmigración.

Fosdick, Raymond. 1915. The passing of the Bertillon system of identification. *Journal of the American Institute of Criminal Law* 6: 364.

Gerbi, Antonello. 1973. *The Dispute of the New World: The History of a Polemic, 1750–1900.* Pittsburgh: University of Pittsburgh Press.

Gibson, Mary. 2002. *Born to Crime: Cesare Lombroso and the Origins of Biological Criminology.* Westport, CT: Praeger.

Gläser, Elisabeth, and Hermann Wellenreuther, eds. 2002. *Bridging the Atlantic: The Question of American Exceptionalism in Perspective.* Washington, DC: German Historical Institute.

Johnson, Lyman L., ed. 1990. *The Problem of Order in Changing Societies: Essays on Crime and Policing in Argentina and Uruguay, 1750–1940*. Albuquerque: University of New Mexico Press.

Kleiner, Alberto, ed. 1983. *Publicidad official sobre Argentina como pais de inmigración (1903)*. Buenos Aires: Poligono SRL.

Kloppenberg, James. 1986. *Uncertain Victory: Social Democracy and Progressivism in European and American Thought, 1870–1920*. New York: Oxford University Press.

Memoria de la División Técnica. 1917. *Año 1916*. Buenos Aires: Imprenta y Encuadernación de la Policía.

Memoria del Departamento de la Policía. 1909 *Memoria del departamento de la Policía de la Capital, 1906–1909*. Buenos Aires: Imprenta y Encuadernación de la Policía.

Montserrat, Marcelo, ed. 2000. *La ciencia en la Argentina entre siglos*. Buenos Aires: Manantial.

Moya, José. 1998. *Cousins and Strangers: Spanish Immigrants in Buenos Aires, 1850–1930*. Berkeley: University of California Press.

Nunes, Guillermo J. 1892. *Memoria del Departamento de Policía correspondiente al año 1891*. La Plata: Talleres del Museo de La Plata.

Piccato, Pablo. 2001. *City of Suspects: Crime in Mexico City 1900–1930*. Durham: Duke.

Registro Oficial. 1893. La Plata: government of the province of Buenos Aires.

Reyna Almandos, Luis. 1909. *Dactiloscopia argentina. Su historia e influencia en la legislación*. La Plata.

Rhodes, Henry T. F. 1968. *Alphonse Bertillon: Father of Scientific Detection*. New York: Abelard-Schuman.

Rock, David. 1987. *Argentina 1516–1982*. Berkeley: University of California Press.

Rodgers, Daniel T. 1998. *Atlantic Crossings: Social Politics in a Progressive Age*. Cambridge, MA: Harvard University Press.

Rodriguez, Julia. 2006. *Civilizing Argentina: Science, Medicine, and the Modern State*. Chapel Hill: University of North Carolina Press.

Ruggiero, Kristin. 2001. Fingerprinting and the Argentine plan for universal identification in the late nineteenth and early twentieth centuries. In Caplan and Torpey 2001, 185–188.

Salvatore, Ricardo, and Carlos Aguirre, eds. 1996. *The Birth of the Penitentiary in Latin America*. Austin: University of Texas Press.

Salvatore, Ricardo, Carlos Aguirre, and Gilbert M. Joseph, eds. 2001. *Crime and Punishment in Latin America: Law and Society since Late Colonial Times*. Durham: Duke University Press.

Sengoopta, Chandak. 2003. *Imprint of the Raj: How Fingerprinting Was Born in Colonial India*. London: Macmillan.

Solberg, Carl. 1970. *Immigration and Nationalism: Argentina and Chile, 1890–1914*. Austin: University of Texas Press.

Soler, Ricaurte. 1968. *El positivismo argentino*. Buenos Aires: Paidos.

Spirlet, F. 1910. Méthode de classification des empreintes digitales. *Archives internationales de médecine légale* 1: 7–34.

Terán, Oscar. 2000. *Vida intelectual en el Buenos Aires fin-de-siglo (1880–1910). Derivas de la "cultura científica."* Buenos Aires: Fondo de Cultura Económica.

Torres, Bibano S. 1894. *Observaciones sobre las Oficinas de Antropometría é Icnofalangometría*. Buenos Aires.

Vucetich, Juan. 1896. *Instrucciones generales para el sistema de filiación "Provincia de Buenos Aires."* La Plata: Solá, Sesé y Compañía.

Vucetich, Juan. 1904. *Dactiloscopia comparada. El nuevo sistema argentino*. La Plata: Jacobo Preuser.

Vucetich, Juan. 1916. *Comentario al proyecto de ley creando el registro general de identificación*. La Plata: Taller de Impresiones Oficiales.

Vucetich, Juan. 1931. Historia sintética de la identificacíon. *Revista de identificación y ciencias penales* 7.

Wilton, George. 1938. *Fingerprints: History, Law, and Romance*. London: William Hodge.

8 Tropical Assemblage: The Soviet Large Panel in Cuba

Hugo Palmarola and Pedro Ignacio Alonso

This chapter examines the construction and history of one of the largest social housing projects in Latin America, which began in 1963 with the arrival of a Soviet-financed factory in Santiago de Cuba to produce large concrete panels. The introduction of the Soviet large-panel factory was important to the Cuban Revolution, as it demonstrated how industrial technologies could standardize ways of living and advance Fidel Castro's plans for an egalitarian socialist utopia. Although it further involved the Soviet Union in the activities taking place on the island, the large concrete panels produced in Cuba became not merely a Soviet but a hybrid Soviet-Cuban technology as they were redesigned by Cuban architects and engineers to make the system feasible for use on the island. In this essay, we analyze how architects and engineers adapted this technology for local conditions, and the intercultural dialogue this exchange entailed between the Soviet Union and Cuba.

We hold that this adaptation was both material and symbolic. The Cuban government used large-panel technology both as propaganda for its revolutionary goals and as a sign of Cuba's success within the cold war struggle. In this context, prefabrication technology and military discipline came together under the tight principles of the Revolution. Workers in the large-panel factory made many modifications to the production process to keep the factory running during its fifty-year history, and residents in large-panel apartment complexes similarly had to adapt their way of life to these structures. By looking at the specifics of the prefabrication technology, the changing political imaginaries of the Cuban Revolution, and the experiences of residents and workers, we reflect on the significance of how these political, technological, and social adaptations occurred. The resulting history juxtaposes the influence of such heterogeneous persons and factors as Nikita Khrushchev, Fidel Castro, Ernesto Guevara (El Ché), hurricanes, filmmakers, concrete, engineers, standardization, production costs, housing, the Cuban missile crisis, architects, local weather, workers, and inhabitants, a combination we refer to as a tropical assemblage. Like Mateos and Suárez-Díaz (this volume), who adopt a transnational framework to study the history of Mexican nuclearity, we tell the history of

Figure 8.1
Aerial view of the José Martí District. Double-page spread, reprinted from *Revolución en el 72. Construcciones* (1972). Permission to reprint image courtesy of Rogelio París.

large-concrete-panel technology in Cuba by tracing how artifacts, experts, and ideas circulated between Cuba and the Soviet Union.

Large-panel housing blocks would become the most widespread housing model of the twentieth century and are therefore an integral part of understanding the materiality of daily life in many parts of the world. Yet their study has been rather marginal in historiographies of technology and architecture, possibly because the signature of the architect is most often concealed behind anonymous collective work and because of their supposedly dull uniformity and poor design qualities, which means that many historians have not considered them worthy of study. Yet a close analysis of the construction of such housing in Cuba allows us to examine in new and revealing ways how the cold war, including the tension between the superpowers of the United States and the Soviet Union, affected the ideal of the socialist "New Man" and daily life in Cuba.

A Globalized Technology for Socialist Aid

In October 1963, Cuba experienced one of the worst natural disasters in its recent history. Hurricane Flora killed approximately 1,160 people, forced around 150,000 people to evacuate their homes, and left several hundred thousand homeless. It was the first natural disaster confronted by Cuba's revolutionary government and provided Castro with an opportunity to differentiate himself from his overthrown predecessor and demonstrate the power of Cuba's revolutionary government. As he said on national television at the time, "a revolution is a force more powerful than nature" (Castro 1963a). Such sentiments also appeared in footage from Santiago Álvarez's 1963 documentary *Cyclone* (*Ciclón*) produced by the government, which shows Cuban families in the Eastern Province fighting against devastating floods and the rescue operations of the Revolutionary Armed Forces. Within Álvarez's film, as Jorge Sánchez notes, the disaster was portrayed "as if the hurricane was a subject, a monster that stalks and kills" (2010, 104). The film portrayed the hurricane as a duel between the Revolution and nature, in which Flora became an enemy for the Revolution to defeat.

Although the hurricane came at the beginning of Cuba's revolutionary process, direct partnerships between Cuba and the Soviet Union were already in place. When the hurricane struck the island, Cuban engineer Vitervo O'Reilly was in Moscow, having been sent by the Cuban Ministry of Construction (MICONS) to take a course on prefabrication. He recalled that the movie *Cyclone* was screened at the Cuban embassy in Moscow for Nikita Khrushchev and his daughter. After the film, Khrushchev decided to donate a factory to produce large concrete panels for building (O'Reilly 2011). It was a critical moment for Soviet-Cuban relations. Hurricane Flora happened one year after the Cuban missile crisis of 1962, and the two countries were in the process of signing a series of new trade agreements as a way to rebuild trust. Khrushchev's gift of the large-panel factory was not simply a humanitarian act, but allowed the Soviets to accelerate their influence on the island in a way that embodied the renewed spirit of Soviet-Cuban collaboration.

The imported large-panel factory, which traveled from the Soviet Union to Cuba via ship, and the panels themselves, which traveled throughout Cuba via local trucks and ships, represented the Soviet hegemony: a deploying of technological components as control and colonization strategies, what Alexander D'Hooghe (2005) has termed "civilizing devices." Concrete panels had already been used in the Soviet Union and Eastern Europe to cope with housing problems and natural disasters. In fact, following the devastation caused by the Skopje earthquake in Yugoslavia on July 26, 1963, the Soviet Union had also donated a prefabrication factory to help the Yugoslavian city rebuild. Khrushchev was now giving Cubans the same kind of Soviet technological assistance to rebuild the 11,000 dwellings that had been destroyed or damaged by the hurricane.

Inside and outside the Soviet Union, Khrushchev promoted new building technologies for building apartment blocks out of prefabricated panels. These technologies brought a new construction paradigm by transferring building efforts from the construction site to the factory and from manual labor to automated mass production of large standardized pieces to be assembled on site. It also transferred the structure and weight of buildings from walls, beams, and columns to the panels themselves, providing reinforced concrete with new possibilities. Housing blocks—functional, cheap and speedy to build—were the result. In the Soviet Union, Khrushchev replaced the collective dwellings known as *kommunalkas* with these single-family apartments, eliminating the ornamental excess of Joseph Stalin's socialist realism and furthering the de-Stalinization of the country. These new buildings were known as *khrushchovkas* and became an emblem of Khrushchev's technological drive, even outside the Soviet Union. The Central Research Institute for the Experimental Planning of Housing (CNIIEP), founded in Moscow in 1949, supported the construction of large-panel factories in several Soviet cities and in nations such as Yugoslavia, Romania, Hungary, Mongolia, Afghanistan, Iran, Vietnam, Cuba, and Chile.

Weeks before Flora hit the Cuban coast, Castro had given the closing speech at the Seventh Congress of the International Union of Architects (UIA), which was held in Havana. In it, the Cuban leader wondered, "What kind of industrialization can be possible in a country with 70 to 80 percent illiteracy, taking into account that to handle any modern working tool, to build businesses of any type, to handle any modern machinery, technical qualifications are needed? ... How many problems we still have, but above all, how big is our need for factories. ... Because, in order to resolve our housing problems, it must be said that we must first begin by resolving the problem of the construction industry" (Castro 1963b). Flora made the concerns raised in Castro's speech even more pressing and increased interest in new building technologies. Prefabrication emerged as a solution that could help to resolve Cuba's housing shortage. Although it arrived in Cuba with the authority of a global technology proven in different countries, it acquired additional symbolic power in the context of the rhetorical tropes used by the Cuban Revolution.

The Egalitarian Ideology of Standardized Prefabrication

This episode of technological exchange was captured on film as part of Cuban government propaganda. The government had already made housing a central part of its political message, and it figured prominently in media propaganda. In 1959, Cuba released its first documentary on the housing problem one year after the fall of Fulgencio Batista. Produced by the Rebel Army's National Bureau of Culture, the film *Housing* (*La vivienda*) asks, "Why is it that in Cuba, housing has never been built for the poor?" At the end of the film, the narrator announces, "One day something came to transform [Cuba's] streets and houses. That day the streets of my city proudly received the triumph of the Revolution. ... That day, a day that dawned like any other, all the houses in my city finally looked the same" (García-Espinosa 1959). As this film shows, Cuba's revolutionary government promoted the idea of housing equality from its outset. The Urban Reform Law, decreed on October 14, 1960, started to transform the housing situation by redistributing homeownership to popular sectors. However, it was only with the arrival of the large-panel technologies that the utopia of socialist egalitarianism was conceived of as a new way of living. The panels married standardized production to social equality and promoted socialist values through industrial mass production. To this day, "Revolution is building" remains the slogan of the Cuban Ministry of Construction and appears on a giant billboard outside their office buildings in Havana and Santiago de Cuba. At that point, socialist construction was thought of as material as well as philosophical. The building of homes was also the building of socialism.

The 1959 documentary *Cubans Forward* (*Adelante cubanos*; García Cuenca 1959) explicitly makes this point by framing prefabrication as "the ultimate solution to the problem of housing" and also as a flexible construction technique that enabled individual choice, although the film showed only semi-prefabricated systems for single-family houses that allowed for greater customization. Initially called Novoa and then Sandino, these prefabricated systems were the first to be used by the Revolution to eradicate slums and *bohíos* (the typical Cuban huts of indigenous origin). The Cuban government used images of the *bohíos* to reinforce the radical contrast between the old shantytowns of the previous government and the new prefabricated housing of the Revolution.

Housing and *Cubans Forward* were the first of what would be many propaganda films about housing. Others included *Building* (*Construyendo*) (1963); *We Have No Right to Wait* (*No tenemos derecho a esperar*) (1972); *To Build a House* (*Para construir una casa*) (1972); *Minibrigades, a Diary* (*Microbrigadas, un diario*) (1973); *Changing Lives* (*Cambiar la vida*) (1975); *White Hard Hats* (*Cascos blancos*) (1975); *Of Life and Housing* (*De la vida y la vivienda*) (1975); and also one of the most famous Cuban animated cartoons for children, *El Bohío* (1984). *Housing* and *Cubans Forward* are particularly significant because they mark the creation of the Cuban Institute of Cinematographic Art and Industry (ICAIC), which in 1959 was founded as the first cultural institution of the Revolution. A year later, Santiago Álvarez, director of *Cyclone*, started the Latin American ICAIC Newsreel. Modeled on Italian neorealist fiction, it became a "documentary school" of considerable influence in Latin America. These developments in mass media allowed the government to construct the "facts" of Cuban housing and prefabrication and their significance. The arrival of Soviet large-panel technologies in Cuba thus formed part of how the Revolution portrayed itself.

Cuba adopted large-panel systems to produce industrialized and prefabricated housing on a massive scale. As Eduardo Ecenarro, a Cuban architect who was trained in the Soviet Union, recalled, "there was no solution without prefabrication ... and so we developed a technical awareness, and a proper ideology that in some comrades grew quite radical regarding the indispensability of industrializing construction ...

driving it toward proper achievements" (Iglesias 1989, 17–19). Prefabrication gained such importance in Cuba's socialist mission that the Technical Research Department of the Cuban Ministry of Construction became the first scientific organization of the Cuban Revolution. Formed between 1959 and 1960 and led by architect Hugo D'Acosta, the Technical Research Department consisted of a multidisciplinary team that was dedicated to experimental research in modular building systems for schools, housing, hospitals, industries, and urban facilities. It sought to standardize all construction activity, ideally through the deployment of prefabricated materials. In 1961, D'Acosta and Ecenarro led the first Cuban trip to study prefabrication in Europe, the Soviet Union, and China. Additional trips by O'Reilly and other Cuban architects and engineers soon followed.

With support from Commander Ernesto Guevara (El Ché), Ecenarro created the Directorate for Standardization of the Cuban Ministry of Construction in 1963, which became central to the process of industrializing Cuban construction with prefabricated building elements. Convincing architects and builders to adopt these elements was not always easy. According to Germán Bode, the second director of the Directorate for Standardization, "A fundamental ideological problem arose because professionals initially saw standardization and classification as enemies to their creative possibilities. Based on the examples provided by other socialist countries, there was a whole set of negative ideas about monotony and repetition in housing, which clearly opposed our traditions and our national character" (Iglesias 1989, 75). Khrushchev's donation of the entire factory, however, allowed Cubans to make innovative changes to the Soviet large-panel technology and by extension to speed up local innovations in their own large-panel systems.

Cubans created their own series of large-panel types, which they numbered from I to VI. Of this series, Large Panel IV became the first Cuban system to be widely used in the country. Between 1963 and 1967, the government set up twenty-five Large Panel IV factories to produce buildings of up to five stories and between 250 and 500 dwellings a year per factory. According to D'Acosta, the Soviet system helped Cuba "to take a leap in the quality of research," serving "as a starting point" for the local systems (1964, 40). Moreover, the introduction of the Soviet large panel in Cuba made the Soviet model of zoning according to districts and micro-districts one of the most influential urban planning models of the period (Azze 2011).

The Cuban adoption and adaptation of prefabrication, therefore, was not only practical but ideological. Cubans expressed their support for the Revolution by building and occupying prefabricated dwellings. Additionally, the arrival of the Soviet large-panel technology provided opportunities for local technological innovations. As prefabrication became increasingly intertwined with the goals of the Revolution, controversies surrounding prefabrication, such as whether Soviet standardization inhibited local creativity, dissipated. Cuban films further linked Cuban political imaginaries to these construction technologies and disseminated a political vision of revolutionary housing throughout the island.

A Tropical Assemblage

Between 1963 and 1965, Soviet technicians directed the construction of the first large-panel factory in the industrial area of Santiago de Cuba. The factory was primarily intended to supply panels to build the José Martí District in the northwest area of the city. Prior to the arrival of the factory, a Cuban committee traveled to Moscow "to study, debate, and improve the system's design and structural conditions" and to adapt the Soviet panels and buildings to the needs of the island (O'Reilly 2011). Adaptations included thinning the facade panels and perforating them with a concave design. These interventions offered several benefits. The first of these was economic, as the thinner facades allowed for significant savings in concrete, which was urgently needed after the United States strengthened the blockade against Cuba in 1962. The cost of the panels and the need to save concrete were the main concerns Castro expressed about the Soviet large-panel technology when he visited the Santiago factory and in various speeches (Rodríguez 2011; e.g., Castro 1965).

A second benefit was that the new panel design allowed climatic adaptation, introducing the cross-ventilation needed in Santiago de Cuba, the country's hottest region. Similarly, it also facilitated structural adaptation to local geographical contexts. The panels were now designed to withstand the tropical hurricanes and seismic events of Cuba instead of the snow loads in the Soviet Union. (Cuba sits above the Bartlett-Caimán Fault, one of the most critical faults in the Caribbean Plate.) The Cubans also modified the Soviet model for transporting the panels, moving them by boat and truck, as opposed to the exclusive use of ground transportation in the Soviet Union. Boats were especially important in moving panels to the far-off town of Pilón located on Cuba's southern coast.

And lastly, these innovations also introduced cultural and aesthetic adaptations, moving away from the severe Soviet design. According to O'Reilly, "the Cubans considered the [Soviet] buildings a bit cold and dull in expression ... but it was possible to modify the facade panels" (O'Reilly 2011). D'Acosta saw the panels as a kind of sculpture that was light and had a more "plastic" appeal but that at the same time maintained the aesthetic expression of machine-made objects, such as the regular curves that made it possible to reproduce them endlessly.

These Cuban solutions were similar to the large-panel systems found in the Muslim Soviet republics such as Uzbekistan and Kazakhstan, where local culture was preserved through ornamental perforations to the facade panels. Sometimes these perforations also made the buildings better suited to warm weather conditions. The most densely populated cities in those republics, Alma-Ata and Tashkent, became important testing

Figure 8.2
Soviet large-panel facades with Cuban design adaptations. This picture shows the transport process in Chile, which produced concrete panels identical to those in Cuba. Photo by Nolberto Salinas, 1972. Permission to reprint image courtesy of Nolberto Salinas.

grounds for the CNIIEP's new building technologies, especially regarding earthquake-resistant structures. The CNIIEP shared the experience it had gained in those cities with Cuban engineers sent to Alma-Ata to be trained in construction and assembly process of the buildings (Cuba Construye 1965, 48).

The similarity between Cuba's adaptations and those made in some of the Soviet republics suggests that we cannot see Cuba's variations as strictly local, but rather as solutions and reinterpretations that emerged from the complex combination of imported technologies, local traditions and situations, and transnational exchanges. As Bruno Latour asks, "How long can a social connection be followed without objects taking the relay?" (2005, 78). The modified panel is itself a tangible relay, or a tropical assemblage, of the translation and intercultural dialogue between Cuba and the Soviet Union.

The Cuban adaptation of the large panel was well received back in the Soviet Union, and its perforated image was widely published (e.g., Lipovskaya 1974). Moreover, the ability to produce perforated panels was incorporated into a second factory that the Soviets gave Chile during the government of Salvador Allende (a gift made to help the country recover from a 1971 earthquake). Unlike the Cubans, the Chileans did not make substantial modifications to the large-panel technology, although the factory was immediately incorporated into the domestic political controversies of Allende's democratic socialist experiment (Alonso and Palmarola 2013). A high level of adaptation was necessary to make the system function in the Cuban context, adaptation that flowed from the specifics of Cuban political, economic, and cultural history. The Cuban case therefore shows that Latin Americans did not simply accept large-panel technology as a form of imported magic but adapted the technology to their needs. These modifications then traveled back to the Soviet Union and altered Soviet ideas about prefabricated housing.

New Men: Workers and Residents in the Communist City of the Future

Besides accounts related to the emergency of Hurricane Flora, stories in the Cuban media presented the José Martí District as an example of the successful overcoming of "the old society" (Sierra Maestra 1967a). The José Martí District was built on the site of San Pedrito, eliminating the most miserable quarter of the city. The provincial Party head, Commander Armando Acosta, declared that "in this place the Communist city of the future will be created" (Sierra Maestra 1967c). The Cuban government's hopes for this unprecedented project, as the first and largest concrete panel district in Latin America, were similar to the expectations set by the Soviets for Cheryomushki, the most famous large-panel district in Moscow, about which a musical film was released in 1963 (Rappaport 1963), based on an operetta by Dmitri Shostakovich. Each character in the operetta and film represented a housing problem in the context of transition from the *kommunalkas* to *khrushchovkas*: the old and the new way of living. A similar sort of socialist rhetoric operated in Cuba as part of the construction of the imaginary about the new José Martí District, especially in the way that new production and material values were associated with the new social values of their workers and residents.

In January 1964, a committee of Soviet engineers and architects arrived in Cuba to begin plans for building the large-panel factory, and construction began three months later. The Soviet CNIIEP sent a set of Russian-Spanish bilingual documents to facilitate the construction process (Anonymous 1964). Acosta and the head of the Soviet economic mission inaugurated the factory on October 17, 1965. Rolando Duarte, one of the factory directors who had been trained in the Soviet Union, remembers, "while I worked in the factory, I tried to be like the Soviets: disciplined!" (Duarte 2012). From its beginning, the factory workers attempted to assimilate the imported Soviet spirit of rational order.

The factory employed over 230 workers and was expected to produce 1,700 apartments a year, the equivalent of 70,000 square meters of living space by 1968.[1] Duarte remembers that the factory's workers were proud of the results. For example, every year they took a mockup of a Soviet large-panel building to the International Worker's Day parade, where, according to Duarte, "It was carried by four of us ... we marched with it" (2012). The buildings became a symbol of political and technological achievement.

The factory thus became a space representative of cold war projections and anxieties. According to one factory worker, "the revolutionary fervor made us alert because at any moment the Americans would attack or sabotage the factory ... it could have been sabotaged just for being Soviet" (Rodríguez 2011). Newspapers like *Sierra Maestra* encouraged these fears, saying, for instance, "Each work center is already a battle trench on alert. Next to the machine or working tool, the *iron* [gun] is ready to fire his song to the enemy" (Sierra Maestra 1966). They also published cartoons of Uncle Sam hanging from a Soviet large-panel crane (Sierra Maestra 1967d) and an article claiming that their Large Panel IV factory was only "a few kilometers from the Yankees" (Sierra Maestra 1968a). This kind of rhetoric militarized the civilian building technologies.

The construction of the José Martí District started near the factory in December 1965. Architect Edmund Azze and engineer Angel Macias led the construction team, which included the Cubans who had been trained in Alma-Ata. At the outset, they constructed five different building models that could accommodate different family compositions. The buildings also had different facade combinations and numbers of stories. However, the government later adopted a single building model, arguably making real the earlier standardization concerns expressed by large-panel critics. The first part of the José Martí District was inaugurated on July 26, 1967, as part of the events commemorating the assault on the Cuartel Moncada in Santiago de Cuba (where in 1953 the struggle against Batista had begun). The first apartments were assigned to veteran heroes of the Moncada assault and to the most needy people from the San Pedrito and Santa Elena slums. In this sense, the technology became the locus for the symbolic confluence here of the Father of the Nation (the nineteenth-century patriot José Martí, who is buried in the city and for whom the district is named), the beginnings of the Revolution, and Hurricane Flora. Most of the factory's 700 assembly workers waited enthusiastically on the eve of the inauguration for construction to finish, some of the workers saying with conviction, "Death rather than failure!" (Sierra Maestra 1967b). Many people volunteered to work in construction brigades to help finish the buildings and even established contests to see who could assemble their panels the fastest. These volunteer brigades were praised by the construction sector and the media as examples of the socialist worker and "the most genuine expression of the New Man" (Sierra Maestra 1968b). Writer Rafael Castro's experience as a

Figure 8.3
Cartoon of Uncle Sam hanging from a Soviet large-panel crane. In the background are a Soviet large-panel building and stacks of panels loaded onto transporter trucks. Cartoon by Angel Quintana. Reprinted from the article "Un presente para la historia futura" (1967). Permission to reprint image courtesy of *Sierra Maestra* newspaper.

volunteer later inspired his novel *Rajayoga* (Castro 1986), which he named after another Soviet large-panel district in Santiago de Cuba.

Early plans for the José Martí District called for the creation of six micro-districts, 8,786 apartments, and 290 buildings in eight years and three months. The district would house 46,571 inhabitants and cost 101 million Cuban pesos (US $101 million in 1965, US $745 million in 2013) (Urbanización Distrito José Martí 1965; Sierra Maestra 1965).[2] Between 500 and 1,000 workers assembled the buildings, the number fluctuating depending on the period.[3] In addition to the José Martí District, the government constructed Soviet large-panel housing projects in other parts of Santiago de Cuba and in the eastern provinces of Guantánamo, Holguín, Granma, and Santiago. Some residents became emotional when they moved to the new apartments in the José Martí District. Maira Vals remembers that when visiting the apartments as a nine-year-old girl, she and her family "had never seen painted, beautiful, and furnished buildings ... [and] areas for children. ... To come here was like an excursion ... everybody wanted to see the buildings, they were a novelty ... we went crazy" (Vals 2012). When her family was randomly selected as one of the first families to occupy an apartment in the José Martí District, her mother cried out of happiness (Noriega 2012).

Thus, the factory and the Soviet large-panel building sites were ideological spaces that led Cuban workers to strongly identify with the Soviet Union and the Cuban Revolution. In the context of the cold war, workers viewed the United States as a real threat to their housing project at the same time as they saw how Soviet principles of standardization and military discipline could improve Cuban life. The José Martí District transformed living conditions for segments of the Cuban population. New panel technologies made possible ways of working and living that fit the ideal of the "New Man" embedded in cold war geopolitics.

Media Adaptations to the Images of Military Discipline

In 1972, Cuba took further steps to adopt the Soviet model of development and became a member of the Council for Mutual Economic Assistance, a Soviet-led economic and trade organization that consisted of the Eastern Bloc and socialist countries in other parts of the world. That same year, the Cuban government released the documentary *We Have No Right to Wait* (*No tenemos derecho a esperar*), one of the biggest propaganda efforts to promote the new infrastructures, in which the Soviet large-panel factory and the José Martí District were presented as protagonists.[4]

Castro proposed making the documentary in a February 1972 speech (Castro 1972), which, according to the director of the film, Rogelio París, became almost "a military order" (París 2012). Paris was given only a few months to produce the film, so that it would be ready in July when Castro returned from a tour to Africa, Eastern Europe, and the Soviet Union. The movie was filmed in 199 locations (including industrial facilities, hospitals, dams, dairies, schools, other buildings) and required the coordinated efforts of the Cuban Institute of Cinematographic Art and Industry, the Cuban Ministry of Construction, and the Cuban Air Force; the famous Cuban singer-songwriters Pablo Milanés, Silvio Rodríguez, and Eduardo Ramos composed original music for the film, which was also released as a musical album. The film created a counterpoint between the socialist man (seen in intimate interviews) and the results of his work (seen in large zoom-out shots). París recalled, "I was not directly interested in technological facts, which I certainly had to register, but I wanted to humanize the discourse ... the focus of the film is man, not technology" (París 2012). Although he thus sought to distance himself from what was perceived as the capitalist and technocratic approach to film, the film, instead of separating the technical and the human, ended up blurring these boundaries, presenting the technology as human and political. The film portrayed construction work as analogous to being in the army, describing workers as "organized into real combat units" that made up a working-class "avant-garde" supported by a "rear guard" of the building materials and prefabricated concrete industries. Moreover, the film adopted a hopeful stance toward prefabricated technologies and praised how their use accelerated the process of building.

The longest unedited sequences in the film were dedicated to the José Martí District, to demonstrate that Cuba was home to the largest prefabricated social housing district in Latin America and present an image of a new Cuba (París 2012). París recalls shooting from a plane so that "I filmed buildings and yet more buildings getting lost into the frame, and more buildings and more buildings and more buildings ... as identical little boxes" (París 2012). Such architectural magnitude, and perhaps monotony, of the José Martí District is also found in the large book of photography that complemented the film (Comisión de Orientación Revolucionaria 1972). The magnificent aerial photography in the film and book, suggesting the rather ascetic "military discipline" of large-panel construction, contradicted the aesthetic ideal of "diversity" portrayed in the same works: "to find variety and beauty within productivity and savings" (film) and to "present the building technique as a flexible instrument" (book). Paradoxically, a different ideal was already evident in the visual appearance of the building blocks from the ground.

The official release of the film took place on July 26 at the Yara Cinema, the most important movie theater in Havana. According to París, "the first thing that Fidel said to me in the cinema was, 'I've seen it four times. ... I have told Raul Roa [the influential Minister of Foreign Affairs] ... that he must show it to everyone because in that way ... we give an image of what is being done in this country. ... I have said he must screen this film to all important visitors'" (París 2012). The documentary was so successful and widely distributed that a special prize given within the construction sector was named after it, and a large placard bearing the title of the film was carried by workers in the 1975 May Day parade in Havana's Revolution Square. It is still considered the most important film on the Cuban construction industry.[5]

This combination of a documentary film, book, and record provided visual, textual, and audio evidence of Cuba's new reality. The Soviet large-panel factory and building sites such as the José Martí District established an ideological connection between the Cuban Revolution and new construction technologies. The documentary was a conscious effort to explain the social meaning of technology, especially by trying to reconcile and soften the revolutionary relationship between New Man and new technique, showing the subordination and moral control involved. Moreover, the film reinforced the military metaphors of the Revolution and translated them into the factory setting at the same time as it unconsciously portrayed the ruthless standardization of large-panel construction.

Adapting for Productivity and Everyday Survival

In 1974, the magazine *Cuba Internacional* published a photographic essay on the José Martí District in which it commented that Santiago's old colonial details "have been replaced by a huge and bright domino game framed by symmetric avenues … really an offbeat city" (Fernández 1974). However, it was precisely this offbeat uniformity of dominoes that triggered a crisis at this time in the Soviet Union and the Eastern Bloc, which had lost their earlier optimistic vision of standardized housing. The Soviet film *The Irony of Fate* (Ryazanov 1975) staged a cartoon parody of large-panel construction that chided the elimination of all ornamentation from the prefabricated buildings, transformed the inspired architect into an authoritarian "Napoleon" who ruled over building-block construction, and showed prefabricated buildings populating the entire world regardless of differences in culture or geography. The cartoon depicted buildings coming alive, marching one after the other in military fashion. The film stressed uniformity and repetition and the sense of being lost in the identical environments of concrete panel neighborhoods. The criticism was even more explicit in Eastern Bloc films like *Panelstory* in Czechoslovakia (Chytilová 1979), *Block* in Poland (Neumann 1982), and *The Prefab People* in Hungary (Tarr 1982) that depicted scenes of everyday life in panel buildings as if they were forgotten ruins of the socialist project set in a threatening environment of anxiety.

David Edgerton (1998; 2007) reminds us that studying its use is a very important part of studying technology. This idea is pertinent to Latin America in light of its reliance on imported products, which it then copies, adapts, or mixes with local techniques to form hybrids. In the case of the Soviet large-panel technology in Cuba, white-collar professionals such as engineers and architects made some of the early design adaptations, although workers and residents also played an important role in adapting the technology to fit the Cuban environment.

In 1977, the Soviet large-panel factory began to face serious problems in Cuba. The Soviets discontinued making the equipment used by the factory, creating a lack of spare parts that made it difficult to keep the factory running. Cuban technicians were forced to modify the machines to attain large savings in fuel, electricity, supplies, spare parts, and concrete while still maintaining production rates. In 1984, the government decided to use the factory to create new models of Cuban-designed buildings that allowed more customization and design diversity. This new model was called the rationalized large panel and demonstrated that Cubans were exhausted with the idea of the universal inhabitant and frustrated by the mismatch between residents' everyday lives and standardized environments. This modified form of the Soviet large panel reflected the growing controversy about ways of living and the mounting struggle to maintain factory production.

The dissolution of the Soviet Union in 1991 and the strengthening of the US blockade in 1992 triggered an economic crisis on the island. This Special Period, as it was called in Cuba, characterized the 1990s on the island. It further exacerbated the decline of the factory and the Soviet large-panel housing districts and resulted in calls to dismantle the factory. In 1992, a ten-year-old girl named Lucía moved with her family to a Soviet large-panel apartment in the Abel Santamaría micro-district. Unlike Maira Vals's mother, who years before had cried for happiness, Lucía's mother "began to cry when she saw the new apartment … she did not like it at all" (Lucía 2012). That micro-district had been recently built by the forced labor of inmates, and it had a rough finish that was different from the "majestic" buildings in the José Martí District in the 1960s (which were now also in a state of decline). The apartments produced and used in the Special Period present a very different image of the Soviet large-panel housing project. Lucía, for example, remembered that it was not a comfortable place to live. "Whenever the neighbors had a shower, all the water [would] run out through the outlet into my room. I could even take stock of the type of soap they used, and the toilet slab leaked and so I had to go there with an umbrella" (Lucía 2012). Because Cuba was in the midst of an energy crisis, Lucía's family cooked with kerosene, which "gave off a dim light and soot blackening the entire house." More dramatically, "people began to cook in the apartments with coal and firewood. This gave an even sadder mood than the blackness caused by the smoke and the constant humidity. And when there was nothing else to light the fire, many took off the wooden windows for cooking" (Lucía 2012). This lack of adequate plumbing and energy made it hard for residents in these buildings to fulfill their basic needs. Crime rates increased. Thieves used the perforated facade panels as ladders to climb into the apartments they intended to rob. In response, residents filled in the holes of the panels and blocked the windows. Many balconies were also transformed into interior spaces when residents extended the apartments to what used to be the exterior. Consequently, the appearance of the original Cuban design changed.

Figure 8.4
A resident of the José Martí District in her apartment. In the background is a building modified by its residents during the economic crisis known as the Special Period. The building was recently rehabilitated by the provincial government. Photo by Hugo Palmarola, May 2012.

During the Special Period, the factory management office told Duarte and the workers to dismantle the factory, an order they ignored. It had been their "battle trench" throughout the cold war, and even after the Soviet Union collapsed they managed to keep the factory open. Today it produces a variant of the Soviet large panel called the modified large panel. In 2009, Santiago de Cuba's provincial government initiated several rehabilitation projects in the various Soviet large-panel districts, improving roofs, repainting walls, and rehabilitating public areas. Today, more than 65,000 inhabitants live in 384 large-panel buildings that contain 11,741 apartments, most of which were built using the Soviet large-panel system (Sierra Maestra 2009). The dwellings in the José Martí District still retain their virtues, described by their inhabitants as "eternal construction," "made for war," and "rustic, but durable, strong, the best!" (Quevedo 2012). In Marta Noriega's words, "they are dull but of great quality" (2012). Furthermore, when their buildings withstood the earthquake that hit

Haiti with such devastating effects in 2010, residents remarked, "Nobody thought this would fall, nobody!" (Arce 2012).

Many factory workers and residents continue to identity with the Soviet large-panel buildings, perhaps because this tropical assemblage is linked in their minds to a number of survival experiences, such as battles of production, natural disasters, the cold war, and economic crises. This has motivated workers to protect the old factory and adapt production processes so it can continue. The present restoration of the Soviet large-panel factory and the apartment buildings is also part of a process to reevaluate the importance of these industrial and architectural elements in the collective identity of the city and the country.

Conclusion

A characteristic aspect of large-panel concrete technologies was the transfer of the structural weight and technical demands of the building to the panel itself. In the case of the Cuban version of the Soviet large panel, an even greater ideological weight was placed on the panel as a symbol and agent of social transformation. The Cuban interventions and alterations to the integrity of the original Soviet panel in the form of perforations, recesses, and indentations are particularly revealing if we see them as in opposition to the tabula rasa initially offered by an imported product that had pretended to be neutral, extemporal, and global.

Moreover, the study of Cuban Soviet-style large panels reveals how visual controversies may also become technological controversies as a foreign building component is altered according to local aesthetic sensibilities. The history of the technology traced in this chapter also shows that propaganda operated as an important means of supporting technological assimilation. The earliest scientific and cultural institutions of the Cuban Revolution[6] produced a particular partnership strategy between prefabrication and film that was intended to build an imaginary assembly between ideology and technology. An examination of that process allows us to observe the manner in which one of the most influential Latin American film traditions also served as a fundamental agent in the production of narratives for technological assimilation. Through film, these technologies were either humanized or framed within a military rhetoric to eliminate any possible controversy (e.g., foreign influence, creative limitations, technocracy, monotony), thereby becoming technological "facts" as indicators of the social country's new reality.

This history also demonstrates that military technologies were not the only technologies to have an active role in shaping the effects of the cold war in Latin America. Although the Soviets' donation of the factory was originally motivated by a desire to restore the weakened relationship between Cuba and the Soviet Union after the missile crisis, its implementation perfectly matched the Revolution's military and political objectives. The study of the Soviet large-panel technology thus permits us to widen the scope within which cold war military metaphors affected the understanding of the identity of some civilian technologies such as those associated with the construction of housing. In fact, in the Latin American context, this process has been seen as a positive one, at least regarding certain ideals of social transformation.

In sum, this examination of the history of the Soviet-style large panel shows how the social value of egalitarianism and the technological value of standardized prefabrication reinforced one another in Cuba. In this tropical assemblage, the redesign of the panel became integral to the Cuban Revolution, to the mix of the social and the technological, the people and the machines, at the core of revolutionary discourse. Building socialism is not only ideological but also material, which we can especially see in the building of large-panel housing projects. Not surprisingly, in Ché Guevara's "El socialismo y el hombre en Cuba," an essay published in 1965, he stated, "We can see, in this period of socialist construction, the birth of a New Man," that is, "a man more complete, in spite of his apparent standardization" (Guevara 1965, 15). At a certain level, the logic of prefabrication and assemblage encapsulated in the single panel is not so very different from the rhetoric underpinning the formation of such a socialist man, understood as a vital piece of the revolutionary process. Like standardized technologies, the New Man is more complete as part of a standardized collective. In this context, standardized prefabrication paralleled the idea of the Cuban New Man and contributed to the government's ability to promote this concept. Prefabricated housing is therefore both a reflection of and a means of producing Cuban socialist ideology, the large panel itself becoming a metaphor for the New Man.

Notes

1. However, the final production rates achieved only about 1,200 apartments annually (equivalent to a whole micro-district) because the factory's labor shifts never worked properly.

2. The first micro-district included 44 buildings, 1,200 apartments, and 6,300 inhabitants, with an estimated cost per district of 13 million of Cuban pesos (US $13 million in 1965, US $95 million in 2013). The Soviet theoretical model of urban distribution of population defined units as primary (2,000 inhabitants); micro-district (6,000); district (36,000); and city unit (50,000).

3. This kind of fluctuation is normal in the world of construction.

4. The film was released in a period when Cuban documentaries had already become a very important influence on the so-called new Latin American film, a movement that since 1967 had sought to build a "social reality" common to the region, one that was clearly opposed to the United States.

5. *No tenemos derecho a esperar* was screened at the Second National Conference of Cuban Architecture, October 1–5, 2011, Camagüey.

6. That is, the Technical Research Department of the Cuban Ministry of Construction and the Cuban Institute of Cinematographic Art and Industry.

References

Alonso, Pedro, and Hugo Palmarola. 2013. A panel's tale: The Soviet I-464 system and the politics of assemblage. In *Latin American Modern Architectures*, ed. Helen Gyger and Patricio del Real, 153–169. New York: Routledge.

Álvarez, Santiago, dir. 1963. *Ciclón*. Havana: ICAIC.

Anonymous. 1964. *Viviendas de grandes paneles para construcción en la República de Cuba*. Album 1, Part 1. Moscow: CNIIEP.

Arce, Raúl. 2012. Interview by Hugo Palmarola, May 13, Santiago de Cuba.

Azze, Edmundo. 2011. Interview by Hugo Palmarola, March 28, Havana.

Castro, Fidel. 1963a. Comparecencia ante las cámaras de televisión cubana para informar al pueblo sobre varios tópicos de interés general. *El Mundo*, October 22, 2–4.

Castro, Fidel. 1963b. Speech at the close of the VII Congress of the International Union of Architects, CTC Theater, October 3. Available at http://www.cuba.cu/gobierno/discursos/1963/esp/f031063e.html.

Castro, Fidel. 1965. Conclusiones sobre el Poder Local. *Cuba Socialista* 51 (November):14–15.

Castro, Fidel. 1972. Speech at the findings of the National Plenary of DESA, CTC Theater, February 13. Available at http://www.cuba.cu/gobierno/discursos/1972/esp/f130272e.html.

Castro, Rafael. 1986. *Rajayoga*. Santiago de Cuba: Editorial Oriente.

Chytilová, Věra, dir. 1979. *Panelstory (Panelstory aneb Jak se rodí sídliště)*. Prague: Film Studio Barrandov.

Comisión de Orientación Revolucionaria. 1972. *Revolución en el 72. Construcciones*. Havana: Comisión de Orientación Revolucionaria del Comité Central del Partido Comunista de Cuba y el Sector de la Construcción.

Cuba Construye. 1965. Planta soviética de prefabricado. *Cuba Construye* 1 (January-February):48.

D'Acosta, Hugo. 1964. La investigación y el desarrollo técnico en las construcciones de Cuba, después de la Revolución. *Arquitectura Cuba* 332 (April-May-June):37–57.

D'Hooghe, Alexander. 2005. Siberia as analogous territory: Soviet planning and the development of science towns. *AA Files* 51 (Winter):14–27.

Duarte, Rolando. 2012. Interview by Hugo Palmarola, May 11, Santiago de Cuba.

Edgerton, David. 1998. De l'innovation aux usages. Dix thèses éclectiques sur l'histoire des techniques. *Annales HSS* 4–5 (July-October):815–837.

Edgerton, David. 2007. *The Shock of the Old: Technology and Global History since 1900*. Oxford: Oxford University Press.

Fernández, Olga. 1974. Santiago la nueva. *Cuba Internacional* (February):29.

García Cuenca, José Antonio, dir. 1959. *Adelante cubanos*. Havana: Professional Films of Cuba.

García-Espinosa, Julio, dir. 1959. *La vivienda*. Havana: National Bureau of Culture of the Rebel Army.

Guevara, Ernesto. 1965. El socialismo y el hombre en Cuba. *Marcha* 1246 (March):14.

Iglesias, Israel. 1989. *Primer decenio; Las obras y los hombres*. Havana: Comisión de Historia del Comité del Partido Comunista de Cuba del Ministerio de la Construcción.

Latour, Bruno. 2005. *Reassembling the Social: An Introduction to Actor-Network Theory*. Oxford: Oxford University Press.

Lipovskaya, Natalia. 1974. Архитектура Кубы, первой социалистической страны Америки. Архитектура СССР 4:28–40.

Lucía [pen name]. 2012. Narrative paper sent by email to Hugo Palmarola on June 15.

Neumann, Hieronim, dir. 1982. *Block (Blok)*. Łódź: Semafor.

Noriega, Marta. 2012. Interview by Hugo Palmarola, May 13, Santiago de Cuba.

O'Reilly, Vitervo. 2011. Interview by Hugo Palmarola, March 15, Havana.

París, Rogelio, dir. 1972. *No tenemos derecho a esperar*. Havana: ICAIC.

París, Rogelio. 2012. Interview by Hugo Palmarola, May 4, Havana.

Quevedo, Roberto. 2012. Interview by Hugo Palmarola, May 13, Santiago de Cuba.

Rappaport, Gerbert, dir. 1963. *Cheryomushki* (Черемушки). Leningrad: Lenfilm.

Rodríguez, Víctor. 2011. Interview by Hugo Palmarola, March 13, Havana.

Ryazanov, Eldar, dir. 1975. *The Irony of Fate* (Ирония судьбы). Moscow: Mosfilm.

Sánchez, Jorge. 2010. *Romper la tensión del arco. Movimiento cubano de cine documental*. Havana: Ediciones ICAIC.

Sierra Maestra. 1965. Iniciada la construcción del Distrito José Martí. *Sierra Maestra*, December 19, 5.

Sierra Maestra. 1966. Alertas y produciendo. *Sierra Maestra*, June 1, 6.

Sierra Maestra. 1967a. Ayer: Barrio San Pedrito. Hoy: Distrito José Martí. *Sierra Maestra*, May 25, 7.

Sierra Maestra. 1967b. Rostros y frases del Distrito José Martí. *Sierra Maestra*, June 18, 5.

Sierra Maestra. 1967c. Será el Distrito José Martí nuestra ciudad del futuro. *Sierra Maestra*, February 17, 1.

Sierra Maestra. 1967d. Un presente para la historia futura. *Sierra Maestra*, July 9, 5.

Sierra Maestra. 1968a. Gran Panel. A unos kilómetros de los yanquis. *Sierra Maestra*, July 12, 6.

Sierra Maestra. 1968b. Hombres heroicos todos los días. *Sierra Maestra*, December 29.

Sierra Maestra. 2009. Los apartamentos vuelven a ser como nuevos. *Sierra Maestra*, November 27, 4.

Tarr, Béla, dir. 1982. *The Prefab People (Panelkapcsolat)*. Budapest: Mafilm.

Urbanización Distrito José Martí. 1965. *Programaciones-Memoria*. Havana: MICONS.

Vals, Maira. 2012. Interview by Hugo Palmarola, May 13, Santiago de Cuba.

9 Balancing Design: OLPC Engineers and ICT Translations at the Periphery

Anita Say Chan

A visit to the Peruvian Ministry of Education's website for its national One Laptop per Child (OLPC) initiative makes plain, for even a first-time visitor, the transformational promise of technological innovation.[1] After entering the site via a page that features a single profile of a young student with her gaze fixed into the screen of her XO laptop—the signature green-and-white computer of the infamous MIT-launched information and communication technologies for development (ICT4D) project—visitors can click a "Testimonials" link leading to accounts of newly minted XO users. Featured are nine photos taken in rural contexts, with quotes labeled as "testimonials" (although none is attributed to any named speaker) alongside each. Beside one photo of two girls working with an XO on a grassy pasture, as a stray goat grazes nearby, appears the quote, "Today we dream of a future, tomorrow we'll achieve it." Beside another of a student sketching lined graphics on her XO appears another quote, "Today [I'm building with] lines; tomorrow it will be a grand building." Others read "Today I discover, tomorrow I'll innovate" or "Today I'll take photos of my town, tomorrow I'll photograph the world." And beside a final image, where seven smiling students raise XOs over their heads, appears: "Today we have our OLPC, tomorrow we'll be prepared for the future." The page of anonymous quotations supplies no further information about the photos, or about what other parties, from local teachers to volunteer-based global engineers, were centrally involved with the program's dispersed deployments—a curiosity, given the program's heavy financial investments in Peru, where some US $82 million was spent to fund 900,000 XOs in public schools nationwide, as the centerpiece of the nation's digital education initiative (Talbot 2008). Neither does the Ministry's website make any reference to two recent Inter-American Development Bank (IDB) studies of the OLPC deployment in Peru (Santiago et al. 2010; Cristia et al. 2012)—the first major studies of the largest and among the oldest of all of OLPC's global partnerships.[2] While Latin American and Caribbean deployments of XOs account for 80 percent of OLPC's global distributions (across some 46 nations), roughly a third of the nearly 3 million XOs are in Peru alone (Cristia 2012).

Figure 9.1
Screenshot of the "Testimonials" page on the Ministry of Education's Una Laptop por Niño website, taken June 17, 2013, from http://www.perueduca.edu.pe/olpc/OLPC_testi.html. The captions read: "Today we dream of the future, tomorrow we'll achieve it" and "Today I'm learning to write, tomorrow I'll be a great author."

The 2012 IDB report—the first large-scale, randomized evaluation of OLPC's pedagogical impacts based on data collected over 15 months in 319 primary schools in rural Peru—notably placed the program in a critical light.[3] It found that although OLPC deployments in Peru generally increased student access to computing resources in schools and homes, they failed to measurably raise math or reading scores. As the report stated, "OLPC aims to improve learning in the poorest regions of the world. … The investments entailed are significant given that each laptop costs around $200, compared with $48 spent yearly per primary student in low-income countries and $555 in middle-income countries. Nonetheless, there is little solid evidence regarding the effectiveness of this program." Elsewhere, it was more pointed: "Although many countries are aggressively implementing the OLPC program ... [n]o evidence is found of effects on enrollment and test scores in Math and Language" (Cristia et al. 2012).[4] Among its findings in Peru were that schools using XOs evidenced little change in time allotments for classroom activities compared to pre-XO use; that there was minimal impact on student attendance rates or time spent on homework; that there was no evidence of improved reading habits;[5] and that there was little evidence that the quality of classroom instruction had improved. As one of the report's authors wrote in his IDB blog summarizing its findings, "The Jury Is Back: One Laptop Per Child Is Not Enough" (Ibarrarán 2012).

Such sober findings notwithstanding, it's clear that investment in ICTs and the ideas of elite global engineers and digital entrepreneurs continue to fundamentally impact national policies, particularly in relation to education. Educational technology policy analysts and reporters note not only the continuing expansion of XO counts in Peru and abroad, but also the broader growth in programs that aim to provide students with their own laptops (Bajak 2012; Cristia 2012; Cristia et al. 2012; Oppenheimer 2012; Severin and Capota 2011; Trucano 2012). A whole range of nations—17 at least are documented in Latin America as of 2011—had launched large-scale ICT for education (ICT4E) programs that generally claim to connect rural communities and provinces and draw them into global circuits of exchange (Oppenheimer 2012; Severin and Capota 2011). Tech industry and education policy analysts indeed credit OLPC with promoting such global trends, with private IT firms like Intel adding some 7 million of their own student-tailored Classmate PC netbooks to public schools across the region (Cristia 2012; Severin and Capota 2011),[6] and new developments in low-cost, student-centered computing products triggered in public and private sectors globally.[7] As IDB policy analyst Julián Cristia underscored after the release of the 2012 report, by that year OLPC had become "just the tip of the iceberg" (2012).

Indeed, there's little doubt that growing investments in new digital education projects will continue—likely faster than researchers can provide studies of broadly scaled plans, and seemingly despite the critical findings of related studies. While the latest reports might have provided significant opportunity for project leaders in OLPC and partnering governments to learn from past deployments and improve future ones, project leaders' reaction to the IDB's report in Peru (as we'll see in this chapter) has reflected to date a rejection of its main research findings, and a reiteration of original project framings that stressed the innovation and functionality of the XO's technical design, and framed it as the necessary and most expedient solution to a broad range of entrenched education-related problems in the "developing world" (Becerra 2012a; Becerra 2012b; Negroponte 2012). These ranged from underskilled teachers to "substandard school facilities" and the presumed absence of student "optimism about their future" (Becerra 2012a; Becerra 2102b). These rationales continued to be dispersed by both OLPC and Peruvian Ministry of Education officials in websites and presentations—often made before global audiences—where the prospect of rural change was linked to repeated images of a rural child's contact with a glowing XO laptop, set against a provincial landscape devoid of other human agents.[8]

Such framings—much like the rapid expansion of digital education projects like OLPC's across the region more generally—readily demonstrate the persistent power that the notion of technology and technological expertise as forms of "imported magic," critically highlighted by this volume, maintains in contemporary regional policies. Such a framing reads science and technology as sources of "magical" effects that are construed as "highly effective, universal, sometimes mysterious, and always as coming from somewhere else" (introduction to this volume; and see Marques 2005). Morgan Ames's chapter in this volume on the "charismatic power" of the XO laptop in Paraguay extends such a critical lens to the OLPC project's spread across the continent. Technology thus is projected as a powerful and necessarily transformational artifact, ideally drawn from centers of advanced innovation into peripheries. Key, then, to the spread of such seductions has been the expectation too of "converting" marginal actors—as philosopher of science Isabelle Stengers has characterized the modern will to reform—into new, future-enabled subjects (Stengers 2007). Implied under such a framing, as this chapter highlights, is not only the notion of innovation centers (understood as Western, urban, and globally discursive) as the definitive and discrete sources of magical solutions that can be diffused outward, but the notion of culturally and geographically peripheral sites as in definitive need of magical solutions. Such global diffusion narratives thus imagine technology as moving not only "from the West to the rest," but from the urban to the rural, from the cosmopolitan to the local, and from the globally networked to the remotely disconnected—in efforts to instantly remake and "convert" peripheries under the single guiding force and image of modern innovation centers. As Peruvian communication scholar Eduardo Villanueva describes it: "OLPC still believes in the power of one technological solution, transforming realities as varied as Afghanistan or Uruguay" (Villanueva 2011).

While OLPC's project leaders, following recent critical evaluations, demonstrated pointed resistance to rethinking project deployment processes or departing from the modes of center-periphery relations reinforced by imported-magic frameworks, other technicians centrally involved in Peru's local deployments developed other orientations altogether. Interviews with free and open source software (FLOSS) engineers—both Peruvian and US-based—volunteering with the project demonstrate how alternative deployment models and resources from those that OLPC and Ministry leadership emphasized could be identified through sustained engagement with local sites. Rather than presuming (or defending) the given "operability" of a foreign digital technology that could function in disparate contexts irrespective of other local factors, locally engaged engineers were able to identify social and technical aspects of the project that were in need of further development, as well as identify underrecognized potentials for leveraging the social expertise of local agents, rural teachers, and provincial community members (Chan, 2013).

This chapter emerges, then, from ethnographic fieldwork I conducted during spring 2011 and winter and spring 2012 while studying rural deployments of OLPC in Peru, with a particular focus on those in the southern Andean province of Puno.[9] While the region is best known as the nation's folklore capital, where large Aymara- and Quechua-speaking indigenous communities reside, it unexpectedly became a center of FLOSS activism in Peru following OLPC deployments there, generating the first teacher's manual for the OLPC, multiple XO-related conferences that mixed regional and international participants, XO software translation projects into Quechua and Aymara, and global media attention for rural teachers' independent efforts to reform OLPC deployments. During those periods, I interviewed some three dozen actors involved in local deployments—including rural teachers, Quechua- and Aymara-speaking educators and language activists, provincially based FLOSS coders and open technology advocates, as well as those from cities across Latin America, the United States, and Europe, Peruvian Ministry of Education officials, and US-based OLPC representatives.[10] The results of this fieldwork, and a focus on the dual engagements and critiques of rural teachers and indigenous-language activists, are compiled in a book on the competing visions of digital culture in information age Peru, as imagined and enacted through government and citizen networks.[11]

This chapter attends explicitly to the experience of engineers volunteering with OLPC, whose eventual investments in local deployments and engagements with diverse translocal agencies provided a means for them to develop their own alternatives to the imported-magic frameworks that still powerfully operate in regional technology-related policies.[12] While I met more than two dozen FLOSS advocates—both old and new to the project and based in scattered rural and urban sites across the United States, Latin America, and Europe—during my fieldwork, I focus here on the experiences of two specific FLOSS engineers volunteering with the project, one based in OLPC's site of origin in Cambridge, Massachusetts, and the other in Puno, Peru, whose ongoing personal histories with OLPC are among the longest possible with the program in their respective sites.[13] Yet the perspectives they came to develop around their roles as technical experts departed from the dominant framework of importing magical solutions for rural users, and instead underscored the potentials in reconfiguring engineers' prescribed relations with local deployment sites and the models of global relations they sustained. They stressed the productive possibilities of defining engineers' roles instead not just in processes of technological design, but in those of multidisciplinary collaboration and "translation" (Latour 2005). As STS scholars have observed, framings of local translation recognize the movement and transfer of technoscientific objects as always contingent processes, necessarily involving transformations that presume not the simple diffusion of science and technology from enlightened centers to peripheries, but technoscientific objects as indeed in constant states of change (Callon 1986; Latour 1987; Latour 2005). Likewise, locally engaged engineers' accounts highlight a framing of "the peripheral" and "rural" less as simple sites in need of conversion than as productive zones from which new understandings of deployment complexities could be cultivated. Given the continued expansion of new ICT4E initiatives well beyond OLPC (Nieusma and Riley 2010), such emerging accounts from locally engaged engineers offer constructive guides for identifying new and underrecognized resources that help address the growing challenge of extending and sustaining deployments in increasingly diverse global sites.

Scripting Reform

That new technologies could allow rural students to transcend local limitations and access a "future" they'd otherwise be denied is a promise Peru's Ministry of Education stressed since its OLPC program began in 2007. The project existed then as a pilot program in Arahuay, an agriculturally based Andean village of 740 residents, chosen—as the Ministry website explained—to connect the most disadvantaged students: "The OLPC program responds to the demand for educational quality and equity through the integration of information and communication technologies … especially in areas with the highest poverty [and] illiteracy … in rural areas."

Oscar Becerra, the Ministry's former chief educational technology officer and head of OLPC deployments, spoke to MIT's *Technology Review* magazine in the months before the pilot expanded exponentially to national scale. Although no official studies had yet been completed to evaluate the impacts of the pilot, and although even some early press accounts expressed concern that it was unclear what concrete results had been achieved, within a year some 400,000 additional XOs were deployed in 9,000 public schools (Economist 2008; Paul 2008; Paul 2009; Vota 2009). And Becerra, who had worked for 22 years at IBM before entering government,[14] stressed rural children's "future" as the stake: "Our hope for [a rural student] is that he will *have hope*. We are giving them the chance to look for a different future. … These children who didn't have any expectation about life, other than to become farmers, now can think about being engineers, designing computers, being teachers—as any other child should, worldwide" (Talbot 2008).

Ministry officials made clear, too, that problems they saw as anchored within local and rural sites were central motivators for its new initiatives (digital and otherwise). The launch of OLPC also coincided with the Ministry's promotion of the highly controversial Teacher Career Law (Ley de Carrera Pública Magisterial) that based teacher promotions and retentions on new national exams—sparking protests among 60,000 Peruvian teachers based in the provinces (Andean Currents 2007; Peru Support Group 2007).[15] Becerra pointedly referenced the shared objectives between Peru's OLPC project and the Teacher Career Law, stressing the stakes involved for Peru's global standing: "The low quality that public education has [in Peru is known]. ... The fundamental problem is that an education system can't improve if the teachers don't. The leading countries in education have teachers that perform best on international exams. This is what Peru will improve via its new Teacher Career Law. But this process will take 10 years. And in the meantime, what will we do with the children?" (Comercio 2011).

Such framings of local and rural teachers as an issue of national concern—and as the primary obstacle to education reform and Peru's future—that required globally expedient solutions like the XO were likewise echoed by OLPC leadership from the project's start. At the XO's debut at the 2005 World Information Society Summit in Tunis, Nicholas Negroponte, the chairman of the OLPC Foundation and founder of MIT's Media Lab, boldly predicted that 100 to 150 million XOs could be distributed worldwide by 2008 as global education solutions. Endorsing such ambitions, Kofi Annan hailed the XO as "an expression of global solidarity"—praise that made international headlines after Annan demonstrated the XO's powering by wind-up crank before a rapt audience of policy makers (Twist 2005).

Such claims invited scrutiny when, at the end of 2010, only 2 million XOs had been distributed (Negroponte 2010a). Although they were implemented across several dozen nations, only two—Peru and Uruguay—committed to more than 250,000 units (Andersen 2010; Warschauer and Ames 2010).[16] Explanations for the project's sobering performance emerged, with press accounts noting the failure to meet the US $100 benchmark unit price; unreliable XO features (including the mouse, keypad, and hand crank); and the US-based OLPC Foundation's inability to work earnestly with national government and local partners (Derndorfer 2010; Paul 2010; Toyama 2010; Toyama 2011). Academic studies especially cited Negroponte's unflinching "techno-determinism" as resulting in an overvaluation of technological promise and a refusal to learn lessons from local implementations (Santiago et al. 2010; Villanueva 2011; Warschauer and Ames 2010). Negroponte, however, still insisted that computers could only enhance a child's capacity to self-educate without being hampered by local factors, including schools or teachers (Negroponte 2010b). A long-time adherent to Seymour Papert's constructionist learning model, which viewed children as radically individualistic, self-directed learners and innate self-teachers, he flatly dismissed the need for project evaluation, stating in a September 2009 IDB forum speech: "That somebody in the room would say the impact [of the XO] is unclear is to me amazing—unbelievably amazing. ... There's only one question on the table and that's how to afford it. ... There is no other question" (Negroponte 2009; Warschauer and Ames 2010).

Such insistence that laptop access was "the only" relevant question suggested that existing education-centered actors—from teachers to local communities—were virtually negligible and could be regarded as external to deployments.[17] That such actors were effectively framed as impediments to "real" learning was reflected further in OLPC's five "core principles," which specified in detail the conditions under which laptops should be adopted: (i) children should be laptop owners, (ii) beneficiaries should be aged 6 to 12, (iii) every child and teacher should receive a laptop, (iv) connectivity should be through a local network or the Internet, and (v) software should be open source and free. Other local actors' roles in project deployment, however, remained conspicuously unaddressed—a point that local teachers critiqued in OLPC's early years in Peru's rural zones (Chan 2011). Their concerns were echoed by IDB reports that similarly cited the stark imbalance between OLPC's abundant instruction on student XO adoption and the neglect of other "essential" education factors: OLPC's "underlying vision is that students will improve their education by using the laptop and through collaboration with their peers. However, the OLPC portal provides limited information about how to integrate the computers provided into regular pedagogical practices, the role of the teachers and other components essential for the successful implementation of the model" (Cristia et al. 2012, 6).

The 2012 IDB evaluation of OLPC deployments in Peru added to the growing academic concerns about the program leadership's efficacy claims and deployment designs, even while similar digital education policies across the region continued to grow. Following the study's release, Oscar Becerra only reiterated the objectives he'd given in early rationales for Peru's heightened OLPC investments (Becerra 2012a, 2012b). Notably refraining from recognizing the IDB's larger concerns, he repeated the program's urgency based on underperforming public school teachers: "The children of Peru, and our country as whole, cannot afford to wait the 10 years required to substantially improve teacher quality. The government is investing in educational interventions, such as OLPC, that will help our children make progress now. ICT skills are key ingredients for success in the 21st century" (2012b). Negroponte likewise reacted to the IDB report by reiterating that granting a poor child a laptop is the key to "giving that child hope, self-esteem, and an opportunity to learn outside, as well as inside school," and dismissing again the need for studies like the IDB's, which he read as equivalent to "applying traditional 19th Century testing to modern learning" (Negroponte 2012).[18] OLPC likewise continued to promote its efficacy in "remote" contexts by reiterating imported-magic narratives for press reports—like coverage in *Wired* magazine's June 2012 issue that described "dropping" XOs into a "zero-literacy" Ethiopian village and enabling "kids who had never seen a computer before [to] learn quickly," and the claim in MIT's *Technology Review* that "given tablets but no teachers, Ethiopian children teach themselves" (Talbot 2012).

STS scholars have cautioned against the means by which such projections of idealized technological interactions by technical experts can foster a neglect of nontechnological elements involved in new design deployments. They observe how professional engineering norms conventionally encourage strict boundary management between local user populations and innovation experts, framing lay users as the design expert's other who must be managed, "configured," and "scripted" for effective design executions (Agre 1995; Akrich 1992; Grint and Woolgar 1997; Oudshoorn and Pinch 2003; Suchman 2007; Woolgar 1991). Emphasizing the means by which engineers strive for a perfect "configuring" and "scripting" of users, entailing detailed prescriptions for the generic user's interaction with products, STS critiques have shown how such models can risk cultivating exclusively technocentric approaches to design (Akrich 1992; Grint and Woolgar 1997; Woolgar 1991). Noting how such practices represent not merely ideal technological design but ideal *social world* design, Madeline Akrich writes: "Designers define actors with specific tastes, competences, motives, aspirations, political prejudices, and the rest, and they assume that morality, technology, science, and economy will evolve in particular ways. A key part of the work of innovation is that of inscribing this vision of (or prediction about) the world in the technical content of the new object" (Akrich 1992, 208). That OLPC could leave a breadth of local actors—from rural teachers to communities—unaccounted for in their scripted projections not only minimizes the role such actors play in actual (and idealized) education practices; it advances prescriptions for social worlds (and global models more broadly) in which these actors could be virtually omitted, and where the only crucial factor would be that of perfectly designed and operating foreign technology—worlds, in short, effectively defined by imported magic.[19]

Such an orientation, however, is not altogether inevitable. STS scholars have recognized technical experts as actors themselves embedded within cultural realms, who are managed by distinct interests (especially those of employers, whether corporate or state) and thus negotiate power dynamics. Feminist science studies scholar and Xerox Parc ethnographer Lucy Suchman argues, for instance, that despite pressures discouraging engineers' engagement with users, dominant design approaches don't necessarily reproduce themselves. Suchman underscores how design practices can also open opportunities for relational reconfigurations when recognized by self-conscious practitioners as explicitly cultural processes. While she notes that standard approaches to scripting user interactions indeed often lead to design as a realization of "cognitive control structures that universally precede and determine actions," she points to the possibility of designers' building other forms of self-awareness (Suchman 2007, 13). To cultivate consciousness of planning as a culturally "situated activity," and of plans as shared "cultural resources," encourages designers to reduce their privileging of ideals of design purity and the perfect configuration of users, and fosters new orientations around *reconfiguring design relations* themselves. Design could then be redefined as an "imaginative and discursive practice" between unpredictable and "unscripted mediators" and other actors involved in deployments, who could together consider design's implications—and translations—in local contexts (Latour 2005, 39). To attend to processes of translation as locally contingent, dynamically unscripted negotiations is to see the object of change as more than that *into which* technoscientific artifacts get channeled; to look beyond to a spectrum of other actors and interests. Bruno Latour emphasized this in reading translation as a "relation that does not transport causality" (2005, 108) from centers of planning ... and calculation to peripheries, but that gets negotiated by unpredictable actors—mediators—who refuse to "follow scripts, [and instead] transform, translate, distort, and modify meanings they are supposed to carry" (2005, 39). Distinct from diffusion models that focus on "the local" as lone object of change, translations open up a continuum of alternative reforms and unscripted possibilities in their process. The following accounts of OLPC engineers underscore then how, in developing alternatives to dominant frameworks of new technologies as imported magic, the participants found themselves reconfiguring both professional and user identities and the global relations they involved.

Rescripting Design

Bernie Innocenti, one of the longest-working engineers behind the XO computer's FLOSS-based operating system Sugar—who, with former MIT Media Lab director Walter Bender, later went on to cofound Sugar Labs to manage the project—makes no secret of his enthusiasm for coding. Of the numerous photos scattered through his website, codewiz.com, nearly all feature him interacting with computers or laptops, including OLPC's recognizable green and white XO laptops. In some, he's seen coding with concentration. In others, he's simply seen in an affable pose, smiling broadly and surrounded by computing equipment and fellow engineers, or proudly gesturing toward the products of his coding efforts.

While he works full-time as a software engineer for Google, coding as part of the search engine team in the company's Cambridge office, most of his personal website's content details his *unpaid* work, dating back more than a decade, as a volunteer programmer with various FLOSS projects. Such investments of volunteered labor are not uncommon for FLOSS users, many of whom are also skilled programmers and many of whom consider themselves participants in a global code-producing network that's frequently identified as a "community," "movement," and even "public" (Coleman 2011; Coleman 2013; Kelty 2005; Kelty 2008). Within the already exceptional volunteer-centered software projects and investments of the FLOSS world, though, Innocenti's include some of the most high-profile. These include his work as the senior system administrator for the Free Software Foundation, contributing to the Linux kernel and the GCC code compiler, and of course his work as an operating systems engineer for OLPC, which began shortly after he left his home country of Italy for the United States in 2007. In geek speak, that is, he's indisputably earned his "geek cred."

He continues to invest much of his work in OLPC on a volunteer basis, coding for Sugar and funding visits to international deployment sites across Asia, Africa, and

Latin America. Such commitments distinguish him among OLPC's US-based engineers, most of whom have never visited a single deployment site. But speaking to me in 2011,[20] he told me he wasn't always interested in deployment sites. Although he moved from Cambridge in the late 2000s for the opportunity to work as a paid developer for OLPC, his excitement with the project was anchored on his identification with Western hacker cultures. He recalled how the prospect of globally extending hacker ethics and its values of self-driven learning and problem solving through tinkering especially motivated his dedication to OLPC. Speaking abstractly for hacker collectives, he explained: "The project gave a lot of visibility to hacker culture, and what we believe is *real* learning versus what a traditional school defines as learning. We completely reject the notion of memorized knowledge; we completely reject the notion that you go to a school to get filled with knowledge. We do subscribe to the fact that everyone is a learner, and that everyone learns from each other, peer to peer." Pausing, he specified how OLPC allowed him to identify a language to describe the education-centered philosophy of hackers, adding simply: "This project embodied all our beliefs. Including beliefs like constructionism [on self-directed learning] that I didn't even know there were words for."[21]

Innocenti underscored the multiple identity-centered processes layered within design, making plain that as much as engineering for OLPC allowed him to "script" users—in the image of constructionism's self-directed learner, in this case—it also afforded him the opportunity to reflect upon his own relation to hacker identity. He was clear, though, that OLPC's construction of a workplace culture that drew talent from engineers professionally and academically trained at top universities, and that proudly privileged the values of engineering enthusiasts, gave little encouragement to considering other values. And it left little space to consider the possible benefits of engagement with nonengineers or local deployment sites. While he remembered hearing about Nigerian and Brazilian deployments in his early years with OLPC, for instance, he confessed he never felt the need to know more about them: "I was totally uninterested in deployments back then, and that was in part because my technical background prevented me from seeing that [local] deployment was actually where the action was—where you apply the things you developed … to see if it actually did work. … As engineers, we felt our place was in Boston, sitting at a desk, and we weren't interested in going to the field."

Innocenti's initial lack of curiosity about deployment contexts, and his single-minded embrace of dominant engineering norms, was something he recalled as reflected by leadership at various stages of OLPC's development. He recalled that such an engineering-centric vision ran deep enough that even minimal means of engaging with local users and contexts were evaded: "For a project that was founded by academics, we were incredibly *not* data driven. … Regrettably, all of Sugar was designed without any children using it—[with] almost zero input from children or teachers.

It was built by people who were visionaries. Typically when you build any project ... you really need a bunch of loops, of iterations, and interactions with users so that you can figure out what works. And we never did that. To date, there are few cases where we went to the field, found the problem, and then fixed it. ... We are lucky that we at least [came] close."

As project complications began to hinder growth, he specified that several pedagogical experts were hired to correct what became an increasingly technocentric, non-educationally-focused design process. Even with such organizational adjustments, however, project leadership refrained from pushing for coordination between "Learning Team" members and project engineers within the OLPC offices, or cultivating more interest in international deployment sites. There was a persistent resistance, he recalled, to collaborating with non-technically-centered actors: "On one hand, none of us engineers had any interest in testing deployments, or just seeing how things ran in the hands of real users. On the other hand, [we thought] there were other people assigned to do this, [since now] there was a Learning Team. ... [So] almost nothing came out of it—no content, no lesson plans, no learning materials—[and] only small feedback from the field." He mentioned that part of the communication difficulties had to do with the fact that project engineers used distinct means of coordination—online IRC channels—independently of the Learning Team. And there remained little organizational pressure or encouragement to change routines: "I'm really not sure what [the Learning Team] used. Maybe we were wrong and should have used more [paper] documents or something else. But the end result was that there was not enough communication and interchange."

Such a structured separation not just from XO users but from pedagogical and nonengineering experts working on the project expressed a failure to recognize the value of multidisciplinary collaboration and translocal knowledge exchange. By Innocenti's account, it was such an exclusive focus on engineering expertise, and unwillingness to recognize the limitations of project configurations, that eventually compromised technical functionality: "For a while, we were all blinded in the same way. ... Most of us were going on about how great the project was ... when in fact many schools were rejecting the software or hardware or both because of problems we never envisioned. For example, the networking function we did wouldn't scale to more than 6 laptops. ... Because the test cases we designed were trivial things that wouldn't show this limitation. But of course, [had we tested] any small deployment, there'd have been enough children to show this problem."

He noted that more recent organizational adjustments were made to improve design and development issues, and to encourage more exchange among OLPC's pedagogical experts and engineers. Still, the challenges of interweaving multiple expertise sets within the organization, working across geographically dispersed sites, and attending to the feedback of teachers and communities globally continue to face

OLPC. As he put it: "The problem that still exists is that we speak different languages. And I don't mean the Spanish and English issue, although there is that as well. But that's a minor issue. The actual issue is that engineers and teachers speak really two different jargons, and [as an engineer] I'm not necessarily interested in the same topics. And we don't communicate well unless we have some people bridging between us." He noted that however much the technical challenges afforded by OLPC might have drawn him into the initiative, the challenge that remains is that of improving communication channels between diverse communities of use and development.

While difficult, he noted that such balanced, multidisciplinary exchanges were not ones he saw as impossible to achieve, even across diversely situated global sites. He pointed, for instance, to ongoing collaborations among local teachers and engineers based in Puno whom he first visited in late 2011. That year had been the third consecutive year that regional teachers, engineers, FLOSS advocates, and indigenous-language activists had organized an international workshop to discuss XO integration into local schools. Such activity resulted in gatherings that drew more than 400 teachers from across the villages and towns of Puno in the 2011 workshop, and the recent launching of a local civic association, Escuelab Puno, that supported local educational technology use via a mixed team of five local teachers and FLOSS advocates (Chan 2011; Chan 2013). Innocenti's 2011 visit with Escuelab Puno allowed him to observe the young, multidisciplinary collective's early work coordinating pedagogical and engineering support for local schools with XOs. Impressed by their work—which interwove distinct expertise sets as OLPC leadership in the United States had found challenging to do, and which also extended volunteer networks to include elder and youth indigenous-language activists from the provinces as well as contributions from global open technology activists—he made a modest $2,500 donation to the effort and encouraged Google to do the same. Explaining the decision, he underscored a key point he had become increasingly conscious of in his own work with OLPC, one he repeated several times as we spoke. The key involved what he called "good engineering," which is about more than either engineering genius or design purity. As he put it, "Good engineering is about finding balance—between a lot of different things."

Critiquing Conversion

Puno, Peru's southernmost province and "folklore capital," is perhaps best known to the outside world for its large indigenous populations, cold stretches of Andean altiplano, and quinoa, potato, and alpaca wool production. Indeed, with the indigenous languages of Aymara and Quechua still widely spoken in the region[22] and a high-altitude climate that contrasts with Lima's winterless coastal climate, Puno often figures as the remote other to Lima's modern, mestizo, urban cosmopolitanism (Jacobsen 1993). But even prior to Innocenti's visit, efforts there had begun to earn some unexpected global visibility for educational technology projects and local XO deployments—even while such efforts lacked official support or endorsements from considerably better-resourced national or global institutions, including either Peru's Ministry of Education or OLPC leadership in the United States. It was from Puno that the first XO-based user's manual—a 100-page, teacher-centered text distributed online and translated from Spanish into English, French, and Arabic—was published (Salas 2009). It was there, too, in parallel with some of the largest conferences for rural teachers' XO use, that workshops for translating XO software into Quechua and Aymara began, organized with indigenous-language activists and elders, and aiming to be among the XO's first indigenous-language localizations.

Among the five founding members of Escuelab Puno (who represent an almost even mix of trained engineers and teachers from the region), however, there has been little detectable sense of exhilaration at this global distinction.[23] Whatever the claims expressed by Peru's Ministry of Education or the US-based OLPC leadership about what IT will bring to "peripheral" sites like Puno, Escuelab Puno's founders, since I first began fieldwork in the region in 2010, have shown little sense of collective confidence in such claims. This became clear not long after I returned to Puno in mid-2011 and interviewed Neyder Achahuanco, the young systems engineer and part-time teacher who helped cofound Escuelab Puno with fellow Puno FLOSS advocates. His investments in FLOSS projects started when he was a high school student volunteering to help local teachers and fellow FLOSS advocates migrate schools and universities from Windows to FLOSS platforms. Although only twenty-four years old when we spoke, his observations on Peru's "digital education" initiatives—and the growing interactions they've entailed between education and engineering specialists—already spanned various cases at local, national, and now (with OLPC) global scales.[24] With some 30,000 XOs deployed in Puno, he made little attempt to veil the critique he had formed after repeatedly witnessing the unfolding of diffusion-centered models that privileged engineering expertise and marginalized teachers' input. As he stated flatly: "This model no longer works—having an engineer in front of someone [in training workshops], at the front of a room with a group of teachers just listening, just does not work. It doesn't work because engineers weren't interested in understanding teachers, and teachers are not interested in becoming engineers."

He was clear too that part of Escuelab Puno's mission was to caution against the risks of imported-magic orientations that reify the "self-evident" need for rural transformation without envisioning parallel reform potentials in other sites, and that amplify the discrete solution-making capacities of IT and urban technology experts. He specified that among the factors that distinguish Escuelab Puno's deployment approach was its recognition that such techno-conversionist urges and their dependence on narratives of imported magic are part of the problem, rather than the solution: "We've become very critical of this idea embraced by many technology projects that the only thing that will save education, [or] improve society, is to throw technology at it. To say, 'Here, take this technology, your magic wand to escape from poverty! Here are your green laptops, your magic wand to improve students. ... [We forget] there's a huge, complex, diverse, and highly multidisciplinary process in what we call education."

He explained that part of Escuelab Puno's philosophy is reflected in its aims to foster spaces of dialogue between engineering and pedagogical experts by balancing such representation in its leadership. And he explained that one of the primary objectives that inspired its founding was to facilitate multidisciplinary collaborations around XO deployments—specifically through a "Partnership Program" the collective proposed to regional and national ministry officials—that would create one-to-one teacher-engineer partnerships in classrooms to collaboratively design local XO instruction techniques and materials. The idea, he explained, was one that emerged only through the involvement of nonengineers in Escuelab Puno's projects: "The idea of the 'Partnership Project' didn't suddenly just dawn on us overnight. It was the outcome of almost a year of collaboration with educators, sociologists, Aymara and Quechua representatives—discussions around multiple interests, daily work, and work in rural sites." He insisted, moreover, that the partnership model they've promoted would operate as much to the benefit of engineers, and their means of deriving solutions through technologies, as to that of local teachers: "Our goal, we realized, was to improve a social problem [rather than a technological one]. Under the direction of just engineers, [we would never] have seen that the problem could be engineers. [Since] engineers always think of themselves as bringing solutions. ... [But] through having multidisciplinary input ... we realized that the problem was with us, in how we thought, and in what we said was the miracle solution and magic wand that would resolve everything. We realized we were creating more problems than solutions for teachers." Pausing for a moment, he underscored then how the process of technological translation might in fact operate as much to decenter and reform the technologist' consciousness as to localize technological artifacts, adding: "But we only achieved this after we sat around the table with everyone together. Only then could we *really see* what we were all doing ... as engineers, teachers, sociologists, linguists, or ordinary people."

Attending to distinct local voices of engineers involved in OLPC deployment sites in Puno reveals the awareness technical experts can develop of the limitations of diffusion models, and of the need to reconfigure design relations for processes of translation instead. Despite OLPC leadership's insistence on adhering to dominant practices of closed-world design and reaffirmation of imported-magic narratives, critical responses emerged from locally engaged engineers who came to observe this approach's incompatibilities with actual deployments' need to learn, grow, and achieve sustainability. Achahuanco thus diagnosed the limitations of applying locally disengaged diffusion models, and reframed local actors and nonengineering experts—from rural teachers to indigenous-language leaders—not as risks or irrelevancies that designers should aim to reform, but as potential collaborators who could diversify expertise sets involved in the deployments. More, that is, than exposing the need to transform local sites, rural zones, and their "underdeveloped" users, such critical perspectives exposed the potentials for engineers' own professional realignments—and a recognition of the "local" as a constructive resource that could aid longer-term sustainability, remaking deployments as multidisciplinary engagements.

Conclusion: Countering Fundamentalisms

A question emerges about the persistent power and regional dominance of the image of science and technology as forms of imported magic. This volume opens a much-needed path toward analyzing the underexamined seductions of technoscience as source of "magical" effects, reforms, and solutions (introduction to this volume; Marques 2005). This chapter asks that we consider the persuasive constructions that have been at work, too, around the projected "other" of technology's imported magic. That projects like OLPC continue to expand internationally, despite expressed concerns about whether technological access may actually improve student learning (as the stated goals of OLPC leadership in the United States and of government officials from participating national partners promise) evidences a twenty-first-century faith—what technology studies scholars have cautioned against as a "techno-fundamentalism" (de la Peña 2006; Vaidhyanathan 2006)—in the *irrefutably* transformative power of information and communication technologies. Coupled with equally entrenched convictions about "the local" and "the rural" as *irrefutably* worthy objects of conversion, the mix takes on potent contours indeed. Particularly when they have been echoed by distinct channels of national state officials, global policy makers, development experts, and engineering professionals, such fundamentalisms prove to result in the aggressive and rapid pursuit of large-scale national initiatives around new ICT integrations, regardless of the hefty expenditures and economic resources they demand (and can draw away from other education-based projects)—and often in spite of research on their questionable pedagogical benefits (Cristia et al. 2012).

Such minimizing of critical perspectives and cautionary adoption practices in ICT for education programs are concerning indeed; as are renewed insistences by OLPC leaders like Negroponte that technology access alone, and simply "dropping" XOs into remote villages, can result in children "teaching themselves" (Hachman 2011; Talbot 2012). It's no surprise either that there have been recent contrary calls to designate OLPC deployments "a failure" altogether (Economist 2012; Hachman 2012; Ibarrarán 2012; Kessler 2012; Watters 2012). But attending to the voices of the local technicians involved with deployments of OLPC in Peru reveals ways we might constructively move beyond contemporary "techno-fundamentalisms"—as either a uniform acceptance or rejection of newly proposed technological "solutions." What OLPC engineers like Innocenti underscore as necessary for the productive deployment and sustainability of new technologies are "balanced" approaches that recognize technological design as a negotiation between diverse interests and forces—in which engineering expertise is but one factor. Engagements with local deployment sites offer opportunities for such a "balanced" consciousness to be cultivated—where change may be a shared objective, but where the engineer is framed less as the lone agent responsible than as a key collaborators among diverse actors invested in deployments, and is enrolled as an actor whose *own* technical practice can likewise be productively transformed.[25]

Innocenti's and Achahuanco's reflections on the value of multidisciplinary collaborations underscore how encounters with nonengineers and local deployment sites could cultivate an engaged self-awareness that allowed technicians not only to newly "see" their practice in relation to other local experts—including rural teachers and local communities (and to recognize the limitations of their own prescribed projections)—but also to identify new potential resources to aid deployments' future sustainability. Rather than framing contact with local users, nonengineers, and rural communities as a liability to the essential work of design (as professional norms and state-centered framings encouraged them to do), both Innocenti and Achahuanco came to read local collaborations as value-adding checks on dominant frameworks. Such collaborative modes reconfigure norms of technological deployment and innovation around processes of multidisciplinary relating and multidirectional agency which recognize rural sites as more than passive zones subject to the reforms of magical imports. Their accounts indeed promote an understanding of innovation that feminist and Latin American STS scholars have stressed as a process constructed through and *for* new social and technological configurations (de Greiff and Maldonado 2011; Kreimer 2011, Suchman 2007). As de Greiff and Maldonado write: "Innovation understood as the effective social incorporation of knowledge in the solution of problems, or in the establishment of *new relations*, [calls for] the interaction between groups and social cultures of experts and nonexperts, which obligates not only a *new social contract*, but the transformation of the means and ends of scientific communication" (2011, 247).

That the global emerges more and more as the desired end of key realms of technoscientific communication today thus presents both new challenges and new opportunities for innovation cultures. Among the assertions this volume underscores is how diverse the landscape of voices, actors, and processes of localization are when technoscientific initiatives unfold on such globally extensive scales as that of OLPC.[26] And attending to technoscience's movement among global sites provides powerful reminders of how the ambition of universal projects has long been that of making certain "local interests global" (Latour 2005, 178). Such ambitions are never frictionless, but always require distinct means and strategies of localization as they come into contact with diverse local spaces. Knowledge and new technologies, that is, never simply diffuse automatically from centers of production, planning, and calculation into remote peripheries—like forms of imported magic—but depend on processes of local translation among involved actors who negotiate transformation processes as multidirectional potentials. What the strategically positioned voices of technical experts like Innocenti and Achahuanco also remind us, however, is how much they rely on diverse processes of exchange—and how such exchanges may be used to productively balance the risks of asserted fundamentalisms, including those around magical imports, in technological deployments. Their accounts make visible how local voices, critiques, and interests may at once travel and trickle *up* to create value in exchanges with centers—moving beyond the standard technology narrative of center-to-periphery transfers and highlighting new potential value for future designs and deployments.

Notes

1. See the website for the Ministry of Education's OLPC Project: http://www.perueduca.edu.pe/olpc/OLPC_Home.html.

2. A status the country has maintained since collaborations with the US-based, MIT-launched computing project began in 2007.

3. While the 2010 IDB report presented preliminary findings of short-term observations that likewise placed the program in a critical light, the report authors acknowledged that longer durations were required for verification of their findings, which should be completed in future studies. The 2012 report that followed indicated that "OLPC in Peru dramatically increased access to computers. There were 1.18 computers per student in the treatment group, compared with 0.12 in control schools. … Eighty-two percent of treatment students reported using a computer at school in the previous week compared with 26 percent in the control group. Effects on home computer use are also large: 42 percent of treatment students report using a computer at home in the previous week versus 4 percent in the control group. The majority of treatment students showed general competence in operating the laptops in tasks related to operating core applications (for example, a word processor) and searching for information on the computer. Internet use was limited because hardly any schools in the study sample had access. Turning to educational outcomes, we find no evidence that the program increased learning in Math or Language. The estimated effect on the average Math and Language score is 0.003 standard deviations" (Cristia et al. 2012, 2).

4. In Peru, educational spending per student was estimated at $438 in 2008.

5. As the IDB report stated, "this is perhaps surprising given that the program substantially affected the availability of books to students. The laptops came loaded with 200 books, and only 26 percent of students in the control group had more than five books in their homes" (Cristia et al. 2012).

6. Peru's own annual deployments of XOs demonstrate a steady climb. By the end of 2008, some 44,700 laptops had been distributed to public schools; by the end of 2009, another 136,600 were added; by the end of 2010, another 109,300 had been added; by the end of 2011, the total was incremented by some 425,400 computers. Although the Ministry of Education made no commitment in 2012 to purchase more XO units itself, it has been supporting new efforts by regional and local governments to expand their XO deployments. These include thousands of new XOs that were purchased by the regional governments of Amazonas and Cuzco, which expected to add another 96,700 XOs to Peru's count by the end of 2012.

7. The website lilputing.com tracks this new market for low-cost, ultraportable machines. It reported that by 2008, some two dozen models were in development; today it estimates that there are hundreds that would fall into the category. See http://lilputing.com/2008/04/over-past-six-months-or-so-asus-everex_24.html.

8. This is the familiar image of the computer-ready child that modern audiences have come to recognize as indicative of future-readiness, that is, except for the rural setting.

9. This also builds on fieldwork I conducted between 2003 and 2011 on free software networks and digital innovation programs in Peru, which overlapped with the OLPC initiative. During that time, I conducted more than 100 interviews with student activists, hobbyist programmers, nonprofit organization personnel, government workers, provincially based teachers, rural producers and communities, representatives from transnational IT companies, and young entrepreneurs involved in FLOSS advocacy in Peru and Latin America more broadly. I also interviewed government officials working under the Ministry of Education and Ministry of Tourism and Trade who were stewards of innovation programs that targeted rural areas in Peru, including the deployment of digital education programs with the XO laptop.

10. See Chan (2013). This longer study represents fieldwork conducted between 2003 and 2012, and covers in part the engagements of free and open source software communities in Peru, the first national site of the United Nations Conference on the use of FLOSS by Latin American and Caribbean governments in 2003, and one of the first nations where legislation to mandate the use of FLOSS in public offices was proposed in 2001. The book also discusses OLPC-related interventions and critiques issued by FLOSS advocates in Peru after the Ministry of Education's national distributions of XOs began in 2007.

11. For a more focused study on the forms of engagement by Quechua and Aymara indigenous leaders, teachers, and language activists, see Chan (2013).

12. These include in Peru and Latin America: Neyder Achahuanco, Irma Alvarez, Francisco Ancco Rodríguez, Vanessa Bosquez, Vladimir Castro, Aymar Ccopaccaty, Koke Contreras, David Cruz, Andres Mauro Cruz Layme, Pilar Saenz, Walter Oswaldo Flores, Alfredo Gutiérrez, Raul Hugo, Illariy T'ika Kispi, Kiko Mayoraga, Ruben Monroy, Eleazar Pacho, Sdenka Salas, Sebastian Silva, Laura Silva, Walter Zamalloa, and Sandra Barragán. And in the United States and Europe: Chris Ball, Walter Bender, Barbara Barry, Pia Campiasino, Luiz Damiao, Bernie Innocenti, Samuel S. J. Klein, Chris Leonard, Alexsy Lim, Daniel Miracle.

13. The project was launched in 2005 from Cambridge, Massachusetts. It arrived in 2007 in Peru. Bernie Innocenti, the Cambridge-based engineer, volunteered with OLPC starting in 2005. Neyder Achahuanco, the Puno-based engineer, marked his first volunteer efforts with OLPC in 2009, when the program arrived in Puno.

14. Becerra worked at IBM as their Latin America education segment manager.

15. Peru had an estimated 313,000 public school teachers in 2012. See http://www.andina.com.pe/espanol/noticia-mas-488-mil-docentes-del-peru-celebran-manana-dia-del-maestro-419089.aspx#.UtSp33krj1o.

16. The minimum number initially required by OLPC. Only Peru came close to approaching the 1 million units originally demanded as the minimum when the project launched in 2005.

17. That such actors were effectively framed as impediments to "real" learning was reflected further in OLPC's mission statement, which contrasted images of existing classroom experiences to the change the XO experience offered: "[We aim] to create educational opportunities for the world's poorest children by providing each child with a rugged, low-cost, low-power, connected laptop with content and software designed for collaborative, joyful, *self-empowered* learning. When children have *access to this type of tool* they get engaged in their own education. They learn, share, create, and collaborate. They *become* connected to each other, to the world and to a brighter future." (Italics added.) See http://www.olpc.in/themission_mission.html.

18. The passage reads in full: "Think of the logic behind applying traditional 19th Century testing to modern learning, especially at early ages. High test scores come from rote learning, and do not evaluate critical and creative thinking, initiative and discovery, let alone peer to peer teaching. It is like using a pedometer to measure the speed of a car. Error." (Negroponte 2012.)

19. STS scholars have indicated too how professional norms that frequently distance professional engineers from users can further reinforce technocentric visions of design. Research has pointed to commercial logics and concerns around intellectual property protection as prescribing a strict separation between engineers and users in actual contexts of technological adoption (Agre 1995; Suchman 2007). Such obligations to remain isolated during innovation cycles, and minimize outside contact, can lead to abstracted concepts of "the user," or what Philip Agre characterizes as a "closed world" of systems development. Under such conditions, he notes: "it becomes difficult to imagine the perspective of someone who does not view a computer system as a logical anatomy, an ontology made of data structures, a set of formal relationships and constraints, and a network of paths for data to move along" (1995, 73). He adds such tendencies can be particularly strong within corporations, whose "culture of programming has historically had an element of monastic withdrawal … [that can develop] to a point of us-and-them hostility to the distant and demanding user" (1995, 73). Under contexts of fast-paced, high-intensity development cycles, the tendency to read local users as mere distractions from the real work of programming, or as nontechnical lay people who have little likelihood of providing relevant feedback, can further flourish—reinforcing technological diffusion models.

20. Most of the cited portions of my interview with Bernie Innocenti were drawn from over four hours of open-ended interviewing in Cambridge, Massachusetts, recorded on January 5 and January 6, 2011. The cited portions of my open-ended interview with Neyder Achahuanco were drawn from a three-hour recorded interview, conducted in Spanish, on May 24, 2011. This was conducted while I was conducting a month's work of fieldwork in Puno, and had relocated to the site. (All translations from Spanish to English are mine.) I conducted additional interviews with Achahuanco in 2010 and 2012 that touched on related issues of Puno's free software community, technology activism in rural Peru, local IT projects, and their connection to broader IT initiatives across the nation, region, continent, and globally.

21. Innocenti also specifies that part of the personal reward from joining OLPC's development team was that it pushed his technical skills, a familiar source of reward among hacker communities: "When I joined OLPC, my technical expertise was a little old, since I had been working as a manager [at an IT company I founded in Italy] for the past three or four years, and was not undertaking enough technical work. So my technical skills had fallen behind. But at OLPC I could learn and relearn new things; and I refreshed my knowledge on low-level Linux kernel [programming and various] system-level details."

22. Aymara is spoken by an estimated 41 percent of Puno's 1.3 million inhabitants, and Quechua spoken by some 30 percent.

23. The founding engineers in alphabetical order are Neyder Achahuanco, David Cruz, and Ruben Monroy, although Achahuanco also works part time as a computing teacher at an elementary school in Puno. The founding teachers are Eleazar Mamani Pacho and Walter Zamalloa.

24. Plan Huascaran, the national digital education initiative launched under Alejandro Toledo's administration in 2001, preceded the state's OLPC investments.

25. Rather than invoking the common designer's mantra of users "not always knowing what they want," a distinct awareness of the limits of engineers' own conscious projections and pre-scripted objectives in design could come to light.

26. There is a larger question of why institutional actors across Latin America have been so uniquely receptive to the project of OLPC, with well over 60 percent of the 3 million XOs in global circulation distributed in this region. Addressing such a question, while enormously worthwhile, is beyond the scope and methodology of this chapter.

References

Agre, Philip. 1995. Conceptions of the user in computer systems design. In *Social and Interactional Dimensions of Human-Computer Interfaces*, ed. Peter Thomas, 67–106. Cambridge: Cambridge University Press.

Akrich, Madeleine. 1992. The de-scription of technical objects. In *Shaping Technology/Building Society: Studies in Sociotechnical Change*, ed. Wiebe E. Bijker and John Law. Cambridge, MA: MIT Press.

Andean Currents. 2007. "Peruvian teachers protest turns violent." *Andean Currents*, July 13. Available at http://www.andeancurrents.com/2007/07/peruvian-teachers-protest-turns-violent.html (accessed June 1, 2011).

Andersen, Lars Bo. 2011. A travelogue of a hundred laptops. Master's thesis, Aarhus University.

Bajak, Frank. 2012. Peru's ambitious laptop program gets mixed grades. Associated Press. Available at http://bigstory.ap.org/article/perus-ambitious-laptop-program-gets-mixed-grades (accessed May 1, 2012).

Becerra, Oscar. 2012a. An alternative reading of the IADB study on Peru's OLPC implementation. *OLPC News*, April 18. Available at http://www.olpcnews.com/countries/peru/an_alternative_reading_of_the_iadb_study_on_peru_olpc_implementation.html (accessed May 1, 2013).

Becerra, Oscar. 2012b. Letter to the editors of the *Economist*. *OLPC News*, April 16. Available at http://www.olpcnews.com/countries/peru/oscar_becerras_letter_to_the_editors_of_the_economist.html (accessed May 1, 2013).

Callon, Michel. 1986. Some elements of a sociology of translation: Domestication of the scallops and the fishermen of St Brieuc Bay. In *Power, Action and Belief: A New Sociology of Knowledge*, ed. John Law, 196–233. London: Routledge Press.

Chan, Anita Say. 2011. Peripheral interrupt: Hacking digital universalism in the Andes. Presentation at the Society for the Social Studies of Science conference, Cleveland, November 3.

Chan, Anita Say. 2013. *Networking Peripheries: Technological Futures and the Myth of Digital Universalism*. Cambridge, MA: MIT Press.

Coleman, Gabriella. 2011. Anonymous: From the lulz to collective action. *The New Everyday*, March 2011. Available at http://mediacommons.futureofthebook.org/tne/pieces/anonymous-lulz-collective-action (accessed February 1, 2012).

Coleman, Gabriella. 2013. *Coding Freedom: Hacker Pleasure and the Ethics of Free and Open Source Software*. Princeton: Princeton University Press.

Comercio. 2011. Una Laptop por Niño: Por una educación de calidad con equidad. *El Comercio*, January 17.

Cristia, Julián. 2012. One Laptop per Child in Peru: Findings and the road forward. *IDB Blogs*, February 28. Available at http://blogs.iadb.org/education/2013/02/28/one-laptop-per-child-in-peru-findings-and-the-road-forward/ (accessed May 1, 2013).

Cristia, Julián, Pablo Ibarrarán, Santiago Cueto, Ana Santiago, and Eugenio Severin. 2012. *Evidence from the One Laptop per Child Program*. Washington, DC: Inter-American Development Bank.

de Greiff, Alexis, and Oscar Javier Maldonado. 2011. Apropiación fuerte del conocimiento: Una propuesta para construir políticas inclusivas de ciencia, tecnología, e innovación en América Latina. In *Estudio social de la ciencia y la tecnología desde América Latina*, ed. Antonio Arellano Hernández and Pablo Kreimer, 209–262. Bogotá: Siglo de Hombre Editores.

de la Peña, Carolyn. 2006. "Slow and low progress," or why American studies should do technology. *American Quarterly* 58 (3):915–941.

Derndorfer, Cristoph. 2010. OLPC in Peru: A problematic Una Laptop por Niño program. *Edutech Debate*, October 28. Available at http://edutechdebate.org/olpc-in-south-america/olpc-in-peru-one-laptop-per-child-problems/ (accessed May 1, 2011).

Economist. 2008. One clunky laptop per child. *Economist*, January 4.

Economist. 2012. Education in Peru: Error message. *Economist*, April 7.

Grint, Keith, and Steve Woolgar. 1997. *The Machine at Work: Technology, Work and Organization.* New York: Polity Press.

Hachman, Mark. 2011. We'll throw OLPCs out of helicopters to teach kids to read. *PC Magazine*, November 2. Available at http://www.pcmag.com/article2/0,2817,2395763,00.asp (accessed May 1, 2012).

Hachman, Mark. 2012. OLPC fails students as tool for education. *PC Magazine*, April 9. Available at http://www.pcmag.com/article2/0,2817,2402756,00.asp (accessed January 13, 2014).

Ibarrarán, Pablo. 2012. And the jury is back: One Laptop per Child is not enough. *IDB Blogs*, March 6. Available at http://blogs.iadb.org/desarrolloefectivo_en/2012/03/06/and-the-jury-is-back-one-laptop-per-child-is-not-enough/ (accessed May 1, 2013).

Jacobsen, Nils. 1993. *Mirages of Transition: The Peruvian Altiplano, 1780–1930.* Berkeley: University of California Press.

Kelty, Chris. 2005. Free science. In *Perspectives on Free and Open Source Software*, ed. Joseph Feller, Brian Fitzgerald, Scott A. Hissam, and Karim R. Lakhani, 415–430. Cambridge, MA: MIT Press.

Kelty, Chris. 2008. *Two Bits: The Cultural Significance of Free Software.* Durham: Duke University Press.

Kessler, Sarah. 2012. 2.5 million laptops later, One Laptop per Child doesn't improve test scores. *Mashable*, April 9. Available at http://mashable.com/2012/04/09/one-laptop-per-child-study/ (accessed January 13, 2014).

Kreimer, Pablo. 2011. Desarmando ficciones: Problemas sociales–problemas de conocimiento en América Latina. In *Estudio social de la ciencia y la tecnología desde América Latina*, ed. Antonio Arellano Hernández and Pablo Kreimer, 127–165. Bogotá: Siglo de Hombre Editores.

Latour, Bruno. 1987. *Science in Action.* Cambridge, MA: Harvard University Press.

Latour, Bruno. 2005. *Reassembling the Social: An Introduction to Actor-Network-Theory.* Oxford: Oxford University Press.

Marques, Ivan da Costa. 2005. Cloning computers: From rights of possession to rights of creation. *Science as Culture* 14 (2): 139–160.

Negroponte, Nicholas. 2009. Lessons learned and future challenges. Speech at "Reinventing the Classroom: Social and Educational Impact of Information and Communication Technologies in Education" forum, Washington, DC, September 15. Available at http://www.olpctalks.com/nicholas_negroponte/ (accessed May 1, 2011).

Negroponte, Nicholas. 2010a. Open letter to India. Published July 29. Available at http://laptop.org/en/vision/essays/35-tablet.shtml (accessed June 1, 2011).

Negroponte, Nicholas. 2010b. Laptops work. *Boston Review* 35 (6) (November/December). Available at http://bostonreview.net/BR35.6/ndf_technology.php (accessed May 1, 2011).

Negroponte, Nicholas. 2012. Nicholas on Peru. Published April 13. Available at http://blog.laptop.org/2012/04/13/nicholas-on-peru/#.Ubatret1FSe (accessed May 1, 2013).

Nieusma, Dean, and Donna Riley. 2010. Designs on development: engineering, globalization, and social justice. *English Studies* 2 (1):29–59.

Oppenheimer, Andres. 2012. Region's One Laptop Per Child plan has a future. *Miami Herald*, April 30. Available at http://www.recordonline.com/apps/pbcs.dll/article?AID=/20120430/OPINION/204300311&cid=sitesearch (accessed May 1, 2013).

Oudshoorn, Nelly, and Trevor Pinch. 2003. *How Users Matter: The Co-Construction of User and Technology*. Cambridge, MA: MIT Press.

Paul, Ryan. 2008. Classmate PC gets a boost with million-unit Venezuelan order. *Ars Technica*, September 29. Available at http://arstechnica.com/gadgets/2008/09/classmate-pc-gets-a-boost-with-million-unit-venezuelan-order/.

Paul, Ryan. 2009. Behind the OLPC layoffs: G1G1 failure and reduced sponsorship. *Ars Technica*, January 30. Available at http://arstechnica.com/gadgets/2009/01/g1g1-failure-and-reduced-sponsorship-behind-the-olpc-layoffs/.

Paul, Ryan. 2010. OLPC's Negroponte offers to help India realize $35 tablet. *Ars Technica*, August. Accessed June 1, 2011. Available at http://arstechnica.com/gadgets/news/2010/08/olpcs-negroponte-offers-to-help-india-realize-35-tablet.

Peru Support Group. 2007. Teachers protest over the government's new education law. Peru Support Group Publication, 122.

Salas, Sdenka. 2009. *La laptop XO en el aula*. Puno, Peru: Sagitario Impresores.

Santiago, Ana, Eugenio Severin, Julián Cristia, Pablo Ibarrarán, Jennelle Thompson, and Santiago Cueto. 2010. Evaluacíon experimental del programa "Una Laptop por Niño" en Perú. Washington, DC: Inter-American Development Bank. Available at http://www.iadb.org/document.cfm?id=35370099 (accessed June 1, 2011).

Severin, Eugenio, and Christine Capota. 2011. One-to-one laptop programs in Latin America and the Caribbean. Washington, DC: Inter-American Development Bank. Available at http://idbdocs.iadb.org/wsdocs/getdocument.aspx?docnum=35989594 (accessed May 1, 2013).

Stengers, Isabelle. 2007. *Cosmopolitics*. Minneapolis: University of Minnesota Press.

Suchman, Lucy. 2007. *Human-Machine Reconfigurations: Plans and Situated Actions*. Cambridge: Cambridge University Press.

Talbot, David. 2008. Una laptop por niño. *Technology Review* 111 (3):60–66.

Talbot, David. 2012. Given tablets but no teachers, Ethiopian children teach themselves. *Technology Review*. Available at http://www.technologyreview.com/news/506466/given-tablets-but-no-teachers-ethiopian-children-teach-themselves/ (accessed January 13, 2014).

Toyama, Kentaro. 2010. Can technology end poverty? *Boston Review* 35 (6) (November/December):12–18, 28–29.

Toyama, Kentaro. 2011. There are no technology shortcuts to good education. *Edutech Debate*. Available at http://edutechdebate.org/ict-in-schools/there-are-no-technology-shortcuts-to-good-education/ (accessed May 1, 2011).

Trucano, Michael. 2012. Evaluating One Laptop per Child (OLPC) in Peru. *World Bank Blogs*, March 23. Available at http://blogs.worldbank.org/edutech/olpc-peru2 (accessed May 1, 2013).

Twist, Jo. 2005. UN debut for $100 laptop for poor. BBC News, November 17. Available at http://news.bbc.co.uk/2/hi/technology/4445060.stm (accessed May 1, 2011).

Vaidhyanathan, Siva. 2006. Rewiring the "nation": The place of technology in American studies. *American Quarterly* 58 (3):555–567.

Villanueva, Eduardo. 2011. The importance of being local. *European Magazine*, April 28. Available at http://www.theeuropean-magazine.com/256-villanueva-masilla-eduardo/255-ict-in-development-cooperation (accessed June 1, 2011).

Vota, Wayan. 2009. If & when schools invest in ICT, teachers first. *Edutech Debate*. Available at http://edutechdebate.org/ict-in-education/if-when-schools-invest-in-ict-teachers-first/ (accessed June 1, 2011).

Warschauer, Mark, and Morgan Ames. 2010. Can One Laptop per Child save the world's poor? *Journal of International Affairs* 64 (1):33–51.

Watters, Audrey. 2012. The failure of One Laptop per Child. *Hackeducation*, April 9. Available at http://www.hackeducation.com/2012/04/09/the-failure-of-olpc/ (accessed January 13, 2014).

Woolgar, Steve. 1991. Configuring the user: The case of usability trials. In *A Sociology of Monsters: Essays on Power, Technology and Domination*, ed. John Law, 57–99. London: Routledge.

10 Translating Magic: The Charisma of One Laptop per Child's XO Laptop in Paraguay

Morgan G. Ames

In April 2008 liberal candidate Fernando Lugo won the Paraguayan presidency with 41 percent of the vote. It was the first time Paraguay's conservative Colorado party had relinquished power in sixty-one years and one of the few peaceful transfers of power between parties in the country's nearly two-hundred-year history (Nickson 2009). Known by his supporters as the "Bishop of the Poor" for his humanitarian service as a Catholic priest in one of Paraguay's poorest districts, Lugo centered his campaign on reducing the nation's bleak social inequalities and widespread corruption (Economist 2008; Nickson 2009). Lugo's messages of hope and reform lasted beyond his election. During my fieldwork in 2010, many told me about their newfound hopes for Paraguay's future on the world stage, even as Lugo became embroiled in scandal and faltered in implementing the dramatic changes he had promised (Economist 2009; Prensa Internacional 2009).

Also in 2008, two elite, well-connected Paraguayans, both fresh out of college (one from Tufts University in Massachusetts, one from the Catholic University in Asunción), took up the banner of hope and steered it toward education. Captivated by the promises of the One Laptop per Child (OLPC) project—and one motivated by nostalgia for his own experiences with computers as a child, which inspired him to become a skilled software engineer—the two formed a nongovernmental organization (NGO) called Paraguay Educa to bring OLPC's distinctive laptops to their country. In September 2008, one month after Lugo took office, they explained their motivations for the project in an article in *ABC Color*, one of Paraguay's two major newspapers and a staunch supporter of the project (ABC Color 2008). Their dream of producing technologically savvy, passionate learners by giving them their own laptops was inspiring to many who became involved with the project, but it was not unique. In fact, much of the vision expressed in that 2008 news article was drawn directly from OLPC's promotional materials, from the hyperbolic list of benefits including customized learning software, rugged construction, and a state-of-the-art screen (OLPC 2011) to the project's Five Core Principles of child ownership, low ages, saturation, connection, and free/open source (OLPC 2012a). The vision that Paraguay Educa and Cambridge,

Massachusetts–based OLPC shared was that their laptop, called the XO, had the power to change a learning culture through the individual child's interactions with it (Papert 1980, 9, 37), making up for a lack of local educational infrastructure and producing children adept at the kind of individualistic mathematical thinking valued in computer engineering cultures (Papert 1993; Negroponte and Bender 2007).

Paraguay Educa's faithful adherence to One Laptop per Child's vision could be seen as problematic: though independent of OLPC, Paraguay Educa nonetheless uncritically adopted a set of ideals largely developed at an elite institution, the Massachusetts Institute of Technology (MIT), in the United States, a country with a history of both military and cultural imperialism in the region. However, this chapter will demonstrate that the hopes held by those using the laptops day to day often did not coincide with those of either OLPC or Paraguay Educa, instead symbolizing many other possible futures for Paraguay.

To explain the laptop's symbolic power, I develop the idea of a *charismatic object*. Charisma as a sociological construct was theorized by Max Weber (1947) to describe the exceptional, even magical, authority that religious leaders seem to have over followers. Though *charisma* usually refers to the power of humans, not objects, the word has been applied to nonhumans as well. Maria Stavrinaki describes the Bauhaus African Chair as a charismatic object based on its portrayal within the Bauhaus community, though she does not dwell long on the idea of charisma itself (Stavrinaki 2010). Relatedly, Anna Tsing (2000) discusses how the *idea* of globalization has been charismatic to some academics who uncritically naturalize or even reinforce globalist agendas. Tsing's model of charisma—a destabilizing force that not only elicits excitement but can produce material effects in the world (even if those material effects differ from those that were promised)—is at play in this chapter as well.[1]

Distinct from fetishism, which theorizes the fixation on the materiality of the presumably passive object itself as a source of power, a charismatic object derives its power experientially and symbolically through the possibility or promise of *action*— what is important is not what the object *is* but what it promises to *do*. As McIntosh (1970) explains, "charisma is not so much a quality as an experience. The charismatic object or person is *experienced* as possessed by and transmitting an uncanny and compelling force" (emphasis added). Charisma implies a persistence of this compelling force even when an object's actions do not match its promises—hence the magical element of charisma.

As I argue elsewhere (Ames 2013; Ames and Rosner 2014), a key component of the XO laptop's charisma among many of OLPC's developers and donors is its ability to evoke the nostalgic and often individualist stories that many in the technology community tell about their own childhood experiences with computers. It is this charisma that Paraguay Educa initially adopted and promoted. However, we will see here that the charismatic authority the XO had among many of its day-to-day users in Paraguay often took different forms, less tied to the ideologies of OLPC than to more general dreams of an information-rich, Internet-enabled future.

The charisma of OLPC's XO laptop takes on a different valence in light of the social history of technological adoption, adaptation, and innovation in Latin America. Projects in ten Latin American countries account for 85 percent of the XOs in use around the world (Klein and Holt 2012; Warschauer and Ames 2010).[2] The project's tensions around how technology is adopted, and on whose terms, have previously surfaced in other projects across Latin America: indeed, Lemon and Medina's historical review in this volume points to the histories of other such projects. Elsewhere in this volume, Chan discusses the resurrection of one of the largest and most troubled OLPC projects in Peru, particularly the negotiations between local and international developers in Puno, Peru, to reappropriate the laptops as tools for local pride and activism. Further afield, Medina's (2011) account of the negotiations between local actors and international "experts" over the utopian visions of a cybernetic society in Chile bears resemblance to the negotiations between Paraguay Educa and teachers. Likewise, Hagood's description of Argentina's dreams for nuclear energy (this volume) and Kreimer and Zabala's (2007) discussion of the compelling "fictions" around Chagas disease in Argentina have parallels in the charismatic stories that circulate about OLPC's laptop in Paraguay and across Latin America. More broadly, the differences between "imported" and locally developed meanings of OLPC's laptop are yet another instantiation of Vessuri's (1987) discussion of development versus dependency in early Latin American science and technology studies.

In the second half of 2008, as Paraguay's new president settled into his duties, Paraguay Educa's founders used their family connections and a captivating story about the promise of OLPC laptops to secure financial, political, and infrastructural support from a number of local and international sources. With the help of a donation from the philanthropic arm of Swift Group, a European banking conglomerate, they purchased 4,000 first-generation XO laptops from OLPC, which arrived in early 2009 (ABC Color 2008). For Phase I of the project in April 2009, Paraguay Educa distributed these XOs to all students in grades one through six (ages six through twelve) and their teachers in ten schools in the small municipality of Caacupé, fifty kilometers east of the capital Asunción. In May 2011, they purchased and gave an additional 5,000 second-generation XO laptops to all primary-school students and teachers in the other twenty-six schools in the municipality in Phase II of the project. Throughout these deployments, Paraguay Educa remained committed to OLPC's ideological roots even as others developed divergent visions, as we will see.

Ideological Worker or Toy? Diverse Translations of the XO

The remainder of the chapter draws on six months of ethnographic fieldwork in the Phase I and Phase II schools in Paraguay to explore whether and how meanings of the XO laptop preferred by the project's creators and sponsors were taken up, resisted, or selectively shaped by local actors in the laptop's use.[3] OLPC and Paraguay Educa promoted meanings that revolved around the tenets of *constructionism*, a learning theory developed over some forty years by former MIT professor and OLPC advisor Seymour Papert. Constructionism, the opposite of "instructionism" (or passive listening and repeating with a teacher), valorizes self-directed learning through active creation using a computer (Papert 1980, 7, 31–32, 115).[4] OLPC's leadership and mission statement openly acknowledge the project's constructionist roots (Negroponte and Bender 2007), asserting moreover that constructionism is the *only* path to really learning how to learn. As OLPC's mission statement says, "constructionism emphasizes what Papert calls 'learning learning' as the fundamental educational experience. A computer uniquely fosters learning learning by allowing children to 'think about thinking,' in ways that are otherwise impossible" (OLPC 2012b). Though claiming to tap into children's innate desires to learn rather than promoting a particular type of knowledge, constructionism ultimately centers on students' learning to "think like a machine," culminating in learning to program the machine itself (Ames 2013).

Papert further argued that a computer was an exemplary apparatus for the ideological work of constructionism. *Ideological work* is described by Bennett Berger (1981) not as state-sponsored dogma but as the more informal—though still powerful—negotiations between ideals and real life on late-1970s "hippie" communes. Here, the ideological work of the project is constructionist learning, and according to constructionism itself, the best ideological worker was a computer. Papert described computers as "the Proteus of machines," appealing to many audiences (Papert 1980, viii) and, like videogames, exerting a charismatic "holding power" over children that most teacher-led classroom learning lacked (Papert 1993, 3–5). He further claimed that specially designed "learning machines" could "be carriers of powerful ideas and of the seeds of cultural change" (Papert 1980, 4) and would naturally replace traditional "instructionist" classrooms in the marketplace of educational innovation (Papert 1993, 8–9).

In practice, how independent could OLPC's laptop be as an ideological worker for constructionism? From the accounts of researchers and visiting journalists, the deployments of 15,000 laptops in Birmingham, Alabama (Warschauer, Cotten, and Ames 2012), and 980,000 laptops in Peru (Derndorfer 2010a; Cristia et al. 2012) have struggled from the lack of social and infrastructural investment beyond handing out the laptops themselves. In many schools in these areas, the laptops are not used at all, suggesting that laptops alone are insufficient in promoting constructionist learning.

Paraguay Educa chose a different path.[5] Shortly after its founding in 2008, the NGO's staff, following the advice of Paraguay's new president (ABC Color 2008), decided to encourage classroom laptop use with teacher training. The employees of Paraguay Educa, still numbering just around half a dozen and lacking much teaching experience themselves, recruited volunteers to lead training sessions for the teachers in the ten Phase I schools. They spent October 2008 familiarizing these trainers with the laptop's functions (Paraguay Educa 2009), and for two weeks in December 2008 these trainers taught all of the teachers from the ten Phase I schools what they had learned (Paraguay Educa 2008).

The focus of this training session, however, was on basic operations such as using the web browser, camera, and word processing, rather than more concrete lessons on how the laptop could be integrated with existing curricula. Moreover, in 2010 interviews, teachers reported that there was little follow-up with them in the months after this training session. Even teachers who would later become the strongest advocates for the program told me that they were left with little idea how to use the laptop pedagogically in that first year, and many struggled enough with basic operations like searching the web or checking their email that they hardly used the laptops at all. One fifth-grade teacher explained, "When I did the teacher training, it was one week and only about the Internet, and Paint and Write. We did not know what to do later."[6]

Some of this difficulty was due to context: before receiving an XO, only one in four teachers had had a computer at home, only one in eight had Internet access, and less than half had *any* access to or experience with a computer (though all had mobile phones and nearly all had televisions). This lack of experience with computers in general compounded the difficulty the teachers had in using a laptop scaled for children, with unfamiliar software and sometimes unreliable hardware (Warschauer and Ames 2010).

As a result, the meanings that teachers and students developed about the XO laptop that first year diverged significantly from Paraguay Educa's and OLPC's. Many told me that they had viewed the XO laptop more as a brightly colored plastic plaything than as a useful tool for learning. One school director said that she and her colleagues "thought the laptops were just a toy for games," especially when they saw their students and their own children using them as media consumption devices for videogames, videos, and music. And rather than thinking of it as a *useful* toy, the teachers made clear from their tone that seeing the XO as a toy was equivalent to writing it off entirely.

Indeed, the ideological work that the laptop appeared to accomplish on its own—and the charisma it held (and continues to hold) for its child users—did not consist of students learning to "think like a machine" in the ways described by constructionism, where students develop an intuitive understanding of mathematics, but of students using the machine as a media device. Even during my fieldwork in 2010 (nearly two years into the project), almost all of the many hours of children's unstructured laptop use I observed—before and after school, during recess, and often at home—was still focused on videogames and media consumption. While constructionism embraces play, the kind of play I witnessed was not the mathematically oriented version described by constructionism, but more like using an interactive television, where the focus was on the products of large media corporations. Paraguay Educa, lacking day-to-day contact with students and teachers throughout 2009, was not present often enough to reinforce a different view, and the laptop by itself was not inspiring the kind of proto-programming exploration for which it was designed.

Seeing the laptop as a toy also meant that the children, parents, and teachers who were involved with the project that first year did not see a particular need to care for it. Though the laptop was designed ruggedly for child ownership, it still did not stand up to the actual roughness of children: one year and three months after Paraguay Educa's Phase I laptops were handed out, 15 percent of the laptops and 10 percent of the AC adaptors were unusably broken, with little recourse for repair.[7] As described in more detail in Rosner and Ames (2014), the material realities of the laptop itself destabilized the meanings that OLPC and Paraguay Educa attached to it: the vision of universal access to XOs as a social leveling force was undermined by how laptops were broken and who could afford the few repair parts available, thus reifying rather than mitigating socioeconomic differences.

In short, the laptop was *not* immediately charismatic to teachers in the ways it was for Paraguay Educa and OLPC—as an educational machine—or in the ways it was to children—as a media machine. As we will see, it was not until Paraguay Educa encouraged teachers to develop alternate meanings of the laptop as a device for education that it started to become charismatic to them.

Teaching Charisma: The *Formadores* Program
When it became clear that OLPC's ideology was not automatically transferred via day-to-day laptop use, Paraguay Educa filled the gap by creating a new type of ideological worker to do the job instead: a teacher trainer. It was not long after laptops were handed out, in fact, that Paraguay Educa staff began to notice their lack of use. Several interviewees explained to me during my fieldwork in August 2010 that one year before, just a few months after the laptop handout, the founders had enlisted the help of a local education expert and family friend to consult on the project, and a month or two later, brought her on full-time as the director of Paraguay Educa's newly formed educational division. One fifth-grade teacher recounted an August 2009 visit from this education expert, where both realized that Paraguay Educa and the teachers needed to take a more active role in shaping the uses and meanings of the XO laptop if the object was to be seen as something more than a toy:

[The educational director] came one day [in 2009] … and said in a strong tone, "You're using the [paper] notebook [and not the laptop]." I didn't know what to say—I only knew how to turn the laptop on and use the Internet, because that's what I was taught, but I didn't know how to use it pedagogically. I told her I can't teach what I don't know.

Realizing that the laptop was not by itself sparking the kinds of exploration they had hoped for, Paraguay Educa hired and trained residents of Caacupé with experience in both technology and education, and with excitement about the project, to become full-time trainers (*formadores*) after the 2009 school year ended in November. When the 2010 school year started in late February, these trainers began working daily in the ten Phase I schools, serving as local mouthpieces for Asunción-based Paraguay Educa by promoting the idea of the laptop as a learning device and giving concrete suggestions for lessons that incorporated the XO. I heard from many sources that their work resulted in a large increase in laptop use. In the words of one third/fourth-grade teacher, "It's so much easier with the trainers. Last year we didn't use the laptop at all—it was impossible without them. They give support and help in class with activities I still don't know well and have a hard time with." Teachers from all ten schools recounted versions of this story: having someone whose job it was to educate the school staff about the laptop and its learning theories transformed the project from one that was marginalized in the classroom to one that, though not as central as OLPC or Paraguay Educa might hope, was still much more integrated.

What Kinds of Charisma? Negotiating Types of XO Use

The trainers were able to stimulate laptop use, but were they also able to enforce constructionist meanings? This question can in part be answered by examining *how* laptops were used. Unique in the ecosystem of classroom laptops, OLPC's XO laptops included not only custom-designed hardware but an educational software suite with several specifically constructionist activities that, according to Papert, would capture the attention and imagination of children, leading them to self-motivated learning.

In fact, the laptop's charisma did appear to have some material effects. I observed and discussed a number of differences between students at schools with laptops and those without: whether their schools were urban or rural, large or small, public or private, the students with laptops seemed much more outgoing and confident than those at otherwise similar neighboring schools without laptops. The difference was especially strong at rural schools, where students were otherwise much shier with visitors like me. One trainer, who himself grew up in a rural household and described a painful transition between rural and city life in secondary education, thought that this could be one of the larger effects of the project—to give rural children contact and common ground with their urban counterparts and more opportunities than they would otherwise have:

I see that all children are given the same opportunity, even if they're in a rural school. Technology is something that's a dream for them. ... There are no secondary schools in the [rural] places I go—they have to come to downtown urban Caacupé and there they'll discover technologies like computers. [When they do,] they won't have such stage fright, where they would be afraid because they are more humble and have less money—they won't be afraid. The project ...

demonstrates to children that they can still progress even if they don't live in the city center or have many resources—if someone is dedicated, they will have the opportunity to benefit.[8]

Teachers and school directors reported that school attendance had improved, and that some students who had been held back in third grade because they could not read or write Spanish were finally motivated to learn because of the laptop and passed to fourth grade, though I was only able to locate one such student. Better reading skills were also reflected in the results of a cognitive exam that I administered with Paraguay Educa in November 2010, which showed a small, but statistically significant, difference of 5 percent in third- and sixth-grade reading comprehension scores between students with laptops and students in nearby schools without them. (However, results in mathematics were mixed, despite constructionism's focus on mathematical literacy.) Even if some of these changes—or belief in these changes—were due to a placebo effect motivated by the laptop's charismatic authority, the result of teachers and trainers *treating* students as more outgoing and creative and believing that the laptops can enact changes may in itself be significant.

These differences, however, do not tell us which of the laptop's activities were used in the classroom, and how. Were these changes due to extensive use of constructionist activities, as OLPC would like to believe, or from laptop use more generally? Or, as one trainer mused, might the changes be less about the laptops at all and more about increased awareness of child-centered pedagogy from Paraguay Educa's trainers and international visitors?

My fieldnotes from an August 2010 school visit depict many commonly recurring features of classroom laptop use that I witnessed. The fourteen students in the fourth-grade morning session I was observing that day haphazardly faced the board in the classroom's sixteen small wooden desks, well-worn and full of carved graffiti. The teacher's slightly larger desk sat in the back right corner, stacked high with students' paper notebooks.

These students were already halfway through their four-hour school day this morning; twelve different students would attend in the afternoon. The teacher had written most of the day's lessons on the chalkboard before the morning session started: one section of the board for language, one section for mathematics, one section for natural science, and one section for a rotating lesson on a fourth subject, as required by the countrywide curriculum set by Paraguay's Ministerio de Educación y Cultura (MEC). We had already completed the mathematics and language sections of the board, using the most common tools for education in Paraguay: pencils and slim paper notebooks, both of which can be bought at the school cantina. The third lesson was on health, and we would be using XOs.

Forty-five minutes had elapsed since the teacher asked students to take out their XOs. Only ten of the fourteen had working laptops at the beginning, and of those, only one had the software the teacher wanted to use, Tux Paint, installed. With the help of the trainer (but losing four more laptops to discharged batteries in the process), the teacher had installed the software on all six remaining laptops and was finally ready to start her lesson. She writes "Alimentos" (foods) at the top of a blank panel of the chalkboard, and then "según su origen" (according to their origin) underneath. She asks, "What are some foods?" The class brainstorms together—first the categories of animals, vegetables, and minerals, and then examples of each (cow, lettuce, salt)—for a few minutes. The girls participate more; the two boys with laptops intently stare at their screens and ignore the teacher. One has closed Tux Paint and has the browse activity open with a search for *chistes* (jokes). Another laptop runs out of battery power.

On the teacher's instruction, students start drawing examples of foods in their laptops or paper notebooks. Those with laptops draw shaky lines to divide the categories using their trackpads, often holding down the keys at the four corners of their keyboards to reset the trackpad hardware when it gets jittery or unresponsive. They then use Tux Paint's built-in image "stamps" to generate examples in the three categories, and soon the room is filled with the sounds associated with those stamps. Students without XOs write out examples and draw pictures in their notebooks instead. After about five minutes of working, the teacher announces that the rest will be homework for tomorrow and asks children to put their laptops away and take out their notebooks to copy a poem in Guaraní from the board in the fifteen minutes before school ends.[9] Most students comply, but the two boys with XOs keep them out, half-open, and peek at them from time to time. I peek at one of them and find a download in progress of *Naruto*, a popular anime series.

This XO lesson illustrates some of the negotiations involved in using XOs in the classroom. One feature of this narrative that may especially strike those familiar with OLPC and its XO laptop is that none of the laptop uses that I noted that day were particularly XO-specific. Despite the XO's ability to rally support from sundry groups with its iconic appearance, in this classroom the laptop did not differ much from other computers, or even from paper and pencil. Furthermore, the activity Tux Paint—though developed for children, relatively easy to understand, and fun to use—was not very constructionist: it did not encourage deep, embodied, passionate exploration any more than any other drawing program. In fact, drawing on the laptop was in several ways inferior to drawing on paper, even after the software was loaded and working. It was beset by the difficulty of using the unreliable trackpad rather than a pencil, the temptation to categorize only the available images instead of thinking of other items and drawing them freehand, and the clarion call of the Internet.[10]

The relatively frequent use of Tux Paint that I witnessed across many classroom observations illustrates a broader issue of which activities were considered best to use in class. The topic was a point of frequent and sometimes heated negotiations between Paraguay Educa and teachers, and a site where different meanings of the machine were particularly evident. The staff of Paraguay Educa, like OLPC, advocated using the most constructionist activities, such as Scratch and Turtle Art. But in both interviews and surveys, teachers said these two activities, along with eToys, were the hardest to learn and the hardest to incorporate into lessons. All have similar interfaces in which command "blocks" are grouped to direct the computer's actions in what amounts to visual programming, and all became even more difficult to manage with a jumpy trackpad. Even in schools that encouraged teachers to use these activities, only a small fraction of teachers were willing to put in the considerable unpaid time needed to learn them adequately. And except in these schools, Scratch, Turtle Art, and eToys were rarely used in the classroom unless a trainer was teaching with them directly.

In contrast, the most commonly used activity by far was Navegar, the web browser. Teachers and students both agreed that it was the easiest activity to use, the easiest to incorporate into existing lessons, and the most compelling. Based on teacher reports and my own observations, teachers used this activity nearly every time they used the laptop, both in class and personally. "The focus is on using the Internet in class," one teacher admitted, because "it's easiest. We need to work more to connect other activities with lesson plans." The reliance on using the web browser was so great that one of the few teachers who had taken up the idea of the laptop as constructionist tool labeled it as one of the project's biggest problems, after breakage—not only were teachers relying on the Internet in the classroom, but they were allowing themselves to be distracted by it at meetings, much as their students were distracted by anime and jokes in the classroom. She said,

I'd teach Phase II teachers to use activities without the Internet, and I'd like them not on the Internet for hours. The Internet should only be a support—train teachers to use only the school server. Let the children learn the activities in Sugar—that would be my ideal. Sugar is educational. Use the XO like a [paper] notebook that has only limited use of the Internet, because very few see the Internet's educational side. Having a foothold with the activities would be good.

We had a meeting with Phase I teachers and everyone had their XOs and was checking their email, everyone looking at their screen and not paying attention. Then we turned off the Internet and the whole room closed their XOs and began to pay attention. Nobody cared that everyone was doing something else.

In short, though its trainers stimulated use of the laptops, Paraguay Educa was less successful in enforcing constructionist meanings. Most teachers struggled to learn how to use the more constructionist activities, and many handed over classroom use of these activities to trainers. In contrast, the most-used activities were the easiest to integrate into existing lesson plans, especially the web browser. To teachers and students, the XO was charismatic not because it could teach programmatic thinking, but because it was a portal to the Internet.

Limits to Charisma: Labor, Conflicting Priorities, and the XO

Other popular activities on the XO were also some of the more general, nonconstructionist ones. After Navegar, the next most-used activities were the word processor and the two drawing programs, all of which straightforwardly replaced parts of blackboard-and-paper lessons. Teachers were clear in their reasons for making only the easiest substitutions: it allowed the one-quarter of students whose laptops were broken to use their paper notebooks instead. Many teachers stated that it was enough work to incorporate the laptops into their lessons at all, much less to use the more complicated activities. In fact, according to the trainers and my own observations, at least one-quarter of the teachers never used the laptops in class unless the trainer taught for them.

These kinds of differences arose from teachers' varied responses to the many demands on their time and loyalties: commands from the school director or coordinator, national curriculum requirements, pressure from Paraguay Educa, teachers' union activities against exploitation and low pay, and personal concerns. Balancing the demands of the MEC's national curriculum against the short school day could be particularly difficult, as both the fieldnotes excerpt earlier in this chapter and the teacher quoted here highlight:

We don't have a lot of time to use the XOs in class. I'd like to use them more—there are so many things to discover and do with it. But the Ministry requires us to complete four lessons a day, and you need at least forty-five minutes for the XO. These aren't activities they'll figure out in twenty minutes, and it takes time to open them [the laptops] up, to type—the children are slow.

The leadership at two of the ten schools in Phase I tried to align at least some of these pressures. One school coordinator who supported Paraguay Educa's mission helped the teachers under her supervision to develop curriculum that used the XO in more constructionist ways. She found, however, that few teachers were committed to investing the considerable time she requested of them—four unpaid training sessions every week—to thoroughly learn the laptop's constructionist activities. Similarly, the director of another school, who had a longstanding interest in a variety of educational reforms, revised her school's mission around constructionist teaching with the XO and encouraged the teachers in her school to volunteer their time for curriculum development and support for fellow teachers, students, and students' parents. Even under her commanding authority, however, some teachers were less committed to the project. In her words, "I can see that we're innovating our old pedagogy with the XO. It's a difficult change. ... If the teacher has training, the XO, and space, but no change of attitude, nothing happens." Her framing of the problem as one of individual "attitude" placed the blame on the teachers themselves rather than on structural issues such as a lack of time or pay for these extra duties.

The other eight schools in the program focused less on promoting constructionist learning in the face of other considerations. The fourth-grade teacher whose classroom is described earlier in this chapter liked Paraguay Educa's mission, but her school's leadership was not so committed and the trainers were not very forceful, making her an anomaly who lacked the influence to rally her colleagues. Similarly, the coordinator at yet another school wanted to promote the project, but both she and the school's trainers lacked the leadership to recruit more teachers to use the XO in the face of an indifferent director. Paraguay Educa found that while the practices of the most committed schools could inform strategies for reaching the others, these strategies would fall flat unless they took into consideration the specific politics and practices already present in each school.

Overall, the changes that schools with laptops experienced appear to be due less to the use of constructionist activities, which was relatively rare, than to laptop use more generally and the attention that the laptops generated. Paraguay Educa continues to champion use of constructionist activities, hosting ongoing teacher training sessions in Scratch and Turtle Art, sponsoring after-school clubs for children interested in learning more about them, and organizing special events such as a local party for the international Scratch Day each May. However, even using the laptop as an optional replacement for a pencil required negotiation between teachers, students, school administrators, and Paraguay Educa's trainers, all of whom may have different visions of how to use the laptop in class. Thus, though constructionism was promoted by certain actors, this ideology was largely absent among the laptop's beneficiaries as they negotiated other uses in the classroom.

Charismatic Objects: The Symbolic Importance of the XO

This chapter has in part been a cautionary tale about entering into an education or development project like OLPC's with too much hubris—illustrating the consequences of overpromising and underdelivering. Unfortunately, both OLPC and Paraguay Educa, along with many NGOs and nonprofits, are caught in a catch-22: they must set lofty goals and show (or at least suggest) great results to continue to attract the interest of investors, even though showing great results could also lead investors to conclude that the project is "done" and does not need more funding. Showy but myopic projects are rewarded either way, even when they rarely produce lasting results. This chapter suggests another course, honestly acknowledging sites of difficulty and conflict as well as the tremendous amount of work—social, infrastructural, and ideological—that must be mobilized to produce even incremental social change. As NGOs take on more functions formerly expected of states in many parts of the world (Ferguson 2006), it becomes ever more important to moderate expectations, promote transparency, and learn from "failures."

This does not mean that such projects are not worth pursuing. To close this chapter, I want to turn back to how the laptop, despite its failings and frustrations, was still a charismatic object. In a country shaped by significant humanitarian intervention, it is remarkable that the XO laptops in particular commanded this charisma, even when other local projects—schools sponsored by American churches, Peace Corps projects, religious mission work, and more—did not. In fact, in six months of fieldwork and hundreds of conversations with teachers, students, and parents, I had trouble finding anyone who was not hopeful about the effects of the laptops, even if their day-to-day experience with them was frustrating or disappointing. Like a semiotic version of a charismatic leader, the laptop inspired hope for change by signifying possible technological futures to those invested in the project. Though their visions often differed, parents, teachers, and trainers alike were unfailingly, unflinchingly hopeful about the laptop's effects. "It's a beautiful project," teachers would remark to me or to each other, even after an hour of venting about how impossible it was to use the XO in the classroom when so many laptops were broken or misconfigured. "It will help the minds of the children develop, and the country will develop too."

While some interviewees were vague on what sort of future the laptops might create for the children of Caacupé, some were quite specific. A number focused on the perceived benefits of technological proficiency and the associated worldliness, similar to the sentiments of the trainer quoted earlier in this chapter who was excited that rural children were finally given the same opportunities as urban children. In the words of one parent, "Children are learning not to have fear of technology—they realize that they can work with it. It's not outside of their realm, or only for really intelligent people. Learning that they can use it too is one step to the technology improving their lives." A teacher likewise stated, "For these kids, nothing will surprise them—teachers can be intimidated even by cash registers, but the kids are prepared for the future, and with luck they'll study more and more and make our city evolve into something great."

Several interviewees linked the processes the XO made possible with the processes of citizenship, comparing Paraguay Educa's project to the historic lack of support for education from Paraguay's government and invoking the importance of initiative and individual responsibility in being a good citizen. During my fieldwork in 2010, as Paraguayans were still adjusting to having a liberal president who came to power in an election seen as fair by the international community for the first time in many of their lifetimes, declarations about technology's role in creating good citizens took on an even more idealistic tone, similar to the way the rhetoric around Peru's open-source software movement produced a vision of its role in civic responsibility (Chan 2004). In the words of one parent, the laptops would enable not only information access, but political thoughtfulness: "We'll have a digital city. These kids won't be shy. Citizenship itself will change—there will be more reflection, and votes will really reflect the best candidate. They won't believe everything they're told because they can investigate anytime and anywhere." One trainer more explicitly implicated the lack of governmental support in the project, but believed that participants would find ways to enact change without it:

In twenty years, Caacupé will be digitized, and the youth will be well-prepared and thoughtful citizens. It will make a profound difference in their cultural background. The government doesn't care that these people will be able to make such a difference—they say it'll create problems for the state, and they won't support the project. But the citizens will be thoughtful and have visions, and will change Caacupé.

It is important to note that the charismatic power that the XO possessed in Paraguay seemed to come not from the messy material realities of the laptop in use, but from the desires that the laptop-as-*symbol* was able to fabricate from theoretical use. The XO laptop seemed most powerful as a prototypical object, its power lying in its ability to produce or connect to certain dreams for a technosocial future. The apparent contradiction between these dreams and the object's much less glamorous material reality did not seem to result in a revision of the dreams—a finding that touches on the moral claims made about new information technologies more generally.

Anthropologist Arjun Appadurai (1991) highlights the social role of fantasy in a world of new media technologies, implicating mass media in allowing more people to "see their lives through the prisms of the possible lives offered"—even when the result is an "ironic compromise between what they could imagine and what social life will permit." Similarly, OLPC's XO laptops revealed social fault lines around the ideals and realities of joining the "information society" that played out not only in the laptop's use but in visions of how it might shape the region's or country's future—from the constructionist learning of Paraguay Educa and OLPC to the portal to the Internet that students and teachers embraced. Since the hopes many Paraguayans had attached to President Lugo were dashed in June 2012, when he was impeached by a Colorado-controlled Congress under suspicious circumstances that international observers have labeled a coup (Guillemi 2012), the dreams that Paraguay Educa's participants have connected to the XO may loom ever more central.

This chapter, like others in this volume, provides an account of the agency of technology users in Latin America, challenging the common story of unidirectional technology transfer from North to South. The specifics of Paraguayan politics and daily life shaped both the use of the XO laptop and the users' visions for it, subverting OLPC's arguably imperialistic aims in favor of uses and meanings that made more sense given the constraints of schools and the interests of students and teachers. In sum, the laptop could not accomplish the ideological work for which it was built, even with significant social intervention. However, many students and teachers using the laptop day to day were captivated instead by the charisma of the XO as a media machine and portal to the Internet. In this way, the hopes held by those using the laptops often did not coincide with those of either OLPC or Paraguay Educa, but instead symbolized a future of better education, better citizenship, a better economy, and better quality of life in Paraguay.

Notes

1. One element of the term *charisma* not at play here, or in Tsing's work, is its connection with Pentecostalism. Though belief in the laptop's powers for change could be characterized as a "faith," it does not resemble the attraction to (or the perceived threat of) Pentecostal religion.

2. Peru and Uruguay host the largest projects (over 500,000 laptops); Argentina, Mexico, and Nicaragua host medium projects (25,000 to 60,000 laptops); and regions in Paraguay, Colombia, Guatemala, Brazil, and Costa Rica host small projects of 1,500 to 10,000 laptops each.

3. I observed classrooms and conducted 133 interviews with students, parents, teachers, and other local actors between late June and mid-December 2010. The bulk of my data comes from fieldwork in the ten Phase I schools. I also visited nearby schools without laptops, many later included in Phase II of the project. Though a volunteer for Paraguay Educa, I often worked independently, and with full access to the schools and the promise of anonymity, I was able to see and hear many local opinions of the project that the Asunción-based NGO could not. I supplemented this fieldwork with quantitative data including breakage reports, attendance records, and exam scores for reading and mathematics, as well as visits to much larger projects in Uruguay and Peru.

4. An overloaded term, *constructionism* is not only distinct from Piaget's constructi*v*ism (Ackermann 2001), but is entirely unrelated to sociological theories of social constructionism.

5. Paraguay's laptop project, though small, has been praised by visitors from the OLPC community for having extensive and ongoing investments in teacher training, student motivation, repairs, and local software (Buderi 2010; Derndorfer 2010b). While it is easy to find accounts of OLPC deployments doing badly, this chapter explores the ongoing complexities and challenges of a deployment that has been described as doing well.

6. Teachers participating in the OLPC program in Paraguay were interviewed in the course of the author's fieldwork in 2010.

7. Of the 1,095 unfixed hardware problems in Paraguay Educa's inventory system in August 2010, 474 involved a broken charger, 403 a broken screen, 139 a broken keyboard or trackpad, and 79 other hardware issues. Uruguay's Plan Ceibal program, even with state-sponsored repairs, has published similar numbers of broken laptops (Derndorfer 2011).

8. This trainer's focus on *opportunity*, and the unspoken assumption that it is an individual's responsibility to take advantage of opportunity, illustrates the individualistic explanations for success I frequently heard from teachers and Paraguay Educa staff alike.

9. Guaraní is a South American indigenous language and one of the two official languages of Paraguay.

10. The first-generation trackpad was especially bad: many who bought XO laptops through the "Give 1, Get 1" program complained about them. OLPC later updated the software to reset automatically when the trackpad started jittering (OLPC 2009).

References

ABC Color. 2008. El proyecto "Una laptop por niño" comenzará a funcionar en Paraguay. *ABC Color*, September 25. Available at http://archivo.abc.com.py/2008-09-25/articulos/453942/en-marcha-el-proyecto-una-laptop-por-nino

Ackermann, Edith. 2001. Piaget's constructivism, Papert's constructionism: What's the difference? *Future of Learning Group Publication* 5 (3):438.

Ames, Morgan. 2013. From MIT to Paraguay: A critical historical and ethnographic analysis of One Laptop per Child. Doctoral dissertation, Stanford University. http://purl.stanford.edu/rc029zm4311.

Ames, Morgan G., and Daniela K. Rosner. 2014. From drills to laptops: Designing modern childhood imaginaries. *Information, Communication and Society*. DOI: 10.1080/1369118X.2013.873067.

Appadurai, Arjun. 1991. Global ethnoscapes. In *Recapturing Anthropology: Working in the Present*, ed. Richard G. Fox, 191–210. Santa Fe, NM: School of American Research Press.

Berger, Bennett. 1981. *The Survival of a Counterculture: Ideological Work and Everyday Life among Rural Communards*. New Brunswick, NJ: Transaction Publishers.

Buderi, Robert. 2010. One ecosystem per child: Walter Bender and OLPC reunite to enhance learning and grow economies in developing nations. *Xconomy*, October 5. Available at http://www.xconomy.com/boston/2010/10/05/one-ecosystem-per-child-walter-bender-and-olpc-reunite-to-enhance-learning-and-grow-economies-in-developing-nations/.

Chan, Anita Say. 2004. Coding free software, coding free states: Free software legislation and the politics of code in Peru. *Anthropological Quarterly* 77 (3):531–545.

Cristia, Julian, Santiago Cueto, Pablo Ibarraran, Ana Santiago, and Eugenio Severin. 2012. Technology and child development: Evidence from the One Laptop per Child Program. IDB Working Paper Series No. IDB-WP-304.

Derndorfer, Christoph. 2010a. OLPC in Peru: A problematic Una Laptop por Niño program. *Education Technology Debate*. Available at http://edutechdebate.org/olpc-in-south-america/olpc-in-peru-one-laptop-per-child-problems/.

Derndorfer, Christoph. 2010b. OLPC in Paraguay: Will Paraguay Educa's XO laptop deployment success scale? *Education Technology Debate*. Available at http://edutechdebate.org/olpc-in-south-america/will-paraguayeduca-scale/.

Derndorfer, Christoph. 2011. Plan Ceibal expands new repair system to address high XO breakage rates. *OLPC News*. December 7. Available at http://www.olpcnews.com/countries/uruguay/plan_ceibal_expands_new_repair_system_to_address_high_breakage_rates.html.

Economist. 2008. The next leftist on the block: Measuring up Fernando Lugo's plans for a misgoverned country. *Economist*, August 7. Available at http://www.economist.com/node/11885681.

Economist. 2009. The boy and the bishop: Paternity claims distract from a struggle for reform. *Economist*, April 30. Available at http://www.economist.com/node/13579202.

Ferguson, James. 2006. *Global Shadows: Africa in the Neoliberal World Order*. Durham: Duke University Press.

Guillemi, Rubén. 2012. Lugo: "La masacre fue montada para derribarme." *La Nación*, July 30.

Klein, S. J., and Adam Holt. 2012. Deployments. *OLPC Wiki*. Available at http://wiki.laptop.org/go/Deployments.

Kreimer, Pablo, and J. P. Zabala. 2007. Chagas disease in Argentina: Reciprocal construction of social and scientific problems. *Science, Technology and Society* 12 (1):49–72.

McIntosh, Donald. 1970. Weber and Freud: On the nature and sources of authority. *American Sociological Review* 35 (5):901–911.

Medina, Eden. 2011. *Cybernetic Revolutionaries: Technology and Politics in Allende's Chile*. Cambridge, MA: MIT Press.

Negroponte, Nicholas, and Walter Bender. 2007. The new $100 computer. *World Bank Group*. Available at http://info.worldbank.org/etools/BSPAN/PresentationView.asp?PID=2070&EID=950.

Nickson, Andrew. 2009. The general election in Paraguay, April 2008. *Electoral Studies* 28 (1) (March): 145–149.

OLPC. 2009. XO-1 Touchpad: Issues. *OLPC Wiki*. Available at http://wiki.laptop.org/go/XO-1/Touchpad/Issues.

OLPC. 2011. Hardware Uniqueness. *OLPC Wiki*. Available at http://wiki.laptop.org/go/Hardware_uniqueness.

OLPC. 2012a. OLPC Principles and Basic Information. *OLPC Wiki*. Available at http://wiki.laptop.org/go/OLPC_Principles_and_Basic_information.

OLPC. 2012b. OLPC's Vision. *OLPC*. Available at http://laptop.org/en/vision/mission/index2.shtml.

Papert, Seymour. 1980. *Mindstorms: Children, Computers, and Powerful Ideas*. New York: Basic Books.

Papert, Seymour. 1993. *The Children's Machine: Rethinking School in the Age of the Computer*. New York: Basic Books.

Paraguay Educa. 2008. Recursos capacitaciones. *Paraguay Educa Wiki*. Available at http://wiki.paraguayeduca.org/index.php/Recursos_Capacitaciones.

Paraguay Educa. 2009. Boletín trimestral: 2008 primavera/verano. *Paraguay Educa Wiki*. Available at http://wiki.paraguayeduca.org/index.php/Boletin_Trimestral.

Prensa Internacional. 2009. Lugo, padre y presidente. *ABC Color*, April 22. Available at http://archivo.abc.com.py/2009-04-22/articulos/514846/lugo-padre-y-presidente.

Rosner, Daniela K., and Morgan G. Ames. 2014. Designing for repair? Infrastructures and materialities of breakdown. In *Proceedings of CSCW 2014, ACM Conference on Computer-Supported Cooperative Work*. New York: ACM Press.

Stavrinaki, Maria. 2010. The African chair or the charismatic object. *Grey Room* 41:88–110.

Tsing, Anna. 2000. The global situation. *Cultural Anthropology* 15 (3):327–360.

Vessuri, Hebe. 1987. The social study of science in Latin America. *Social Studies of Science* 17 (3) (August 1): 519–554.

Warschauer, Mark, and Morgan G. Ames. 2010. Can One Laptop per Child save the world's poor? *Journal of International Affairs* 64 (1):33–51.

Warschauer, Mark, Shelia R. Cotten, and Morgan G. Ames. 2012. One Laptop per Child Birmingham: Case study of a radical experiment. *International Journal of Learning and Media* 3 (2): 61–76.

Weber, Max. 1947. Charismatic authority. In *The Theory of Social and Economic Organization*, ed. Talcott Parsons, 358–392. New York: Free Press.

11 Nanoscience and Nanotechnology: How an Emerging Area on the Scientific Agenda of the Core Countries Has Been Adopted and Transformed in Latin America

Noela Invernizzi, Matthieu Hubert, and Dominique Vinck

One of the main problems facing STS studies in Latin America regards the relevance of public policies in social, cultural, and economic contexts that are considered "peripheral." A particular feature that we analyze in this chapter is the way in which emerging thematic priorities and science policy models circulate among central and peripheral countries. We question the idea of a unidirectional flow of science and technology from developed to less developed countries ("imported magic"), in spite of the apparent isomorphism in policy design in the different countries, and explore alternative views on how scientific ideas and technologies move in the case of an emerging technoscientific field.

Over the past ten years, the majority of Latin American countries have adopted some sort of policy to drive research and development (R&D) in nanoscience and technology (NanoS&T).[1] The rapid engagement of these and other developing countries in NanoS&T research has often been interpreted as a trait of the global character of this emergent technology (Treder 2004), and as a new trait of global production in science (Hassan 2005). In this view, this globalization, unlike with previous technological breakthroughs, would place developing countries (or at least some of them) in a more favorable position to embrace the new technology to promote development and reduce poverty. The rationale underlying this argument is that during a technological paradigm change, new possibilities open for developing countries to catch up with technologically more advanced nations.[2]

In this chapter we analyze NanoS&T programs in Latin America, highlighting opposing forces in play there. On one side, we call attention to the phenomenon of the transference of emerging scientific-technological themes to countries considered as "developing" or "peripheral," and the implications in terms of objectives, instruments, and ways of organizing the research in that field. On the other, we explore how local factors, such as scientific traditions and capabilities, the role of international collaborations, and the relationship with industry, among others, give distinctive traits to the effective development of NanoS&T in those countries, beyond the common rhetoric in their policies. Among these opposing forces there is some space for maneuver that is disputed by diverse actors to influence the development of NanoS&T in the region.

The history of policies supporting NanoS&T, initially in Northern countries, emerged in the context of the end of the cold war and the resulting decline in importance of the technical-military R&D programs spawned by it. The process of globalization and economic neoliberalism in the 1990s, which resulted in an increase in economic competition at a global scale, then became the principal motor driving investment in science and technology (S&T). In particular, NanoS&T is seen by the national governments of the United States and Europe as a key resource in the technological and economic race in which emerging countries of the South might threaten their competitiveness.

NanoS&T has likewise dominated a range of converging discourses regarding the knowledge economy. This discourse is accompanied by the launch of new scientific policies in which the principal objective is an orientation of research toward innovation and economic competitiveness. Such an orientation is heavily promoted by international bodies (mainly the OECD and the World Bank) and the larger research funding agencies. In the decade 2000–2010, large multinational companies started making NanoS&T a priority area for R&D investment. NanoS&T has also been taken up by scientific communities whose research requires ever-increasing funding due to its technological requirements, leading them to support and actively legitimize such policies. However, beyond the converging official discourses regarding competitiveness and the knowledge economy, there are important differences in NanoS&T orientation among countries and geographic zones (Schummer 2007). Furthermore, scientists, industrial firms, and the policies that support NanoS&T are not the only actors involved in the emergence of this scientific-technological complex; other local actors also influence its direction. These divergences reflect the local trajectories of scientific and industrial actors and the concerns of distinct social groups, whether these involve toxicity, or citizens' privacy, or unemployment created by industrial relocation to countries with lower labor costs. This shows, despite the discursive convergence among national NanoS&T policies, that different programs are in fact implemented in light of the scientific, industrial, and societal particularities in distinct local (Robinson, Rip, and Mangematin 2007; Vinck 2014) and national (Jasanoff 2005) contexts. This leads us to reflect on the possibilities of choice for the development of alternative agendas with local focus, and on the need to evaluate in greater detail the complexity of the dynamics that are in play.

We argue that there is a tension in the making of Latin American NanoS&T policies between the isomorphism resulting from external influences and the local and regional particularities. Amid this tension, some spaces for maneuver emerge, opening up possibilities for a more locally contextualized agenda. More precisely, we intend to evaluate to what degree the international dynamics determine scientific policy choices at the national scale, and assess the influence of local dynamics—taking into account characteristic processes in Latin American socioeconomic contexts—in the formulation of those scientific policies. We begin by presenting the discourses, instruments, and objectives of national scientific policies in different Latin American countries. Then we focus on the strategies and orientation of research conducted under that umbrella. In particular, we identify the key mechanisms of convergence and divergence at play, showing that these fundamentally concern the preexisting thematic specialization and industrial capabilities, the central role of international research networks, and the role of particular scientific communities, foundations, or companies.

The Formulation of National NanoS&T Policies in Latin America

Most countries in Latin America have designed programs to drive the development of NanoS&T. Brazil, Mexico, and Argentina, the largest countries in the region and the ones that have historically accumulated greater scientific capability, are in leadership positions.

In Brazil, the Ministry of Science and Technology (MCT) began to coordinate activities toward a strategy for the development of the field at the end of 2000, reckoning the existing human resources in this area and organizing the first four research networks. In 2004, a Program for the Development of Nanoscience and Nanotechnology was included in this ministry's Multiannual Plan for 2004–2008. The strategy adopted since then, which included a goal of decentralizing scientific capabilities, was based on the building of modern multiuser laboratories distributed throughout the country, and the organization of human resources in cooperative research networks that connected researchers from various universities and research centers in order to share infrastructure. The same year, industrial policy began to consider NanoS&T as a strategic area, a "doorway to the future." The National Nanotechnology Program, launched in 2005, highlighted this increased integration of science, technology, and innovation (STI) policy and industrial policy, and considerable funding was made available for the promotion of R&D in the field among companies and for cooperation between universities and companies. Aspects such as risk analysis, social impact, regulation, and public information, although tentatively a part of the 2004 program, only began to be taken up very recently (Invernizzi, Korbes, and Fuck 2012).

In Mexico, the Special Program for S&T for 2001–2006, part of the National Development Plan, highlighted the strategic importance of NanoS&T for the first time, and the STI Special Program for 2008–2012 placed it among the nine priority areas for technoscientific development. However, the country does not have a national program that centrally coordinates NanoS&T activities, with the research groups that are more consolidated leading its development, making use of a variety of public research funds, university-business cooperation, and international cooperation. These groups have succeeded in implementing considerable infrastructure for research, although with large differences between regions. It was only in 2009 that the National Council on S&T contributed to the creation of a national network in NanoS&T, bringing research centers together. Mexico emphasizes the development of technological parks, where business, government, and academia converge, such as the NanoS&T cluster in the city of Monterrey and the micro- and nanosystems field in the clusters located in Chihuahua and Puebla (Foladori and Záyago 2012).

In Argentina, the Secretariat of Science and Technology placed NanoS&T among the priority areas in 2003. Two years later, the Argentinean Nanotechnology Foundation (FAN) was created by the Ministry of Economy and Production, with the aim of stimulating human resource training and the creation of technical infrastructure to promote the development of the field and facilitate its adoption by industry. The National Agency for S&T Promotion financed four cooperation networks among researchers dedicated to the development of NanoS&T in different laboratories and institutions within the country (Andrini and Figueroa 2008; García, Lugones, and Reising 2012). While these programs distinguished funding for the public sector on the one hand (essentially the research networks) and the private sector on the other (projects of the FAN), the nanosectoral fund (FS-NANO), launched in 2010, provided funding for projects that were dedicated to basic and applied science via public-private partnerships (Spivak et al. 2012).

Various middle-size countries in the region, such as Colombia, Chile, and Venezuela, and smaller ones, such as Uruguay and some Central American and Caribbean countries, have also pushed for the development of NanoS&T. In Colombia, it has been defined as a strategic area for the competitive development of the country since 2004 and is associated with a program for the development of centers of excellence. Of note is a national network on materials research. In Chile, S&T policies have driven the formation of NanoS&T research centers since 2004, although the topic had already been the focus of that country's Millennium Institute created some years before. In Venezuela, the National STI Plan (covering the years 2005–2030) highlighted the need to incorporate cutting-edge technologies in the country, among them nanotechnology, but did not implement any specific plans. The Venezuelan Nanotechnology Network (RedVNano) was established in 2010, with members from the productive sector, universities, and some state bodies. Uruguayan researchers created the Uruguay Nanotechnology Group in 2006, and later, in 2010, nanotechnology was included as a transverse priority area in the National Strategic Plan for STI. In Central America, Costa Rica is notable for the LANOTEC facility, a laboratory dedicated to nanomaterials created in 2004, which receives support from the Ministry of Science and Technology, universities, high-technology industries, and the US space agency NASA. In the Dominican Republic, the Strategic Plan for STI for 2008–2018 has a subchapter on nanosciences within its priority areas. In Cuba, the Ministry of Higher Education's Nanotechnology Network was created in 2003, with the Center of Advanced Studies being built in 2010, dedicated to converging technologies and nanotechnologies (Foladori, Invernizzi, and Záyago 2012).

This brief review provides an initial impression of the replication of NanoS&T initiatives in Latin American countries shortly after they had appeared in the more industrialized countries. The bottom section of table 11.1 shows some of the pioneering initiatives leading in this direction. In the middle section, it is of interest to note that apart from the middle of the last decade, a number of countries had designated NanoS&T as an important strategic area in their STI plans. Finally, Brazil and Argentina led this process with the establishment of national policies in the field.

Although they vary in format and exhibit highly unequal levels of funding, the policies adopted in these countries to stimulate NanoS&T share general goals centered on the development of national competitiveness and also share some instruments, such as human resource training, the building of infrastructure, and cooperation between universities and businesses. In table 11.2 we can see a strong policy convergence in the three countries that have more structured policies and activities, noting some differences in the definition of priority areas to be focused upon.

According to these programs, the strategic character of NanoS&T lay in the possibility that it would allow the countries to raise their industrial competitiveness and find a niche in an emerging area within the world market, opening windows of opportunity to catch up with industrialized nations. Thus, for example, in the Brazilian nanotechnology program it is argued that "in an imminent paradigm change brought about by nanoscience and nanotechnology, we are facing a unique opportunity to enter this new era at the same time as the developed countries" (MCT 2004, 8).

Therefore, the policy instruments were designed to foster S&T capabilities—laboratory infrastructure development, human resource training within the country and abroad, facilitation of researcher links via networks—to provide those countries with competitive conditions in global NanoS&T R&D. In this sense, the idea of NanoS&T as a strategic area has been coupled with the establishment of centers of excellence in research, which in general make up the core of research networks. University-business cooperation as a key instrument is expressed in practically all of the calls for nanotechnology research undertaken in the various countries, although the kinds of cooperation between these two actors in funding the research are considerably varied.

This imitation or isomorphism in the formulation of STI policies has been a recurring feature in Latin America (Bastos and Cooper 1995; Dagnino, Davyt, and Thomas 1996; Albornoz 1997; Dagnino and Thomas 1999; Hurtado 2010). These authors emphasize the role played by international bodies in spreading institutional models, as well as the roles attributed to S&T within the import-substitution industrialization

Table 11.1
The Emergence of Nanotechnology Programs in Latin America

	2001	2002	2003	2004
National nanotechnology program established				**Brazil** Program for the Development of NanoS&T
Nanotechnology identified as a strategic area in STI Plans	**Mexico** Special Program for S&T, 2001–2006			**Brazil** Multiannual Plan, MCT, 2004–2008
Initial activities specifically directed to nanotechnology	**Brazil** CNPq: creation of four research networks in nanotechnology		**Cuba** Nanotechnology Network of the Ministry of Higher Education	**Argentina** ANPCYT call for nanotechnology projects (vacant fields) **Colombia** Nanotechnology as a strategic area: convocation for the creation of centers of excellence **Costa Rica** LANOTEC laboratory, financed by the Ministry of S&T, universities, business, and the CR-USA Foundation

Source: Authors' creation.

2005	2006	2007	2008	2009	2010
Argentina Argentinean Nanotechnology Foundation					
Argentina Medium-Term Action Plan in STI, 2005–2015 **Guatemala** National Plan for STI, 2005–2014 **Ecuador** National STI Policy, 2005–2010	**Colombia** Vision Colombia II Centennial, 2019 **Peru** National Strategic Plan of STI for Competitiveness and Human Development, 2006–2021		**Dominican Republic** Strategic Plan for STI, 2008–2018		**Uruguay** National Strategic Plan for STI, 2010
Venezuela Support of FONACYT for human resource training in nanotechnology and the First Nanotechnology School			**Chile** Conycit finances projects and supports research centers in nanotechnology **Venezuela** Ministry of S&T undertakes initial studies on converging technologies and nanomaterials	**Ecuador** Ecuadorian Center for Research and Development in Nanotechnology	

Table 11.2
Objectives, instruments, and priority areas in policies promoting NanoS&T in Argentina, Brazil and Mexico

Country	Objectives	Principal instruments	Priority application areas for the definition of initial nanotechnology policies*
Argentina	Stimulating the development of a human and technical infrastructure that will permit the country to compete at an international level in the application and development of micro- and nanotechnologies that increase the value added to products for internal consumption and export (Fundación Argentina de Nanotecnología, Decreto Presidencial 380/2005)	• Vacant Areas Program (PAV) for the promotion of areas with insufficient development • Promotion of strategic areas: research, human resource training, and international collaboration • Creation of cooperative networks among researchers, with a geographic decentralization • Promotion of the development of sectors with high-impact production, stimulating value chains and productive clusters through public–private consortiums	Nanomaterials; characterization of objects and structures at the nanoscale; design, construction and characterization of nanometric devices; theoretical development and simulation of properties; nanotechnology and bioinformatics
Brazil	The development of new products and processes based on nanotechnology to increase national industrial competitiveness (Plan Plurianual 2004–2007, MCT 2004)	• Formation of cooperative networks of researchers, with business participation and geographic decentralization • Creation of large multiuser laboratories and retrofitting of existing laboratories • High-level human resource training • International cooperation • Funding for R&D in businesses and in cooperation with universities and research centers; support to business incubators	Nanostructured materials; molecular nanotechnology and interfaces; nanobiotechnology; semiconductor nanodevices
Mexico	Developing quality education in nanotechnology; strengthening basic and applied sciences, technological development, and innovation to assist in the improvement of the quality of life in society and achieve greater competitiveness (Program Especial de C&T 2008–2012, CONACYT, 2008)	• Construction of National Laboratories; improvement of existing research infrastructure • Development of technological parks • NanoS&T Network, connecting researchers, technologists, and the private sector • Bi-National Sustainability Laboratory (LBNS) in cooperation with the United States, including incubation of start-ups	Advanced materials; energy and petroleum

Source: Created by the authors based on the national STI programs of the countries indicated.

* The original priority application areas tended to be diversified or specialized over time in the cited countries.

paradigm, and later the economic opening and global market integration policies. Velho (2011) also notes that similar science policy models respond to dominant conceptions of science in different historic moments. In the case of NanoS&T, international bodies play an important role in the formulation of policies (Robles-Belmont 2011; Foladori, Figueroa, et al. 2012), as does the alignment of actors and conditions in the region at the moment that NanoS&T emerged.

Since the 1990s, the World Bank has promoted the integration of countries into the knowledge economy as a path to development, through innovation and the adoption of leading-edge technologies that could facilitate their competitive insertion into the world market (World Bank, 2007). The Millennium Institute project was shaped toward this objective, since the end of the 1990s promoting the creation of centers of excellence in Chile, Brazil, Mexico, and Venezuela to spur research at the frontier of scientific investigation, such as NanoS&T, along with human resource training and links with the productive sector (Macilwain 1998). The Organization of American States, in 2004, called NanoS&T a field that is "crucial for the Americas" and recommended that governments include it as a national priority and allocate the necessary funding (OAS 2004, 55). In 2005 the International Center for Science and High Technology of UNIDO (United Nations Industrial Development Organization) organized the conference "North-South Dialogue on Nanotechnology: Challenges and Opportunities," with the objective of promoting the participation of countries in NanoS&T development (Brahic and Dickson 2005). Finally, the meeting of the Union of South American Nations (UNASUR), held in Argentina in 2010, agreed on a path for the pursuit of STI that considered the field a strategic area (Hirschfeld 2010).[3] The timing of these developments in the promotion of NanoS&T in STI policies of the regions suggests an apparent convergence.

These recommendations found fertile soil among Latin American policy makers, for whom the catching-up rationale had been current since the nineties and had been applied to biotechnology supported by similar rhetoric. Other factors at the time also help explain why this new research area gained legitimacy so quickly. It is interesting to note the confluence of some elements in the regional context when, at the beginning of the 1990s, NanoS&T began to appear as an emerging scientific-technological field in the most industrialized countries. It occurred at a moment in which regional universities began to broadly adopt resources such as the Internet and online databases of scientific journals, which, with wider opportunities for postgraduate studies abroad and, in the case of the Southern Cone, the return from abroad of many scientists after long periods of exile at the end of the 1980s, contributed to newly expanded contact by local scientific communities with more dynamic centers of knowledge production in Northern countries. Furthermore, the systems of scientific evaluation and compensation that privileged publication in journals with international prestige were being increasingly adopted in Latin America, contributing to a strengthening of local ties to global science. Thus, in good measure, Latin American scientists followed the incorporation of NanoS&T in developed countries much more closely than they had probably done in other areas in the past, and thus its inclusion in local agendas occurred very quickly (Invernizzi and Foladori 2012).

Although NanoS&T policies, their instruments, and the field's rapid legitimation among the scientific community in Latin American countries reinforce the idea of the transfer of policies and research agendas, the concrete development of NanoS&T in the region reveals more diverse approaches, as we argue in the following section.

Research Processes: Convergences and Divergences among Latin American Countries

Aspects of the development of NanoS&T in Latin American countries reveal, beyond the identities at the policy design level, that differences and locally important particularities also arose. This section starts by analyzing the network format adopted for research in NanoS&T, which resulted from the stimulus of scientific policies, from international cooperation programs, and from initiatives by researchers. We then show that the weight of traditions and the historically specific traits of scientific development in each country contribute particular details to the scientific communities in each country. In the same way, the different industrial structures influenced local NanoS&T policy interpretations to increase competitiveness. Lastly, we mention some specific actors that played a role in determining the direction taken by NanoS&T development in some national contexts.

The Role of National and International Networks

Regional and global NanoS&T policies, as well as the operation of research funding programs in Latin America by the European Commission and the United States, facilitated a rapid process of integration of local researchers into international networks. For Latin American scientists, the issue is one of maintaining themselves as legitimate scientific actors, at the frontier of science, while gaining access to better scientific working conditions. From the point of view of international agencies from the United States and the European Union, one of the objectives of these policies of cooperation with Latin America is to ensure the availability of qualified personnel to support their scientific-industrial competition with Asia. In this sense, the research networks play an important political role (Vinck 1996; Vessuri 2011). The participation of researchers from developing countries in international scientific networks has generated debate about the conditions in which they are inserted into the international scientific division of labor and about their ability to influence research agendas in accordance with their countries' needs (Kreimer and Levin 2014; Wagner 2008; Hubert and Spivak 2009).

In Latin America, the national and international research networks in NanoS&T were as much spontaneously formed by scientists as they were directed by promotional policies in the field and by programs of international cooperation. In all cases, the networks played a central role in ensuring access to laboratories and highly expensive equipment, as well as bringing together a critical mass of scientists working on a given topic. In Brazil, Argentina, and Mexico, the promotion of national networks also had the objective of decentralizing scientific research, which had traditionally been concentrated in the larger universities and the more economically dynamic regions.

Studies conducted in nine countries of the region (Foladori, Invernizzi, and Záyago 2012) suggest that the link with international research groups was significant in the early phases of NanoS&T development, as seen in coauthored publications with authors from industrialized countries. However, some countries were able to develop national networks much more quickly than others, as soon as their local research conditions were strengthened. This translated into a gradual increase in the number of articles by national coauthors alone. Much less consolidated, in terms of resulting in the creation of joint publications, was the process of regional collaborations within Latin America, although various initiatives in that direction had been pursued.[4]

Let's see a few examples of this. In the Mexican case, publications coauthored with foreigners were predominant until 1998. In the period 2000–2008, publications of national authors represented some 53.6 percent of the total, while those coauthored with foreigners (with the United States being the most represented) were 46.4 percent. Collaborations within the Latin American region remained low (the most significant was with Cuba, at 3.07 percent of the total) (Robles-Belmont and Vinck 2011; Robles-Belmont 2011). In Brazil, over the period 1990–2006, nationally coauthored publications prevailed while those coauthored with foreign researchers accounted for only 36 percent of the total, showing greater integration of domestic researchers, although they maintained active international collaborations. Brazilian researchers' papers coauthored with their counterparts from other countries within Mercosur[5] were few, at 3 percent of the total. In Argentina, in the same period, half of the publications were coauthored with foreigners, while Mercosur collaborations accounted for almost 8 percent (Kay and Shapira 2009). In Chile and Venezuela, collaborations with foreign authors still predominate (Cortés-Lobos 2012; López, Hasmy, and Vessuri 2012). Undoubtedly, the greater research capabilities, in terms of human resources and research laboratories, held by Mexico and Brazil, and to a lesser extent Argentina, strengthened national research networks, while countries with weaker capabilities tended to remain more tightly linked with external networks.

Traditions and Historic Conditions of Scientific and Industrial Development

It is important to note that the research dynamics generated by the stimulus of NanoS&T policies did not challenge or profoundly modify the preexisting positions and scientific strategies of each country. That is to say, each country maintained its specific thematic focus, since it had accumulated scientific capabilities and the scientists in those areas were motivated by the availability of public funds and the construction of NanoS&T policies. Thus, although the interdisciplinary character of nanotechnology is frequently reaffirmed in policy texts, researchers continue to operate within and are guided by their scientific fields (disciplines) rather than in response to problems or challenges that need to be tackled in their respective countries and that would require a more interdisciplinary approach. For example, in Argentina, the four cooperation networks created to give structure to Argentinean research on NanoS&T maintain the disciplinary divisions between physics, chemistry, material sciences, and biomedicine (Spivak et al. 2012). In Mexico, networks were created on the personal initiative of the researchers, with resources obtained for their own projects. The key universities in the country have their own NanoS&T networks; their themes are defined by the specializations within each institution. Physics is the discipline most representative of NanoS&T development (Robles-Belmont 2011), which reflects the long tradition of this scientific branch in the country (Mateos and Suárez-Díaz, this volume). Moreover, the NanoS&T network at the national level organized by the National Council for Science and Technology (CONACYT) has no specifically defined areas. In Brazil, the role of researchers in physics and chemistry, and to a lesser extent in biomedicine, was key to the legitimation of research on nanotechnology in the country (Invernizzi 2008), and this is reflected in the first four research networks formed in 2001, focused on nanostructured materials, molecular nanotechnology, semiconductors and nanoinstruments, and nanobiotechnology. However, the later expansion of the Brazilian networks has allowed for a greater thematic variety, and even a mild refocusing from scientific fields to social problems prompted by the call made in 2010 to develop research areas that could take up the millennial challenges (Invernizzi, Korbes, and Fuck 2012). In Chile, groups of researchers from various universities in materials science, physics, physics-chemistry, and biomedicine have joined in the search for public financing. Their research is oriented to the design of new materials, biomedicine, and environmental applications (Cortés-Lobos 2012). This situation illustrates how scientists have been able to take advantage of opportunities, adapting their research orientations to that of an emerging new field and new funding programs, and in this process the better-established groups in each country have been the more successful.

Another aspect of note is that the common policy discourse that considers NanoS&T an opportunity for the development of industrial competitiveness within an intensive knowledge economy, replicating the policies of the most industrialized countries, is faced with a slack industrial development in many countries of the region and a very frail tradition of innovation. In many cases, this incongruence tends to reduce the development of NanoS&T principally to the area of basic science. In the most advanced countries of the region, where industry is well established, like Brazil, Mexico, and Argentina, the funding of university-business projects and of industrial R&D has contributed to the formation of some companies that develop or apply nanotechnology. In the medium-sized and small countries, the industrial applications of nanotechnology are reduced to a few cases or are nonexistent.

The same phenomena can be seen in the small degree of participation of industry in publications on nanotechnology, which suggests that companies active in nanotechnology are predominantly those involved in cooperative, university-business projects stimulated by public financing. Thus, for example, in Mexico between 1990 and 2008, few articles (149) were published in collaboration with businesses, and most of those were with companies whose headquarters were located abroad (108 articles). Over time, the collaboration with industry has increased, though it is still primarily with foreign companies (Robles-Belmont and Vinck 2011). In Brazil, over the same period, of the 10,304 scientific articles produced on NanoS&T, more than 95 percent came out of universities, while the participation of businesses reached a maximum level of 5 percent in 2008. Over this period, 64 companies had authorship in those publications (Kay, Invernizzi, and Shapira 2008).

The Role of Local Actors
Lastly, it is necessary to draw attention to various Latin American countries that are following their own paths in the development of NanoS&T due to the central role played by some local actors. In Colombia, for example, the Administrative Department of Science, Technology, and Innovation (Colciencias) is heavily pushing an orientation toward nanobiotechnologies, replicating the North American model, supported by agreements signed with the United States. The NanoS&T policy in Argentina, at least in its early years, is explained by the weight afforded to it by the physics community in the organization of research networks and the role of the multinational corporation Bell Labs in the original proposal for the Argentinean Nanotechnology Foundation, which created a negative public reaction, given the presence of such a private, foreign company in the formulation of a national policy (Andrini and Figueroa 2008). In Mexico, FUMEC (the Mexico-United States Science Foundation) played a central part in the promotion of micro- and nanosystems research, the installation of infrastructure, the creation of scientific networks and human resource training, and the transfer of knowledge (Robles-Belmont 2011). Noteworthy in Costa Rica, likewise, are the role of the CR-USA Foundation and even the support of NASA along with the STI ministry in the creation in 2004 of the only laboratory in the country equipped for nanomaterials research (Vega Baudrit and Campos 2012). In Uruguay, it was a group of researchers who pressured the government to include NanoS&T in the National S&T Plan (Chiancone 2012), and similarly, in Venezuela, the RedVNano was organized to show the government the potential of nanotechnology to achieve some of the

socioeconomic development goals contained in the government's plan (López, Hasmy, and Vessuri 2012).

Conclusions

A collection of STS studies on NanoS&T in Latin America problematizes the development of public policies and the dynamic of peripheral societies (scientific communities, businesses, and civil society) in this emerging field in the scientific agenda of the core countries. They examine in what way Latin American countries have become involved in this area, what their orientations have been, whether they have transferred thematic priorities and scientific policy models from the core countries, and to what extent they have produced alternative orientations taking into account the concerns of local and regional actors.

We have shown that there is a strong convergence in Latin American public policies in terms of their chronology, their objectives, and their policy instruments, and have underlined the important role played by international organizations and researchers integrated into global scientific networks in the region's NanoS&T enterprise. In this way, the construction of research policies in Latin America appears to be an imitation of models from the North, with very little adaptation to specific social, cultural, and economic contexts. Even the scarcity of local resources does not appear to decisively affect the definition of those policies, at least at the early stages. Therefore, we can conclude that a significant phenomenon of policy imitation occurred. This suggests a condition of dependency among Latin American countries that reflects a weak autonomy of their scientific communities and scarce local demands on the part of businesses and civil society.

At the same time, the incorporation of the region's scientists into international networks tends to reinforce the idea that they are integrated into global NanoS&T research in a dependent way, since the specific needs of their respective countries do not appear to have significantly influenced the research agenda. The lack of dynamism in national industries regarding R&D endeavors and the inability of civil society to influence scientific research means that neither of those groups have a relevant role to play in the setting of agendas. Thus, the policymakers tend to respond primarily to the demands of researchers, of international organizations, or of multinational corporations, or to proposals from foundations. The question of how NanoS&T agendas are set and what actors and interests are more represented in this process is a subject scantily covered by Latin American STS studies so far.

In that sense, the case of NanoS&T seems to be one of straightforward North-to-South diffusion ("imported magic"), unlike other areas where Latin American countries have been able to affirm their own strategy of research and innovation.

In the field of biomedicine, for instance, clear priorities have been implemented in various Latin American countries in order to solve specific national or regional public health issues, whereas, until now, innovation in NanoS&T has been more dependent on the orientation of multinational companies toward globalized markets. The difference between the two areas may also have a technological explanation. Indeed, biomedical research technologies require a lower investment than those of NanoS&T, especially for the construction of high-tech infrastructures such as clean rooms or synchrotron radiation facilities.

However, the aforementioned trend is not the only force in action, and the transnational circulation of S&T policy models, as well as the diffusion of technological artifacts (Chan, this volume) or scientific ideas (Rodriguez, this volume), implies local contextualization, negotiation, and adaptation. Indeed, we have shown that the development of NanoS&T in Latin America cannot be reduced to an isomorphic copy of what the countries of the North have done, as even in those Northern countries, despite their employing similar policies, there is no homogeneous development of NanoS&T in thematic terms or activities. We can say, then, that there are significant differences between and within the countries of the North, as well as between and within Latin American countries. We have shown that the path to insertion into international, national, and regional networks has given specific forms and focuses to NanoS&T research in the region's countries. Furthermore, the forms of insertion are conditioned by the degree of development of the structure of research and human resources, which are unequally distributed among the countries. Larger countries are, indeed, creating more autonomous capabilities. We have also seen that the historically accumulated capabilities and traditions add a crucial skewing to the form in which new fields are taken up. Scientists adapt to new areas promoted by policy, without deviating from the paths they already follow. So, beyond the rhetoric of NanoS&T's transdisciplinary character, the majority of countries have embraced this field from the disciplines in which they have the most qualified resources. Similarly, existing industry, which is weak in the majority of small and medium-sized countries and shows low innovation performance even in those with a more developed industry, does not constitute either an adequate material infrastructure and knowledge base or a demanding force for NanoS&T to increase its competitiveness. In the majority of countries, it has proven difficult to create a relationship between research and business, which leaves advances in NanoS&T very much tied to basic science.

These characteristics of diversity show that there is a marginal space in which action can be taken to orient the policy and the development of NanoS&T in ways that are suitable to local contexts. Taking advantage of this room for maneuver requires action at various levels: setting up participative instruments that enable enduring interactive processes between local actors (governments, companies, and civil society) that express local needs in the framework of their country-specific projects; developing a context-specific NanoS&T agenda and a capability for autonomous thought in the making of S&T policy objectives (without losing the international connections to identify resources and synergies); designing specific S&T policy instruments in order to reinforce local, national, and regional scientific communities in such a way as to not dissolve within international scientific networks (as can be seen in the increase of publications with national coauthorship); and negotiating at the international level the design of the intellectual property system.

Notes

1. The most widespread definition of nanotechnology refers to the knowledge and methods by which it is possible to create, manipulate, visualize, and use objects (materials or devices) at the nanometric scale (1 nanometer = 1 billionth of a meter), the scale at which the materials exhibit specific properties (RS&RAE 2004; NTSC 2007). However, the emphasis on scale hides political and economic debates regarding different conceptions of nanotechnology, as well as debates on regulation and risks related to it (Bensaude-Vincent 2004). Various actors—industrialists, scientific communities, governments, and civil society organizations—alternatively attempt to limit the definition to exclude regulations or seek to open it up, to convince politicians that something new and interesting is under way or that new risks are involved.

2. For the general concept of "catching up" in a technological paradigm change, see Perez and Soete (1988), and for a study on catching up in nanotechnology, see Niosi (2006).

3. These issues are further developed in Foladori, Figueroa, et al. (2012).

4. For example, Brazil has active cooperation with Argentina, Mexico, and Chile via bilateral centers, and maintains other forms of cooperation with Venezuela, Colombia, Cuba, Peru, Uruguay, and Ecuador (MCT 2008).

5. Mercosur is an economic and political agreement established between Argentina, Brazil, Paraguay, and Uruguay in 1991. In 2012 Venezuela joined the agreement.

References

Albornoz, Mario. 1997. La política científica y tecnológica en América Latina frente al desafío del pensamento único. *Redes* 4 (10):95–115.

Andrini, Leandro, and Santiago Figueroa. 2008. El impulso gubernamental a las nanociencias y nanotecnologías en Argentina. In *Las nanotecnologías en América Latina*, ed. Guillermo Foladori and Noela Invernizzi, 27–39. Mexico City: M. A. Porrúa.

Bastos, Maria Inés, and Charles Cooper. 1995. *Politics of Technology in Latin America*. London: Routledge/UNU Press.

Bensaude-Vincent, Bernardette. 2004. Two cultures of nanotechnology? *International Journal for Philosophy of Chemistry* 10 (2):65–82.

Brahic, Catherine, and David Dickson. 2005. Helping the poor: the real challenge of nanotech. SciDev.Net, February 21, 2005. Available at http://www.scidev.net/content/editorials/eng/helping-the-poor-the-real-challenge-of-nanotech.cfm (accessed September 5, 2012).

Chiancone, Adriana. 2012. Nanociencias y nanotecnologías en Uruguay: Áreas estratégicas y temáticas grupales. In Foladori, Invernizzi, and Záyago 2012, 201–210.

Consejo Nacional de Ciencia y Tecnología (CONACYT). 2008. Programa especial de ciencia, tecnología e innovación 2008–2012. Mexico City: CONACYT. Available at http://www.siicyt.gob.mx/siicyt/docs/contenido/PECiTI.pdf (accessed October 31, 2009).

Cortés-Lobos, Rodrigo. 2012. Nanotecnología en Chile. ¿Qué tan preparado se encuentra el país para desarrollar esta disciplina? In Foladori, Invernizzi, and Záyago 2012, 85–100.

Dagnino, Renato, Amilcar Davyt, and Hernán Thomas. 1996. El pensamiento en ciencia, tecnología y sociedad en Latinoamérica: Una interpretación política de su trayectoria. *Redes* 7 (6):13–51.

Dagnino, Renato, and Hernán Thomas. 1999. La política científica y tecnológica en América Latina. *Redes* 12 (6):49–74.

Decreto Presidencial 380/2005 de la República Argentina. 2005. Autorízase al Ministerio de Economía y Producción a constituir la Fundación Argentina de Nanotecnología. Available at http://www.fan.org.ar/wp-content/uploads/2012/08/estatuto.pdf (accessed January 13, 2014).

Foladori, Guillermo, Santiago Figueroa, Edgar Záyago, and Noela Invernizzi. 2012. Nanotechnology: Distinctive features in Latin America. *Nanotechnology Law and Business* 9 (1): 88–103.

Foladori, Guillermo, Noela Invernizzi, and Edgar Záyago, eds. 2012. *Perspectivas sobre el desarrollo de las nanotecnologías en América Latina*. Mexico City: M. A. Porrúa.

Foladori, Guillermo, and Edgar Záyago. 2012. La política de ciencia y tecnología en México y la incorporación de las nanotecnologías. In Foladori, Invernizzi, and Záyago 2012, 13–32.

García, Marisa, Manuel Lugones, and Ailin María Reising. 2012. Conformación y desarrollo del campo nanotecnocientífico argentino: Una aproximación desde el estudio de los instrumentos de promoción científica y tecnológica. In Foladori, Invernizzi, and Záyago 2012, 13–32.

Hassan, Mohamed. 2005. Nanotechnology: Small things and big changes in the developing world. *Science* 309 (5731):65–66.

Hirschfeld, Daniela. 2010. Latin America defines roadmap for innovation. *SciDevNet*, December 13, 2010. Available at http://merid.org/en/Content/News_Services/Nanotechnology_and_Development_News/Articles/2010/12/15/Latin_America_Defines_Roadmap_for_Innovation.aspx (accessed January 9, 2014).

Hubert, Matthieu, and Ana Spivak. 2009. Integrarse en las redes de cooperación en nanociencias y nanotecnologías. El rol de los dispositivos instrumentales. *Redes* 15 (29):69–91.

Hurtado, Diego. 2010. *La ciencia argentina. Un proyecto inconcluso: 1930–2000*. Buenos Aires: Edhasa.

Invernizzi, Noela. 2008. Visions of Brazilian scientists on nanoscience and nanotechnology. *NanoEthics* 2 (2):133–148.

Invernizzi, Noela, and Guillermo Foladori. 2012. ¿Hacia dónde van las nanotecnologías en América Latina? In Foladori, Invernizzi, and Záyago 2012, 231–235.

Invernizzi, Noela, Cleci Korbes, and Marcos P. Fuck. 2012. Política de nanotecnología en Brasil: A 10 años de las primeras redes. In Foladori, Invernizzi, and Záyago 2012, 55–84.

Jasanoff, Sheila. 2005. *Designs on Nature: Science and Democracy in Europe and the United States*. Princeton: Princeton University Press.

Kay, Luciano, Noela Invernizzi, and Philip Shapira. 2008. The role of Brazilian firms in nanotechnology development. Atlanta Conference on Science and Innovation Policy, 1: 1–8. Atlanta: Institute of Electrical and Electronics Engineers (IEEE).

Kay, Luciano, and Philip Shapira. 2009. Nanotechnology development in Latin America. *Journal of Nanoparticle Research* 11:259–278.

Kreimer, Pablo, and Luciano Levin. 2014. Scientific cooperation between the European Union and Latin American countries: Framework programmes 6 and 7. In *Research Collaborations between Europe and Latin America: Mapping and Understanding Partnership*, ed. Jacques Gaillard and Rigas Arvanitis, 79–106. Paris: Editions des Archives Contemporaines.

López, Sonsiré, Anwar Hasmy, and Hebe Vessuri. 2012. Nanociencia y nanotecnología en Venezuela. In Foladori, Invernizzi, and Záyago 2012, 211–228.

Macilwain, Colin. 1998. World Bank backs Third World centers of excellence plan. *Nature* 396 (711):24–31.

Ministério da Ciência e da Tecnologia (MCT). 2004. Plano Estratégico do MCT 2004–2007.

Ministério da Ciência e da Tecnologia (MCT). 2008. Relatório analítico Programa de C,T&I para Nanotecnologia. Coordenação Geral de Micro e Nanotecnologias. Available at http://www.mct.gov.br/upd_blob/0028/28213.pdf (accessed February 22, 2011).

National Science and Technology Council (NTSC). 2007. *The National Nanotechnology Initiative: Strategic Plan*. Washington, DC: Executive Office of the President, National Science and Technology Council, Nanoscale Science, Engineering, and Technology Subcommittee.

Niosi, Jorge. 2006. Biotechnology and nanotechnology: Science-based enabling technologies as windows of opportunity for LDCs? *World Development* 35 (3):426–488.

Organization of American States (OAS). 2004. Final report of the fourth regular meeting of the Inter-American Committee on Science and Technology (COMCYT) OEA/Ser.W/XIII.3.4. Available at http://www.science.oas.org/COMCYT/english/Resolution.htm (accessed April 4, 2006).

Perez, Carlota, and Luc Soete. 1988. Catching up in technology: Entry barriers and windows of opportunity. In *Technical Change and Economic Theory*, ed. Giovanni Dosi et al., 458–479. London: Pinter.

Robinson, Douglas, Arie Rip, and Vincent Mangematin. 2007. Technological agglomeration and the emergence of clusters and networks in nanotechnology. *Research Policy* 36 (6):871–879.

Robles-Belmont, Eduardo. 2011. Les Fondations, acteurs de l'émergence des nouvelles technologies dans les pays non hégémoniques: Le cas des micro et nanotechnologies au Mexique. Thèse de doctorat, Université de Grenoble.

Robles-Belmont, Eduardo, and Dominique Vinck. 2011. A panorama of nanoscience developments in Mexico based on the comparison and crossing of nanoscience monitoring methods. *Journal of Nanoscience and Nanotechnology* 11 (6):5499–5507.

Royal Society and Royal Academy of Engineering (RS&RAE). 2004. *Nanoscience and Nanotechnologies: Opportunities and Uncertainties*. London: Royal Society and Royal Academy of Engineering.

Schummer, Joachim. 2007. The global institutionalization of nanotechnology research: A bibliometric approach to the assessment of science policy. *Scientometrics* 70 (3):669–692.

Spivak, Ana, Matthieu Hubert, Santiago Figueroa, and Leandro Andrini. 2012. La estructuración de la investigación argentina en nanociencia y nanotecnología: Balances y perspectivas. In Foladori, Invernizzi, and Záyago 2012, 33–53.

Treder, Mike. 2004. Nanotechnology and society: Times of change. Presentation, October 18, São Paulo. Available at http://www.crnano.org/Speech%20-%20Times%20of%20Change.ppt (accessed September 5, 2012).

Vega-Baudrit, José R., and Allan Campos. 2012. Nanotecnología en la región Centroamericana y Panamá: Caso Costa Rica. In Foladori, Invernizzi, and Záyago 2012, 127–134.

Velho, Lea. 2011. Conceitos de ciência e a política de ciência, tecnologia e inovação. *Sociologias* 13 (26):128–153.

Vessuri, Hebe. 2011. La actual internacionalización de las ciencias sociales en América Latina, ¿vino viejo en barricas nuevas? In *Estudio social de la ciencia y la tecnología desde América Latina*, ed. Antonio Arellano Hernández and Pablo Kreimer, 21–55. Bogotá: Siglo del Hombre Editores.

Vinck, Dominique. 1996. The dynamics of scientific intellectuals within the integrative trend in Europe: The case of co-operation networks. In *Internationalism and Science*, ed. Aant Elzinga and Catharina Landström, 162–198. London: Taylor Graham.

Vinck, Dominique. 2014. The local configuration of a science and innovation policy: A city into the nanoworld. In *The Local Configuration of New Research Fields: On Regional and National Diversity*, ed. Martina Merz, Philippe Sormani, and Peter Biniok. Dordrecht: Springer.

Wagner, Caroline. 2008. *The New Invisible College: Science for Development*. Washington, DC: Brookings Institution Press.

World Bank. 2007. Building knowledge economies: Advanced strategies for development. Washington, DC: World Bank. Available at http://siteresources.worldbank.org/KFDLP/Resources/461197-1199907090464/BuildingKEbook.pdf (accessed March 18, 2011).

12 Latin America as Laboratory: The Camera and the Yale Peruvian Expeditions

Amy Cox Hall

Through our lens today we view Hiram Bingham's Yale Peruvian Expeditions (1911, 1912, 1914–1915) to Machu Picchu as tarnished by industry and ego, couched in imperial narratives of progress and development, and perpetuating a racialized understanding of Peru as burdened by its indigenous population. Moreover, recent publications and current politics have further framed the Yale Peruvian Expeditions as a popular narrative of intrepid adventure (Heaney 2010; Adams 2012). Today the archaeological site is remembered for its beauty, its energy, its mystery, its iconicity of national heritage, and its commercial possibilities as a new modern wonder of the world. Quickly in tow is a photograph buttressing such claims.

The fact that these expeditions were initially rooted within a scientific frame has become a forgettable backdrop to the current discourse. Nonetheless, a failure to read Bingham's pursuits as scientific would be to miss a critical aspect of the three Yale expeditions to Peru and the ways in which so-called discoveries were made and publicized. The expeditions were not the superficial adventures of an Indiana Jones, nor did they happen by chance, destiny, or luck. Nor were they simply the conspiratorial deceits of agents of empire (Mould de Pease 2001; Mould de Pease 2003). As the first professor of South American history at Yale University, Bingham, at least initially, sought to develop exciting and relevant questions that built upon previous scholarship before methodically attempting to answer those questions.[1] His main goal was to contribute to something beyond individual experience that would enrich "civilization's" knowledge, and along the way make a name for himself.[2] The frame of science facilitated his expeditions, validated his practices, and legitimized his findings. Bingham's most powerful scientific expeditionary tool, the camera, materialized and made fact the imagined discovery of a lost city in the clouds. In many ways, Bingham's significant and most lasting contribution was, and remains, a photograph of Machu Picchu.[3]

In this chapter, I focus on Bingham's use of the camera to examine the relationship between expeditions and the formation of scientific knowledge about Peru and Machu Picchu. I argue that photography was considered a primary scientific practice conducted on the expeditions, as it allowed the team to bring back material evidence of their explorations and findings. Through the camera, landscapes were negotiated and scenes were framed, creating compelling and popular narratives about Bingham, Yale's discovery, and Peru's Incan past and indigenous present.

Although Bingham conceived of Machu Picchu and the Andean landscape as a laboratory to test Kodak's latest technology, they were not docile subjects of his objectification. Other authors in this volume focus on the translation of technology in the hands of new users (see Ames and Chan, this volume), but Bingham did not share his cameras with Peruvians. The cameras were operated only by Bingham and a few selected team members, although one camera was stolen from his saddlebag during the 1911 expedition. While some locals were perhaps awed by the photographic technology and certainly assisted the expedition in other ways, such as locating sites and excavating remains (Cox Hall 2012), the camera remained in the sole purview of expedition members. This does not mean, however, that Kodak's camera technology "flowed" (Appadurai 1996) to Peru or that images were not negotiated, translated, and transformed in practice. Kodak's cameras were not simply taken out of their case and used. Instead, there was friction in their diffusion in the contact zone not only due to their users but also because of the technology and the Andean climate (Tsing 2005; Pratt 1992). In other words, a variety of actors, human and nonhuman, had to be considered to obtain a photograph. In this chapter, I examine the work exerted through the camera—negotiated practice—to manifest a specific and scientific imagining of Machu Picchu as a "lost city" with Yale as its discoverer.

Photography and Expeditionary Science

By the early twentieth century, a camera was considered indispensable on scientific and anthropological fieldwork. Images were critical evidence for legitimizing not only the anthropological discipline as a whole but also the findings of scientific explorations (Edwards 1988; Edwards 1990; Edwards 1992; Ryan 1997). Professional photographers were hired to accompany scientific expeditions, and guidelines outlining what constituted scientifically valuable photographs were widely distributed to travelers and amateur explorers (Garson and Read 1892; Freire-Marreco and Myres 1912). Photography was considered invaluable because it was thought to capture what words could not, offering a form of mechanical objectivity (Daston and Galison 1992). Moreover, photography's imaging technology could record cultural practices and physiognomic features as a way of documenting peoples who were thought to be fast disappearing (Edwards 1992). Photography thus acted as a documentary technology and evidentiary object and was used to bring faraway places home for future study.[4] As a mechanism for collecting and materializing evidence for the Yale Peruvian Expeditions, photography became a powerful way to stabilize facts about Peru due to its elision of cultural making and its claim to produce an objective reality.

The photograph's status as a collected object or as scientific evidence obtained through a specific type of laboratory technology is often obfuscated when scholars consider the power of photography's imaginings.[5] Scholarship on photography has emphasized the visualized referent, the image itself, and the power of the photograph's reproducibility. Thus, photographs have been examined for their import as a technology of seeing (Crary 1990; Daston and Galison 1992; Dumit 1997; Tucker 1997); for the types of gazes exhibited in their frame (Berger 1972; Barthes 1981; Kuhn 1985; Lutz and Collins 1993); for the creation, expression, and perpetuation of racialist understandings of the Other (Alloula 1986; Chavez 2000; Poole 2004); for their role in state formation (Sekula 1983; Sekula 1986; Tagg 1989); and for the ways in which they create and consolidate powerful conceptualizations of nations (Anderson 1991; Ryan 1997; Poole 1998). Photographs are powerful producers of knowledge, not only because they are indexical imaginaries, but also because they themselves are material objects—a "melding of image and form, both of which are direct products of intention" (Edwards and Hart 2004, 2).

Prior to the materialization of that intention was the camera—a potent device of inscription and technology of extraction (Latour and Woolgar 1986). Although Yale's images were published worldwide, here I focus on the actual use of the camera on the expeditions.[6] I suggest that it is critical to figure camera technology and its deployment into the discussion of expeditionary photography to further explicate the ways in which intention and imaginaries were initially married with facts and to understand how this technology was deployed in the Andes. The photograph did not become a byproduct of intention simply with its insertion into text. It was not only that the images Bingham produced appeared to be divorced from his rendering (Poole 1998), or that they circulated widely in *National Geographic*, museum exhibitions, illustrated lectures, and newspapers. It was also that the specific camera equipment, and its use, initiated the technological framing of the site as an exotic find, a nativist utopia, and a national treasure. Camera technology and its negotiated practice in the laboratory of Latin America was the first step in translating and manifesting Bingham's intention: the discovery of a city, lost and hidden behind the ranges, waiting for him.[7]

Visualizing Science: Kodak and the Panorama Camera

In recounting his "discovery" of Machu Picchu in the *Lost City of the Incas*, Bingham wrote, "Would anyone believe what I had found? Fortunately ... I had a good camera and the sun was shining" (Bingham 1952, 180). Bingham and his team were avid and conscientious photographers, taking over nine thousand photographs on their three expeditions.[8] Many of the images were landscape views that highlighted the spatial organization of the monument. Other photographs documented the location of material objects within the remaining structures of Machu Picchu and other archaeological ruins (Poole 1998; Lopez-Lenci 2004). The few photographs of the Andean peoples who inhabited the site and its environs served either to typologize the customs and dress of Quechua peoples or to highlight medical ailments, such as massive goiters and skin diseases (figure 12.1).[9] Physicians accompanied the last two expeditions, and Bingham felt that with a little training in anthropometry, they might serve a dual purpose: they could care for the health of the team members while at the same time further the expedition's scientific agenda of studying the local indigenous population as a fallen, or degenerated, race of the Incas (Salomon 1985; Thurner 2003; de la Cadena 2000).[10]

Kodak was a significant corporate sponsor for the three expeditions and contributed more to the expedition than any other organization except National Geographic (Salvatore 2003).[11] Bingham wrote Kodak on March 25, 1911, asking if they would like to provide camera equipment "for a scientific expedition in the tropics" (Bingham to Kodak, Kodak Select Correspondence, March 25, 1911, YPEP, box 14, folder 210). The company agreed to provide the expedition with an entire photographic outfit on the condition that Bingham and the other members use the equipment properly, report on their work, and test out the equipment in various conditions (Bingham to George Eastman, Kodak Select Correspondence, May 16, 1911, YPEP, box 14, folder 210). A month before departing for Peru, Bingham wrote to George Eastman, Kodak's founder, confirming the agreement, stating that he would do all in his power to abide by it, including making reports and developing the pictures as soon as possible after exposure. In exchange for outfitting the expedition with cameras, tank developers with solution, and films at no charge, Kodak was "to secure the benefit of our experiments with this material in damp tropical valleys and second, in order that you may have the benefit of this experience for advertising purposes" (Bingham to Eastman, Kodak Select Correspondence, May 16, 1911, YPEP, box 14, folder 210). The 1911 photographic outfit provided by Kodak included four cameras with lens attachments, tripods, three complete tank developers with chemicals, and 284 film cartridges containing 2,608 exposures.[12] Kodak's contribution brought home noteworthy results.

Returning from the 1911 expedition, Bingham wrote to Eastman that the camera outfit supplied by Kodak had worked "splendidly" and "the results certainly justified our efforts." He added: "I don't know how high your Kodaks have been carried before but we took two to the top of Mount Coropuna, which is probably about 22,500 feet above sea-level" (Bingham to Eastman, Kodak Select Correspondence, December 27, 1911, YPEP, box 14, folder 210). In general, the Yale team followed the agreement with Kodak, developing the film a few days after the pictures were taken. When they lacked water, the expedition sealed the tins containing the film, developing it two to three weeks later.

Figure 12.1
Quechua Indians, untitled photograph labeled "F," taken by David Ford, c. 1915, Yale Peruvian Expedition Papers, Manuscripts and Archives, Yale University Library. Six of several images, taken by the 1915 expedition's physician, David Ford, depicting medical ailments and goiters of the surrounding Quechua-speaking public.

Kodak also supplied film and cameras for the 1912 expedition. Although the second expedition continued to conduct survey work and mapmaking, Bingham also emphasized collecting as a primary goal. He was bringing along more men, wanted to take more photographs, and most of all wanted images of the Urubamba canyon. Although Eastman initially offered him a speed Kodak for the 1912 expedition, Bingham declined, saying that a panorama camera would be more useful. This choice reflected the subjects and views Bingham privileged. He had already witnessed the area and was thus able to articulate the specific needs of the 1912 expedition for documenting panoramic vistas, taking anthropometric measurements, and collecting. In addition, National Geographic had become a primary sponsor and also wanted photographs.[13] In his letter to Eastman, Bingham wrote:

> I am almost convinced that it would be extremely advisable to have one Panorama Kodak in the outfit. I do not feel so sure about the Speed Kodak. We do not have to take pictures of the fastest moving objects. There are no automobiles, and athletes and race horses are equally scarce. I have understood, however, that the lens is better than that in the 3A Special, and will give finer details. Can you give me some advice on this? ... In many of the deep canyons where we are expecting to work it needs a Panorama Kodak to show the opposite side of the mountain up to the top. ... If you can give us three new 3A Specials, and one No. 4 Panorama we shall have nine Kodaks in the outfit and ought to be well equipped for the scientific work that lies ahead of us. (Kodak Select Correspondence, April 15, 1912, YPEP, box 14, folder 11)

Bingham's choice of cameras would ultimately shape the ways in which knowledge about Machu Picchu and its history was fashioned. For the ends of the expedition, only the Kodak panorama camera could help Bingham "show the opposite side of the mountain up to the top." Kodak's panorama camera allowed him to conjure a specific framing of Machu Picchu as "lost and hidden behind the ranges," discovered by the intrepid Yale historian. The 3A, in turn, would be useful in capturing the details of archaeological remains. Without the 3A and the panorama camera, Bingham's gaze would never have materialized and circulated. His "discovery" of an imagined lost Incan refuge was first made possible by Kodak technology.

Although documents suggest that the team gave minimal feedback on how the cameras performed in the tropics, Kodak did use the expeditionary images for advertising purposes. The company made thirteen enlarged bromide prints, lantern slides for illustrating lectures, and an exhibition catalog for the Peruvian exploration display in a Kodak exhibition that toured for two years between 1912 and 1914 (E. Craig to Bingham, Kodak Select Correspondence, April 10, 1912, YPEP, box 14, folder 211). Additionally, due to the interest that was shown in that Peruvian exploration display, Kodak requested further loan of the images for their display in the Panama-Pacific Exposition in San Francisco (Craig to Bingham, signed by T. H. Griffin, Kodak Select Correspondence, July 11, 1914, YPEP, box 14, folder 244).

Deborah Poole has suggested that Bingham's vision combined physical achievement with an imperialist's triumphant sense of geography and conquest (Poole 1998). Analyzing one of Bingham's photographs, Poole argues that his aesthetic of discovery began at the horizon and sought to visualize the mysterious and hidden beyond. However, specific camera technology was first needed to manifest Bingham's scientific vision. The 3A camera, fitted with a wide-angle lens, was useful for highlighting details at shorter focal lengths (e.g., the rocky walls of the monument), while the panorama camera could collect wider vantage points.[14] The scientific evidence showing the Andes as an area with wistful panoramas, detailed rock formations, and expansive landscapes devoid of inhabitants was initially made possible because of these two Kodak cameras. Combined with more than 3,500 negatives brought along for the 1912 expedition, the camera materialized and made possible Bingham's imagining of Peru.[15] Technology was thus the first step in the visualization of Machu Picchu as a "lost" city "discovered" by Yale, with Bingham at its helm. The next step was making pictures.

Photographing a Lost City

Although understood as a primary scientific practice on the expeditions, making pictures involved skill, time, and careful translation. The photographs were intended not to be travelers' snapshots but to serve the scientific goals of the expedition: documenting and surveying the landscape and its remains. In this section, I discuss the ways Kodak camera technology was deployed. My goal is to examine the ways in which a specific vision of Peru was communicated to the expeditionary team and to witness the work involved in translating that vision into scientific evidence. The camera was an unstable tool, and Bingham expended much effort to obtain "accurate" images.[16]

Bingham provided directions and instructions to his team members through typed expedition circulars. Organized in black notebooks, the Yale circulars included information and instructions as disparate as packing lists of the boxes, daily meals, vaccinations, care of rifles, rules for hygiene and camp sanitation, and collecting.[17] Throughout the expedition, additional circulars were distributed to party members as a way of communicating new information such as shipping schedules and travel arrangements (e.g., General Orders for 1915, No. 1 and No. 3, YPEP, box 23). Although different, the 1912 and 1914 circulars on photography conveyed the sense that while the camera technology was understood as a tool for obtaining scientific evidence, it was also suspect and unstable (Poole 2005).

The instructions in the 1912 circulars suggest that visual documentation was not sufficient on its own as evidence (Derrida 2002). While the collection of views provided a visual record of the activities of the expedition, additional work was needed to fashion those views into useful data. Labeling, noting color schemes, and connecting the photograph with more detailed drawings formed a minable database indexing itineraries, geographies, practices, and people. The 1912 circulars primarily contained instructions on the use of photography as an aid in documenting collected materials.

Bingham instructed that photographs should be taken before, during, and after excavation.[18] Considered a supplement to sketches, the images needed to record the geographic locality, geological formation, and position and orientation of all human remains (Alfred Bumstead, Journal and Notebook, June-September 1912, YPEP, box 19, folder 17). Specific instructions for labeling the photographs were also included as an attempt to enable the collector to link specimens with locations. A random assortment of images was not useful for future study, and so the photographs had to be tethered to other information. Each photograph was to be marked with the serial number given to it by the quartermaster on the lower right-hand corner of the glossy side of the negative print on the margin of the film so as not to show on the image. On the left side, the photographer needed to write in the month, day, and his initials (see Circular No. 14, 1912, YPEP, box 19, folder 17).

Instructions for coloring the slides were also included in the 1912 circular. Anticipating the image's future use as evidence and illustration for public lectures, expedition members were told to use the Golsch Reference Color chart and number the film accordingly (see Circular No. 19, 1912, YPEP, box 19, folder 17). For example, the foreground could be numbered 41; foliage, 17; poncho, stripes alternation, 8 and 1; small stripes, 24, etc. (figure 12.2). Bingham assured his team members that while marking the image for future coloring took a great deal of time, it was not wasted if the work was done accurately and carefully.[19] Bingham's instructions attempted to stabilize the technology, reducing not only the noise inherent in the capturing of one's intention, but also the black-and-white photograph's fleeting and slippery meaning.

The photographic circulars for the 1914 expedition expanded upon the instructions for the 1912 expedition. Perhaps due in part to Bingham's initial absence during the first year of the expedition, he offered greater detail and direction on expeditionary photography in the 1914 circulars. While the instructions for the 1912 expedition provided careful detail for labeling the images and using photography as a complement to sketches when collecting archaeological remains, the 1914 circulars contained specific direction on the making of pictures. As such, the 1914 circulars provide insight into the evolving needs of the expedition and ways in which making pictures became the expedition's most valued scientific practice.

For the 1914 expedition, Bingham sought to further direct and discipline the gaze of the expedition photographers. He offered explicit guidelines as to which subjects were desired and how to obtain accurate views.[20] The city of Cuzco, for example, was not of interest. Rather, Bingham desired that all film be reserved for use in the field. Moreover, photographs of landscapes or objects connected with other departments of

Figure 12.2
Untitled, Yale Peruvian Expedition Papers, Manuscripts and Archives, Yale University Library. A lantern slide, hand-tinted by W. C. Ives in Mystic, Connecticut, for use in Bingham's illustrated lectures.

the expedition were only to be taken on request of the director or if it was likely that no one else had already taken the picture (see Circular No. 20, YPEP, box 23).

Simply snapping pictures was not enough. Taking pictures in the field required a considerable amount of know-how for the images to become valuable evidence. Accuracy was required. For Bingham, *accuracy* meant materializing original intention. He urged the members of the expedition to understand the camera and to spend time making good pictures:

The more painstaking the process of setting up the tripod, leveling the camera, using a small diaphragm, and accurate time exposure, and exact records of the time, place and subject, the better the results of the Expedition will be. No time is better spent than in studying your negatives after they have been developed and dried, with the record book in hand, seeing whether you got what you desired, whether the exposure was too much or too little, and whether the camera is working according to your expectations. Careless and hasty work in photography means waste of time, loss of opportunities, and poor results. (Circular No. 20, YPEP, box 23)

Team members were instructed to go slow at first, becoming familiar not only with the particular camera but also with the atmospheric conditions, the exposure meter, and the level.[21] The members were encouraged to learn to estimate the distances and focus of the camera, taking test images to assure the reliability of the camera and its operator (see Circular No. 20, YPEP, Hasbrouch, box 23).[22] Snapshots were not desired. If it was not worth setting up the tripod, leveling it, and getting the exposure correct, "the photograph was not worth taking" (Circular No. 20, YPEP, Hasbrouch, box 23). Above all, team members needed to make records immediately after taking the image.[23] Without contextualizing the photograph and marking it according to weather conditions, the image was of little use as an accurate record.

Anyone who has taken photographs can sympathize with Bingham's approach. After a period of time, one forgets the initial meaning of the view. Where was it taken? When was it taken? Why did I take it? The photograph is slippery as a record. Separated from contextualizing clues, the framed image becomes useless as scientific evidence. The expedition members took thousands of photographs. Without attributing information and acquiring some level of skill, they would not produce images that could function the way Bingham intended: as data and record for future study or evidence of discovery. To mobilize photographs as facts, Bingham had to fix intention, disciplining observation through a lens. But, Kodak's technology was not enough. The camera was only as scientifically useful as the hands of its operator.

During the second year of the third expedition, Bingham circulated detailed instructions on how to develop the negatives, including how to fix solutions and bath, tank developing and printing photographs.[24] The considerable number of steps coupled with the conditions in which members developed the films offer a sense of the vulnerability of photographic evidence and, by association, the outcome of the expedition and Bingham's reputation.

First, members were told how to prepare the developing solutions. Expedition photographers were provided two powders that were to be mixed and dissolved in hot water in the developing tanks provided by Kodak. The mixed solution was good for "only one" roll of film.[25] The film had to be wound into a cylinder, which was kept tightly closed and placed in the developing solution. After they were used for developing the film, the tanks then had to be washed with at least three changes of water before establishing the fixing bath. The film was subsequently fixed with an acidifier composed of acidic acid, powdered aluminum, and sulphide of soda. Once that solution was made, it could fix fifteen to twenty-five spools of film. Moreover, prints had to be left in the hypo bath for at least twenty minutes and then washed for about twenty minutes or longer.[26]

The amount of time and effort expended by the members on taking the image, hauling and heating the water, processing the negatives, and labeling and systematizing their records affirms that photography was a critical, though suspect, expeditionary practice. The camera did not necessarily capture intention, and the photograph, as fact, was unstable (Poole 2005). To produce the reality of Bingham's lost city, team members, the landscape, and the camera were negotiated to realize the shot as scientific evidence. Photographers collecting views passed through various trials of strength that necessitated a keen sense of work, effort, dedication, diligence, and negotiation to quiet the silent actors of time, memory, atmosphere, and the camera (Latour 1987; Callon 1999). The photographic extraction from Kodak's cameras of team member's intention was anything but a reflection.

Circulating Evidence, Mobilizing Facts

Returning after the 1911 expedition, Bingham remarked that while he was not permitted to bring a wall of Machu Picchu back to the United States, he did have excellent photographs (New York Herald 1911). Upon returning from the field, Bingham sent the collection of negatives to Eastman Kodak in Rochester, New York, where they were printed in their entirety for a special price (Craig to Bingham, Kodak Select Correspondence, January 28, 1913, YPEP, box 14, folder 214).[27] Bingham selected the specific tone, treatment, and paper for printing the negatives based on the effect that they would lend to the photograph as scientific document.[28] Kodak assembled the photographs, mounting them in chronological order in albums.[29] The donated films and camera equipment provided Bingham and National Geographic with ample evidence for unveiling Machu Picchu to a global public (Kodak Select Correspondence, March 14, 1912, YPEP, box 14, folder 211).

The accumulation of thousands of images affirms the privileged position held by photography as scientific expeditionary practice. Certain members were provided with a camera and film, instructed on how to take pictures, where to take pictures, appropriate subjects for pictures, and how to properly develop the negatives in the field. In addition to being asked to label and contextualize the images for future research and to provide color matches for hand tinting, team members were also asked to make test images assuring the successful capture and inscription of intent. Photographs were then cataloged and organized for future study.

The expedition's most important scientific tool thus required a significant amount of the team's time and attention. Paul Bestor, Bingham's assistant on the expedition, spent much of his time developing and assisting in the taking of photographs. In two journal entries around Bingham's visit to Machu Picchu, he wrote that after he left camp at 7 a.m. on July 25, he took dictation and assisted Bingham in photographing the ruins. The following day Bestor developed six spools of film and typed orders (1912, Jun-Nov Journal of Paul Bestor, YPEP, box 19, folder 14). Bestor's time was centered on creating photographic evidence.

Figure 12.3
"The Ruins of an Ancient Inca Capital, Machu Picchu." Bingham's panoramic photograph, made possible by Kodak's panoramic camera technology, was included as a panoramic insert in the 1913 *National Geographic* article "In the Wonderland of Peru." The image was reprinted in a variety of other publications. Used with permission. Credit: Hiram Bingham/National Geographic Stock.

And yet, once it was visualized as scientific evidence, the photograph instantly obscured the sociotechnical relationships involved.[30] The corporate sponsorship, camera technology, climate, and labor in practice were elided in the eventual reproduction of prints in a variety of media. Although obtaining an image required the photographer's thoughtful negotiation of Kodak's camera technology, Bingham's instructions, and the damp and changing Andean climate, the images were considered direct imprints of reality. Through this successful coupling of displacement and representation, the photographic translation of Andean landscapes came to constitute the facts of the expedition.[31]

In April 1913, National Geographic published "In the Wonderland of Peru," a magazine issue entirely dedicated to Bingham and the Yale Peruvian Expedition (Bingham 1913). In it was a specially printed, hand-folded panorama photograph. The expensive insert was intended to be both visually arresting and exciting for the reader.[32] The panorama not only impressed the magazine's readers but was also picked up as illustration for newspaper copy (figure 12.3). On June 15, 1913, "Lost City in the Clouds Found after Centuries" was published in the *Boston Morning Herald* along with the panorama photograph of Machu Picchu.[33] Although the 1913 National Geographic edition was published nearly two years after the initial sighting of Machu Picchu in 1911, Bingham admitted that he did not realize the magnitude of his findings until his photographs were published in the *National Geographic* issue dedicated to the expedition (Bingham 1989).

Photography became a critical actor not simply as a way to provide evidence of the last or perhaps first place of the Incas, but also as a flexible technology.[34] Photography tested in the tropics allowed the expedition to bring back evidence of its explorations, producing compelling narratives about scientific expeditions generally, but also of Incan discoveries, lost cities, Andean landscapes, and native subjects specifically. However, obtaining images necessitated figuring the camera and expedition members to work toward that end. The expedition's most powerful scientific evidence relied on disciplining a practice that sought to stem the tide of both suspicion and unruliness.

Making pictures in the laboratory of Latin America produced and therefore claimed the Andes as a space dedicated for US scientists. Unlike the other chapters in this section that focus on the ways technologies have been translated and adapted by Latin American users, Bingham retained the camera for Yale alone. This was not a scientific expedition by patriots to help marshal a nation into modernity (see Cukierman, this volume). Rather, this was an expedition reminiscent of an earlier era, where transnational outsiders removed evidentiary objects and local knowledge for reflection and production at home (Pratt 1992). Although Peruvians such as Martin Chambi would eventually use camera technology to frame their own imagining of Machu Picchu, Yale's images, so laboriously produced, brought Machu Picchu and indigenous subjects into a "dominant field of visibility" (Salvatore 2003), enabling a powerful gaze which easily circulated into a variety of spatial and textual fields throughout the United States, Peru, and abroad. While Peruvians helped guide Bingham to Machu Picchu, making its picture was the exclusive privilege of Yale.

Ultimately the team's photographs became collected objects, evidencing and indexing a monumental find for scholars, aficionados, and a global public. Circulating beyond the initial intent as extracted record of and for scientific inquiry, the photographs became potent fixers and fixations of people and history. However, the vulnerability in the camera's capacity as a scientific technology of extraction marks Machu Picchu and the Yale Peruvian Expeditions, promulgating a multiplicity of competing desires around the site that continues today. As the images that resulted from the expeditionary lab work circulated, their meaning or informational content shifted based on their insertion within diverse narrative frames. As Ames (this volume) described the XO laptop, the camera was Bingham's charismatic object: full of magical promise and hopeful futurity, yet elusive and ultimately impossible to stabilize.

Notes

1. One of Bingham's primary sources was the geographer Sir Clements Markham. Writing to a potential sponsor, Bingham suggested that Markham knew more about Peru than any living geographer (Hiram Bingham to Huntington Smith, January 11, 1911, Yale Peruvian Expedition Papers, Manuscripts and Archives, Yale University Library, box 5, folder 2). Future citations will use the abbreviation YPEP.

2. In a course Bingham taught in 1916 entitled "On the opportunities offered to college graduates in exploration and scientific travel," he expressed his belief that much of the world had already been discovered and at least partially explored. Consequently, for the scientific traveler, "the great opportunities of the future lie in intensive, rather than extensive exploration." According to Bingham, such travel would be "much more useful to his fellow citizens" (Bingham's teaching papers, c. 1916, YPEP, box 34, folder 38).

3. For a list of scientific articles stemming from the expedition, see A. Bingham (1989).

4. This practice of photographing "faraway places" was not only conducted between the North and the South, but also within Peru to reconstruct the nation after the disastrous effects of the War of the Pacific (1879–1884). Peruvian photographer Fernando Garreaud was also supported by President Nicolas de Pierola to photograph the Peruvian republic for display at the Universal Exposition of Paris; see Majluff and Wuffarden (2001).

5. Scholarship on collecting in the early twentieth century has tended to privilege and examine the collection of more traditional ethnographic objects. See Gosden and Knowles (2001); Edwards, Gosden, and Phillips (2006); Barringer and Flynn (1998). This emphasis is perhaps due in part to the fact that until quite recently, photographs in museum's collections were not registered as part of the museum's formal collection and were stored in multiple and disparate spaces within the museum (Crimp 1993).

6. For a discussion of the circulation of images from the expedition, see Cox (2010).

7. In *Inca Land* (1922), Bingham wrote about being inspired by Rudyard Kipling's "The Explorer": "Something lost behind the Ranges. Lost and waiting for you. Go!" See also Poole (1998) and Bingham (1989).

8. Citations list the number of images as ranging from 8,000 to 11,000.

9. A mainstay on colonial expeditions, anthropometry sought to order and typologize all of humankind based on a hierarchical scale of evolutionary development. Physical attributes were thought to betray not only one's cognitive development, but also one's moral behavior. See Cox (2007); Wallis (1995); Edwards (1988); Edwards (1990).

10. The anthropometric images of the Yale expedition were not widely circulated. The material compiled by Ferris from the 1912 expedition was eventually published in *Memoirs of the American Anthropological Association*. Ford's images taken on the 1914 expeditions were originally intended for Aleš Hrdlička at the National Museum. In the end the museum did not want the images since Hrdlička was too busy with his own. None of the images appear to have been printed in popular magazines or newspapers.

11. Bingham secured tents and camping equipment at a discount from Abercrombie & Fitch, guns and ammunition from Winchester Rifle, and saddles from Smith Worthington. Waltham Watch donated a set of chronometers and astronomical watches (see YPEP, box 5, folder 4).

12. Specifically, Kodak provided a No. 3A special Kodak with leather case, portrait attachment and color screen, wide-angle lens, and a Kodak metal tripod; a No. 2 1 Bull's-eye tripod, two 3A

folding pocket Kodaks with portrait attachment and wide-angle lens, tripods, and a leather case for packing on a saddle; 200 film cartridges, 10 exposures each, 3½ × 5½, specially packed in tin boxes; 24 cartridges, 12 exposures each, and 40 cartridges, 6 exposures each for a No. 4 model "A" screen focus Kodak; a No. 3 folding pocket Kodak with 20 cartridges of film, 6 exposures per cartridge, and three complete tank developers with necessary chemicals (Bingham to Eastman, Kodak Select Correspondence, May 16, 1911, YPEP, box 14, folder 210).

13. The two other primary monetary sponsors were Yale and Bingham himself. Photography was critical for the goals of the *National Geographic* magazine. In their contract with Bingham, they specified first rights to all expedition photographs to illustrate the initial story (National Geographic Select Correspondence, April 24, 1912, YPEP, box 15, folder 237).

14. Bingham asked Eastman for information on how to take a good photograph. Eastman promptly provided Bingham with copies of the pamphlet "How to Make a Good Picture." The choice of words is interesting: "making" implies human involvement and manipulation, whereas "taking" connotes a direct replication without intervention.

15. Bingham requested 1,000 additional negatives for the 1912 expedition because on the 1911 expedition they had brought 2,500 negatives and had run out. For the 1912 expedition, Kodak sent 250 spools of 10-exposure film for the 3A, 100 spools of six exposures for the 3A, and 100 spools of 10-exposure "bulls-eye." (Kodak Select Correspondence, April 15, 1912, YPEP, box 14, folder 11.)

16. For a discussion on the camera and "accurate" images, see Daston and Galison (2007).

17. For more on collecting and circulars, see Cox Hall (2012). Also included were circulars on mail addresses, itinerary, personal equipment, contracts, drinking water, dentistry, food preparation, passports, care and selection of mules, and treatment of snake bites. General orders about ship times and packing instructions were also included.

18. This was consistent on both the 1912 and 1914 expeditions. See circular No. 17, YPEP, box 23.

19. Hand-tinted lantern slides were made of several of the images in Mystic, Connecticut, by W. C. Ives.

20. Photographic negatives were assigned to each member based on the kind of work he was doing, the number of photographs taken by other members of the expedition, and the individual's skill in taking photographs (Official Circular No. 20, YPEP, Hasbrouch, box 23). So, for example, Ellwood Erdis was given negatives 6000 through 6200, surgeon Ford was given 8201 through 8400, etc. The most negatives were given to the naturalist Edmund Heller (6401–6800) and to the topographer (7505–7999). Negatives 8580–9004 were left vacant for later assignment.

21. Bingham instructed members to use the exposure index found in the back of the red Burroughs Wellcome photographic record book.

22. In the circular, Bingham suggested taking approximately six test images.

23. Developing was to occur "definitely before 3 weeks time."

24. It is unclear from the archive whether this was the only circular providing detailed instruction on developing the negatives. The circulars cited are from expedition members' notebooks and may be incomplete.

25. See notes on developing dated February 16, 1915, J. J. Hasbrouch notebook, YPEP, box 23, folder 55.

26. Two types of paper for printing were also used in the field. One was grade "E hard" for soft effect and "E hard x" for contrast. The hypo bath cleared away the silver bromide.

27. Letter to Mr. Craig, Eastman Kodak, dated January 2, 1913, YPEP, box 14, folder 214. Bingham went on to make research albums for all the photographs.

28. From a series of letters written in January 1913 to Mr. Craig at Kodak Corporation (Kodak Special Correspondence, YPEP, box 14, folder 214). Bingham wanted the pictures mounted on black and slate/dark-gray paper (at the request of National Geographic), printed with a glossy and Velox finish. The photographs with a glossy finish were to be burnished to prevent hindrances for reproduction.

29. Bingham kept these albums as research references, and made copies, donating a set to National Geographic. Evidence also suggests he made an album for the Hispanic Society of the Americas, but the archive currently only has loose images.

30. In her study of the mapping of the human genome, Donna Haraway (1997) argues that the gene is not a thing or a master code, but a signification of a node where many actors meet. The gene obscures and disavows the very sociotechnical relationships that generate both objects and value. Through the process of visualization, the sociotechnical relationships involved in an object's making are obscured.

31. For a discussion on the ways in which early anthropological photographs came to be "facts in themselves," see Edwards (1992); Edwards (2001).

32. Gilbert Grosvenor, director and editor of National Geographic, wanted to spare no expense and to bring the article out in "a striking and emphatic form." The printing of the panorama photograph cost about $2,000. The month's issue was "considerably larger than usual in order to contain the complete series of pictures" (Grosvenor to Bingham, National Geographic Select Correspondence, April 7, 1913, YPEP, box 15, folder 242).

33. The panoramic image continues to circulate today. For the celebration of the 100-year anniversary of Hiram Bingham's initial visit to Machu Picchu, an enlargement of the panoramic photo was included in the exhibition "Machu Picchu 100 years: A Glimpse of the Expedition that Astonished the World." The exhibition was displayed on the fourth floor of the Sumaq Machu Picchu Hotel with images on loan from National Geographic. The panoramic photograph was also included in National Geographic's exhibition "Machu Picchu: A Lost City Uncovered: Photographs from the Hiram Bingham Expeditions 1911–1915," which lasted from July 24, 2011, through September 11, 2011. The panoramic photograph is also displayed on National Geographic's website under the tab Rediscovering Machu Picchu. See http://ngm.nationalgeographic.com/1913/04/machu-picchu/bingham-photography.

34. Bingham originally was looking for the final refuge of the Incas, but in *Harper's* magazine (1913) he quipped that perhaps instead of finding the last capital he had found the first, the birthplace of the Incas.

References

Adams, M. 2012. *Turn Right at Machu Picchu: Rediscovering the Lost City One Step at a Time.* New York: Penguin.

Alloula, M. 1986. *The Colonial Harem.* Minneapolis: University of Minnesota Press.

Anderson, B. 1991. *Imagined Communities: Reflections on the Origin and Spread of Nationalism.* London: Verso.

Appadurai, A. 1996. *Modernity at Large.* Minneapolis: University of Minnesota Press.

Barringer, T., and T. Flynn, eds. 1998. *Colonialism and the Object: Empire, Material Culture and the Museum.* London: Routledge.

Barthes, R. 1981. *Camera Lucida.* London: Jonathan Cape.

Berger, J. 1972. *Ways of Seeing.* Harmondsworth: Penguin.

Bingham, A. 1989. *Portrait of an Explorer: Hiram Bingham, Discoverer of Machu Picchu.* Ames: Iowa State University Press.

Bingham, H. 1913. In the wonderland of Peru. *National Geographic* 24 (4):387–573.

Bingham, H. 1922. *Inca Land.* Washington, D.C.: National Geographic.

Bingham, H. 1952. *Lost City of the Incas: The Story of Machu Picchu and Its Builders.* London: Weidenfeld and Nicolson.

Callon, M. 1999. Some elements of a sociology of translation: Domestication of the scallops and the fishermen of Saint Brieuc Bay. In *The Science Studies Reader*, ed. M. Biagioli, 67–83. New York: Routledge Press.

Chavez, L. 2000. *Covering Immigration: Popular Images and the Politics of the Nation.* Berkeley: University of California Press.

Cox, A. 2007. Purifying bodies, translating race: The lantern slides of Sir Everard Im Thurn. *History of Photography* (October): 348–365.

Cox, A. 2010. Framing Machu Picchu: Science, photography and the making of heritage. Ph.D. dissertation, University of Florida.

Cox Hall, A. 2012. Collecting a "lost city" for science: Huaquero vision and the Yale Peruvian expeditions to Machu Picchu. *Ethnohistory* 59 (2):293–321.

Crary, J. 1990. *Techniques of the Observer: On Vision and Modernity in the Nineteenth Century.* Cambridge, MA: MIT Press.

Crimp, D. 1993. On the museum's ruins. In *The Anti-aesthetic: Essays on Postmodern Culture*, ed. H. Foster. Seattle: Bay Press.

Daston, L., and P. Galison. 1992. The image of objectivity. *Representations* (Berkeley) 40:82–128.

Daston, L., and P. Galison. 2007. *Objectivity*. Brooklyn, NY: Zone Books.

de la Cadena, M. 2000. *Indigenous Mestizos: The Politics of Race and Culture in Cuzco, Peru, 1919–1991*. Durham: Duke University Press.

Derrida, J. 2002. The archive market: Truth, testimony, evidence. In *Echographies of Television*, ed. J. Derrida and B. Steigler, 82–99. Cambridge, UK: Polity.

Dumit, J. 1997. A digital image of the category of the person: PET scanning and objective self-fashioning. In *Cyborg and Citadels: Anthropological Interventions in Emerging Sciences and Technologies*, ed. G. L. Downey and J. Dumit, 83–102. Santa Fe, NM: School of American Research Press.

Edwards, E. 1988. Representation and reality: Science and the visual image. In *Australia in Oxford*, ed. H. Morphy and E. Edwards, 27–45. Oxford: Pitt Rivers Museum.

Edwards, E. 1990. Photographic types: The pursuit of a method. *Visual Anthropology* 3 (23):241–247.

Edwards, E., ed. 1992. *Anthropology and Photography, 1860–1920*. New Haven: Yale University Press in association with the Royal Anthropological Institute.

Edwards, E. 2001. *Raw Histories: Photographs, Anthropology and Museums*. Oxford: Berg.

Edwards, E., C. Gosden, and R. Phillips, eds. 2006. *Sensible Objects: Colonialism, Museums and Material Culture*. Oxford: Berg.

Edwards, E., and J. Hart. 2004. Introduction: Photographs as objects. In *Photographs, Histories, Objects*, ed. E. Edwards, 1–15. New York: Routledge.

Freire-Marreco, B., and J. L. Myres. 1912. *Notes and Queries on Anthropology*. London: Royal Anthropological Institute.

Garson, J., and C. Read. 1892. *Notes and Queries on Anthropology*. London: Royal Anthropological Institute.

Gosden, C., and C. Knowles. 2001. *Collecting Colonialism: Material Culture and Colonial Change*. Oxford: Berg.

Haraway, D. 1997. *Modest_Witness@Second_Millennium. FemaleMan©_Meets_OncoMouse™: Feminism and Technoscience*. New York: Routledge.

Heaney, C. 2010. *Cradle of Gold: The Story of Hiram Bingham, a Real-Life Indiana Jones, and the Search for Machu Picchu*. New York: Palgrave Macmillan.

Kuhn, A. 1985. *The Power of the Image: Essays on Representation and Sexuality*. New York: Routledge.

Latour, B. 1987. *Science in Action*. Cambridge, MA: Harvard University Press.

Latour, B., and S. Woolgar. 1986. *Laboratory Life: The Construction of Scientific Facts*. Princeton: Princeton University Press.

Lopez-Lenci, Y. 2004. *El Cusco, paqarina moderna: Cartografia de una modernidad e identidades en los Andes peruanos (1900–1935)*. Lima: Fondo Editorial Universidad Nacional Mayor San Marcos.

Lutz, C., and J. L. Collins. 1993. *Reading National Geographic*. Chicago: University of Chicago Press.

Majluff, N., and L. E. Wuffarden. 2001. *El primer siglo de la fotografía. Perú 1842–1942*. Lima: Fundación Telefónica y Museo de Arte de Lima.

Mould de Pease, M. 2001. Un día en la vida peruana de Machu Picchu: Avance de historia intercultural. *Revista Complutense de Historia de America* (27): 257–279.

Mould de Pease, M. 2003. *Machu Picchu y el codigo de ética de la Sociedad de Arqueologia Americana*. Lima: Pontificia Universidad Catolica del Peru.

New York Herald. 1911. Explorers find a city that was "snuggled under corn fields." *New York Herald*, December 22.

Poole, D. 1998. Landscape and the imperial subject: US images of the Andes, 1859–1930. In *Close Encounters of Empire*, ed. Gilbert M. Joseph, Catherine LeGrand, and Ricardo D. Salvatore, 107–138. Durham: Duke University Press.

Poole, D. 2004. An image of "our Indian": Type photographs and racial sentiments in Oaxaca, 1920–1940. *Hispanic American Historical Review* 84 (1):37–82.

Poole, D. 2005. An excess of description: Ethnography, race and visual technologies. *Annual Review of Anthropology* 34:159–179.

Pratt, M. L. 1992. *Imperial Eyes: Travel Writing and Transculturation*. London: Routledge.

Ryan, J. 1997. *Picturing Empire: Photography and the Visualization of the British Empire*. London: Reaktion Books.

Salomon, F. 1985. The historical development of Andean ethnology. *Mountain Research and Development* 5 (1):79–98.

Salvatore, R. 2003. Local versus imperial knowledge: Reflections on Hiram Bingham and the Yale Peruvian Expedition. *Nepantla* 4 (1):67–80.

Sekula, A. 1983. The invention of photographic meaning. In *Thinking Photography*, ed. V. Burgin, 84–109. London: Macmillan.

Sekula, A. 1986. The body and the archive. *October* 39 (Winter):3–64.

Tagg, J. 1989. *Burden of Representation: Essays on Photographies and Histories*. Basingstoke, UK: Macmillan.

Thurner, M. 2003. Peruvian genealogies of history and nation. In *After Spanish Rule: Postcolonial Predicaments in the Americas*, ed. M. Thurner and A. Guerrero, 141–175. Durham: Duke University Press.

Tsing, A. 2005. *Friction: An Ethnography of Global Connection*. Princeton: Princeton University Press.

Tucker, J. 1997. Photography as witness, detective and imposter: Visual representation in Victorian science. In *Victorian Science in Context*, ed. B. Lightman, 378–408. Chicago: University of Chicago Press.

Wallis, B. 1995. Black bodies, white science: Louis Agassiz's slave daguerreotypes. *American Art* 9 (2):38–61.

Part III Science, Technology, and Latin American Politics

13 Bottling Atomic Energy: Technology, Politics, and the State in Peronist Argentina

Jonathan Hagood

On March 24, 1951, Argentine president Juan Perón invited members of the nation's press to the Casa Rosada, the presidential palace, to announce the success of Proyecto Huemul, Argentina's atomic fusion research program. With émigré German scientist and project leader Ronald Richter at his side, the president began by explaining why Argentina had pursued research in atomic fusion rather than atomic fission. Perón stated that the United States, Britain, and Russia had undertaken atomic research under the threat of war when the fission of uranium had been the safest and most viable option. In contrast, Perón noted, "The New Argentina decided to take the risk and adopt all measures which might bring the desired approach ... with the application of materials less costly and easier to obtain." He then enumerated the benefits of an atomic research program for Argentina, which included supplying the energy needs of industry, creating radioactive isotopes for scientific research, and, most important, demonstrating that "Argentina wants once again to give an example of the possibility of applying scientific advancements toward the welfare of her people and of humanity" (New York Times 1951b).

The story of Proyecto Huemul began with Richter's arrival in Buenos Aires in August 1948.[1] Richter, a relatively unknown German scientist of Austrian descent, traveled to Argentina at the request of Kurt Tank, a German aviation engineer working on Argentina's fighter jet, the *Pulqui II*. After a brief meeting between Perón and Richter, Perón, who had been elected president of Argentina in February 1946, placed Richter in charge of Argentina's atomic research program in Córdoba, a city in the mountains about four hundred miles inland, where Tank's Instituto Aerotécnico was located. In early 1949, a fire destroyed Richter's research laboratory. Blaming the fire on sabotage, Richter appealed to Perón for a new location removed from espionage and interference, and the scientist began an aerial search of the northern reaches of Patagonia for suitable surroundings. He eventually selected Huemul Island, located on Nahuel Huapi Lake in the Andean foothills. Water isolated Huemul from its immediate surroundings, and sheer distance—nearly nine hundred miles—separated it from Buenos Aires.

Construction began in July 1949 while Richter continued working in Córdoba; in March 1950, he moved the project to Bariloche.

The day after the March 24 press conference, the state awarded Richter the Peronist Medal and an honorary doctorate from the University of Buenos Aires, but in contrast to the honors immediately surrounding the public announcement, the following months were not successful for Richter. Despite the positive reports from visiting politicians and journalists sympathetic to the Peronist regime, anti-Peronist politicians, foreign scientists, and Argentine physicists excluded from the project criticized the secrecy surrounding his work and the lack of reproducible results. Throughout 1952, Richter continued to leave promises unfulfilled, and even Peronist politicians began to worry about the significant amount of money being spent.

Eventually, Perón halted work on Proyecto Huemul after successive legislative and scientific investigations revealed the fraudulent nature of Richter's fusion experiments. The final inspection, led by physicist José A. Balseiro, made use of concealed gamma ray detectors, which failed to register any radiation during Richter's demonstrations of "successful" fusion experiments (Balseiro 1988). On November 22, 1952, while Richter was away in the capital, a military operation occupied the research facility. Navy captain Pedro Iraolagoitía, head of Argentina's National Atomic Energy Commission (CNEA; Comisión Nacional de Energía Atómica), led the group, which discovered that many machines used in Richter's experiments were not connected to one another, a clear indication that his claims were fraudulent. The scientist quietly moved to a suburb of Buenos Aires, where he lived for the next three years, effectively under house arrest and out of the public eye. After a military coup overthrew Perón in September 1955, the case featured prominently in charges of corruption and disorder against the president (Comisión Nacional de Investigaciones 1958, 162–165). Argentina had spent 62.5 million pesos (US $15 million in 1950 and US $143 million in 2012) on Richter's failed research.[2]

The story of Proyecto Huemul might appear to be the simple case of a charlatan hoodwinking the ignorant president of a developing country into investing valuable resources in hopeless research, and most accounts emphasize some part of this interpretation. This case might also figure as an example of the failure of a given technology (controlled atomic fusion) to stabilize within a given set of knowledge and political conditions. Indeed, atomic fusion to this day remains a development just on the edge of the scientific horizon, much as it was in the 1950s. Nevertheless, explanations of failure and the assignment of blame neglect crucial aspects of the story that point to the multiple, sometimes contradictory ways in which the project was both a success and a failure. In addition, revisiting this history offers valuable insight into the relationship between technology, the state, and innovative political rhetoric. To this end, this chapter does not address the question of why or how Argentina's atomic research failed but instead explores the reasons Perón undertook the project in the first place.[3] This perspective renders notions of fact, fiction, success, and failure problematic.

Contemporary Explanations

Attempts to understand Proyecto Huemul began in 1951 with the Austrian physicist Hans Thirring's post-announcement article in *United Nations World* in which he highlighted Richter's invisibility within the physics community and the scientific difficulties associated with controlled atomic fusion. Thirring outlined the likelihood of different "possible explanations" for Richter's "pseudo-technical achievement" as follows:

(a) Perón has fallen victim to a crank suffering from self-delusion. 50%
(b) Perón has been taken in by a sly swindler. 40%
(c) With the aid of Richter, Perón is attempting to bluff the world. 9%
(d) Richter's assertions are true. 1%

The low probability assigned to (c) is based, perhaps, on a too favorable opinion of President Perón. (Thirring 1951, 1–2)

Thirring based his list largely on the opinion that Richter's research was flawed—whether maliciously or not—but left unquestioned the premise that Argentina could undertake such a program in the first place.

Writing after the 1955 coup, the anti-Peronist National Investigation Commission (CNI; Comisión Nacional de Investigaciones) also placed the blame for the failure of Proyecto Huemul squarely on Richter's shoulders.[4] The summary of what the CNI labeled "the Richter Case" stated, "Richter completely lacked the scientific background that would have justified confidence in his promises." Richter was also responsible for an "inadmissible administrative disorder, in which [he] was conferred absolute power." Furthermore, those who did suspect the worst "could not make public their opinions because it was known that they would face the punishment that fell to those who acted against National Security" (Comisión Nacional de Investigaciones 1958, 164–165).

Nearly thirty years after the CNI's report, Mario Mariscotti, former research and development director of Argentina's CNEA, conducted a thorough investigation of Proyecto Huemul, which included interviews with the reclusive Richter. Mariscotti's work is an example of STS literature written by Latin American academics that deserves greater attention from European and North American scholars. In his book *El secreto atómico de Huemul* (1985), Mariscotti argued that Richter bore most of the responsibility for the failure of the project because of his self-delusion, an autocratic and paranoid management style, and the lack of institutional or personal oversight. Mariscotti blamed Perón for the last point and none too subtly suggested that had Perón utilized Argentina's existing community of physicists, he might have averted an embarrassing failure. Yet Perón's dismissal of the Argentine physics community was partly due to the Argentine academy's general distaste for his government and Perón's interest, despite his repeated claims for the peaceful use of atomic energy, in maintaining military control over the project. Perón's problematic relationship with the Argentine intelligentsia in general, most whom never came to his side, was also a factor (Carassai 2010).

These explanations for the failure of atomic research in Argentina require a critical reading. As a scientist firmly established in the field of physics, Thirring was both suspicious of any claims that fell outside accepted scientific knowledge and naive about Perón's political intentions. Thirring's political inexperience is evident in his omission of an equally valid explanation of the project: Perón neither knew nor cared whether Richter was a crank, a fraud, or a genius but sought whatever political benefits he could gain regardless. The historical evidence indicates that Perón's political use of Proyecto Huemul reflected the irrelevance of the technical success or failure of the project in light of the immediate political gains that would accrue to Perón from an announcement of success.

The members of Argentina's CNI also invite skepticism, as their political motives clearly manifest the rabid anti-Peronism of the Argentine state immediately after the 1955 coup. For example, the CNI's report consistently referred to Perón as "the dictator," even though he had twice been voted into office in elections widely considered fair and without the irregularities common in Argentine elections in the fifteen years before his administration. The report begins with a history of Argentina that places Perón within the context of the nation's traditional principles of "liberty, equality, and democracy" and the now triumphant battle against the antidemocratic ideas of Perón and his ilk. However, the section on Proyecto Huemul condemns Richter's role in the project without placing any direct blame on Perón himself other than the implication that Perón should have reined in Richter sooner and the admonition that Perón had "embarrassed the nation" with his false announcement. Perhaps the emphasis on Richter's culpability stemmed from the fact that Perón had removed him from the program while still in power and, although the incident was certainly embarrassing, the Peronist government had handled it in as discreet and straightforward a manner as possible. The scientists who took over the CNEA and the facilities at Huemul avoided taking public positions for or against Peronism, and the Argentine Navy—traditionally not a close ally of Perón—took over institutional management of subsequent atomic research projects. Mariscotti's later investigation affirmed the CNI's findings, and no documents or statements have ever claimed or supported the idea that Perón knowingly supported a failing or doomed project.

Still, Mariscotti's role as a director at the CNEA suggests he would have sympathized with the Argentine physicists left out of Proyecto Huemul. A significant portion of his finely detailed and exhaustively researched book attempts to recover the role played by Argentine physicists in atomic research before Richter entered the scene and to suggest that, with appropriate leadership and funding, the project might have succeeded as research on atomic fission. This is certainly a valid interpretation of events, in that other nations were able to successfully develop a working knowledge of atomic fission and how to apply it, but it leaves unexplored the broader questions as to why Perón would embrace this research, execute it in the manner he did, and use it to achieve a variety of political ends. Mariscotti did note that "the idea of the New Argentina was sustained in these great accomplishments and on them revolved the mystique of the government ... [and] it is for that reason that the scientific fact, in addition to its intrinsic value, was used like a platform for the development of a political action"; however, he did not explore the innovative political rhetoric that made up the political action, rhetoric that translated atomic research as a complex form of knowledge into the distinct conditions of Argentina (Mariscotti 1985, 163).

Why Atomic Research in Argentina?

A better understanding of these issues comes from addressing not the problem of why atomic research in Argentina failed but rather the related questions of why Perón supported it in the first place, and why it was even feasible to embark on such a project. The political motivations that underlay Perón's support of Proyecto Huemul were threefold and involved (1) his very public efforts to support and maintain Argentina's industrial expansion, (2) his primary foreign-policy goal of advocating a "third position" independent of both the United States and the Soviet Union, and (3) the very practical need to distract public attention from some of his regime's more authoritarian and unpopular excesses. While the demand for cheap energy was perhaps itself incentive enough for an atomic research program, the other three motives dramatically influenced the manner in which Perón conceived, executed, and promoted Proyecto Huemul—that is, the way he deployed innovative political rhetoric in order to translate atomic research for Argentine political discourse.

Industrial Expansion

As president, Perón inherited both a booming wartime economy and the fear of a postwar downturn. The worldwide economic depression of the 1930s revived a process of Argentine industrialization that had enjoyed a brief beginning during World War I (Cardenas, Ocampo, and Thorp 2000). Industrial expansion accelerated during World War II because of the worldwide decline in international trade. In addition, a strong wartime demand for beef, wool, and grain on both sides of the conflict created a financial windfall for Argentina's traditional agricultural industries because of the nation's officially neutral status.

As a consequence, the policy debates in the late 1940s did not concern choosing between diversified or selective industrialization but instead focused on sustainability by determining how to conserve the capacity of the new productivity structures to continue industrialization with increasing employment (Gerchunoff and Antúnez 2000, 139). The government feared a repetition of the events after World War I, when the return of commercial imports from Europe bankrupted the local Argentine manufacturers who had prospered during the war. A similar industrial collapse would have created severe unemployment and threatened the Peronist labor alliance, the foundation of Perón's political appeal. Perón's solution was to intensify state involvement with increased import tariffs to encourage the placement of foreign production facilities in Argentina. Perón believed that these policies would fund the expansion of the nation's energy and transportation infrastructure, and his government nationalized industries such as the railroads and telecommunications and modernized government-owned defense industries. These policies saw some initial success. Manufacturing's share in the economy increased by more than 15 percent from 1940 to 1955, and finished consumer goods, which accounted for around 40 percent of total imports in 1930, fell to less than 10 percent by the early 1950s (Rock 1987, 264).

The growing industrial sector placed a great strain on Argentina's energy resources, however, and made clear the need for new sources of energy production. To this end, between 1946 and 1955 the Peronist government funded the construction of thirty-seven hydroelectric plants and an oil pipeline to feed the refineries near Buenos Aires (Rock 1987, 263–264). Although the state-owned petroleum company increased production between 1947 and 1952 by 20 percent, this was not enough to meet rising demand. As a result, Argentina continued to import nearly half of its oil. In addition, despite the opening of coal mines in Patagonia and the construction of a railroad linking them to Atlantic ports, coal production lagged behind the demands of power production. Argentina's electrical system, hindered in large part by these factors, increased its capacity between 1945 and 1955 from 1.29 million kilowatts to 1.62 million, an annual rate of only 2 percent (Rock 1987, 298). Voices within and outside the Peronist state repeatedly identified the insufficiency of oil production as the principal factor explaining Argentina's economic vulnerability (Gerchunoff and Antúnez 2000, 183). A successful atomic energy program hypothetically, at least, would have helped solve this problem.

Foreign Policy

The method and timing of Perón's announcement of Proyecto Huemul's success reveal much about his political agenda for atomic research in Argentina. When questioned about the exclusion of non-Argentine journalists from the press conference, Perón admonished that he was "not interested in what the United States or any other country in the world thinks" and that he had made the announcement "not for other countries but for [his] own." Going further, he stated that he was "avoiding the course followed by politicians and newspapers in other countries of the world who lie consciously, directing their lies to their own people and spreading them abroad. ... They have not yet told the first truth, while I have not yet told the first lie" (Warren 1951, 1). These statements are evidence of the way in which Perón used the rhetoric of international politics to build up his domestic image as a leader dedicated to authenticity, autonomy, and truthfulness.

Perón's pursuit of a foreign policy agenda independent of both the United States and the Soviet Union presents a historical problem for scholars in determining whether his third position was truly the "central axis" of Peronist foreign policy or simply an "independent posture for internal consumption" (Barbero and Rougier 2002, 145). His announcement of successful atomic fusion is representative of this question, as the distribution of an official statement in English counterbalanced the exclusion of foreign reporters. Seen from the perspective of internal consumption, Perón carefully calculated the rhetoric of the announcement to appeal to his domestic audience. Newspapers touted the success of atomic research as a triumph for the nation of Argentina and its people. Beginning with the evening of March 25 and continuing the next morning, all major newspapers—Peronist and otherwise—gave the project's success highly supportive front-page coverage. On the other hand, Perón's statement that "it will interest scientists of foreign countries to know that in the course of our labors with thermonuclear reactors ... we were able to prove that the writings of most authoritative foreign scientists show they are enormously far from their goal" was certainly intended for both domestic and foreign audiences (New York Times 1951b).

Distracting the Public

Political considerations also influenced the timing of the announcement. Although the *New York Times* correspondent in Buenos Aires reported that "after wondering what made such an announcement necessary at this time, most persons went back to the fact the weather was good this Easter Sunday, and continued on from there with their usual desultory conversation" (Warren 1951, 11), others were not so naive. Writing from the United States, the *New York Times* journalist Milton Bracker previewed the Fourth Meeting of Consultation of Ministers of Foreign Affairs of the American States, which took place in Washington, D.C., from March 26 to April 7, 1951, by noting that Perón's "implacable fight against the newspaper *La Prensa* has had worldwide repercussions and touched off a battle of historic proportions for freedom of the press" (Bracker 1951, 122). Making the connection to Perón's announcement, Bracker wrote: "as if it would have taken an atom bomb to distract attention from his suppression of *La Prensa*, General Perón has come up with just that: his own atom bomb" (1951,

122). These contradictory interpretations of Perón's timing also indicate differences between the domestic and international intentions of the announcement. Perón's goal of overshadowing the bad press surrounding the closure of an opposition newspaper was eminently successful in the Argentine media. Although the US press made the connection, the *New York Times* placed Perón's announcement on page 8 and buried Bracker's *La Prensa* piece on page 122.

Nevertheless, Perón had calculated the announcement to further his political goals of claiming international significance for Argentina outside of the US and Soviet orbits and assuring his domestic audience that Proyecto Huemul would also solve the problems of importing necessary raw materials and resolving the growing currency crisis. Ironically, Richter later claimed that the industrial partnership was his idea, as a way to "shift the reactor project from the political and propaganda level to a more reasonable and profitable commercial level." Perón's support of the idea was "much to [Richter's] surprise" (National Archives and Records Administration 1957).

Why Was Atomic Research in Argentina Possible?

Atomic research was feasible in 1940s Argentina due to the worldwide interest in the potential of atomic energy immediately following World War II, the participation of Argentine scientists in the rapidly expanding field of atomic physics in the 1920s and 1930s, and Argentina's profits from the wartime sales of foodstuffs and raw materials. Perón was not alone in seeing the potential advantages of the nascent form of knowledge that was atomic research, and his plans invoked both the scientific and political possibilities that a stable project of atomic fusion promised. After World War II, people throughout the world worked to translate the science behind the atomic bomb into peaceful use. The first atomic reactor to produce energy was completed in the United States in December 1951, but technical and scientific hurdles kept the first commercial atomic power plant from coming on line until 1960. Other nations also developed atomic reactors for commercial power. France started power production in 1959, Canada began in 1962, and the Soviet Union's first commercial reactor went online in 1964.

The application of atomic physics to power production drew upon research that had taken place during the first half of the twentieth century. Argentine physicists participated in atomic research at centers of science in France, Germany, and the United States. For example, Enrique Gaviola, born in Mendoza, Argentina, in 1900, studied in the 1920s at the University of Göttingen in Germany under Albert Einstein and Max Planck. He later worked with Robert Wood at Johns Hopkins University in Baltimore before returning to the University of Buenos Aires in 1930, where he founded the Argentine Physics Association (AFA; Asociación Física Argentina) (Mariscotti 1985, 38–42; Vessuri 1997, 317). By the 1940s, Argentine physicists like Gaviola and his mentor, the German émigré Richard Gans, had obtained "citation visibility" in European scientific publications for their work in Argentine universities. Argentina was the only nation in Latin America whose physicists registered in the citation index for academic publications during this period (Glick 1994, 495). Therefore, although the story of Proyecto Huemul clearly reads as the movement of science and technology from the more developed North (Germany) to the less developed South (Argentina), Argentine scientists were already participating in a multidirectional flow of scientific and technological knowledge, thus establishing the preconditions for atomic research.

Argentina could afford Perón's significant investment in atomic research because of the strength of financial reserves the nation had accumulated through the exportation of foodstuffs and raw materials during the war and the new earnings from exports in the postwar years, the value of which doubled between 1946 and 1948 (Rock 1987, 290). Between 1939 and 1948, Argentina experienced ten consecutive years of a trade surplus. The nation had been saving in excess and disposing of a surplus of currency, and had extended "an invitation to spend, consume, invest, and repatriate debt" (Gerchunoff and Antúnez 2000, 142–143). As Perón's policies demonstrate, state investment in science and technology was heavily tied to economic cycles of boom and bust and the availability of government funds.

Argentina's relationship with Nazi Germany explains, in part, the manner in which Perón undertook the project. The military government had maintained close ties to Germany throughout World War II through Argentina's official neutrality. At the war's end, the former Allies began a heated competition for the scientists behind the German war machine. Perón's political sympathies were well known in Europe, and Argentina positioned itself as an ideologically attractive destination for members of European right-wing groups (Goñi 2002). Tomas Eloy Martinez notes that, in contrast to ideological affinities or concerns, "for the Argentine president, the most important element—according to the statements which we have seen—was his desire to have a group of trusted policemen and technicians on hand to help with Argentina's industrialization, as well as to monitor and control the opposition, especially among the unions" (1984, 15). Indeed, in *Yo, Juan Domingo Perón* (1976), the former president discusses only the transfer of "useful Germans" arranged through Argentine diplomats assigned to Sweden and Switzerland for that purpose (Luca de Tena, Calvo, and Peicovich 1976).

Perón clearly had both the motive and the opportunity to invest in atomic research. From a purely economic perspective, the risk involved in supporting Proyecto Huemul was potentially worth the reward of cheap and seemingly limitless energy. Perón's political future depended on continued industrialization and economic expansion to underwrite the social welfare programs that were the key to his popularity and electoral success. Although this was certainly true for many twentieth-century populist leaders in Latin America like Getúlio Vargas in Brazil, for Perón a postwar budget surplus made such a major investment in industrial expansion possible. These two factors coincided with both the worldwide interest in adapting atomic research to power production and the lack of a proven model for doing so. Although in the late 1940s the atomic power plant had yet to appear as a material artifact, Argentina was just as likely a location as any other for such a development to succeed.

Framing the Local Context

Illuminating the contingent and contextual explanations for Argentina's pursuit of atomic research in the first place requires an examination of the local context. Marcos Cueto (1997) outlines components of scientific success in Latin America—geographic and institutional concentration, utility and social relevance, nationalism and the state, access to technology, and international scientific networks—that constitute an interpretive framework for understanding Proyecto Huemul and its relationship to general strategies for successful research and development in Latin America.

First, due to the small size of scientific communities relative to those in industrialized nations, the concentration of resources and personnel has been a successful strategy for research in Latin America. Proyecto Huemul demonstrates the ineffectiveness of extreme concentration, as it would seem that in the hands of a single scientist in a secret facility, concentration becomes isolation. Perón's placing the entire operation in Richter's hands clearly became a liability when Richter let no one in on his atomic secret and allowed no other scientists—except for his two German assistants—to work on the project. Richter claimed that a larger staff would threaten security and was unnecessary given the time- and labor-saving "electronic brains" that the project had in the form of advanced technical equipment (Clarín 1951, 4).

Second, the survival and success of science in Latin America—if not globally—depends upon scientists' claims of public utility for their work, an issue clearly key to understanding Proyecto Huemul. The project itself focused only on the development of atomic energy as a useful piece of technology without taking account of continued research or the training of atomic engineers. After all, if one believed that Richter already knew the atomic secret and had communicated it to Perón, it was only a matter of working out the details and harnessing the secret for the improvement of society. This idea taps into what Allen Batteau has labeled the "utilitarian discourse" of technology wherein the ability of tools to "allow human groups to harness external sources of energy and bend Nature and fellow humans to their will" is the primary way of understanding technology (2010, 1–2). Richter's "secret" also evoked the myth of a scientific discovery made by an individual separate from the networks of collaboration, knowledge, and technology that make such discoveries possible.

Third, throughout the nineteenth and twentieth centuries, Latin American nations often coupled the pursuit of new areas of scientific research with projects of national prestige and identity. In the case of Proyecto Huemul, Perón linked the promise of atomic research to his vision of a "New Argentina" as an energy-independent and highly industrialized nation—an almost self-evident example of the translation of technology. In addition, Perón was always careful to refer to the "Argentine" scientists responsible for the project's success despite the fact that Richter was a German émigré who had become a citizen of Argentina in early 1950 after less than two years in the country—a process that required bending, if not breaking, several immigration laws (Mariscotti 1985, 109). More important, possession of the "atomic secret," much like possession of battleships before World War II and atomic weapons during the cold war, had a function similar to David Hess's (1995) concept of a "technototem," a technology that embodies a nation's distinctiveness and first-tier status.

Fourth, the difficulty in Latin America of obtaining access to the technology that in the industrialized world makes scientific research efficient and timely meant that the scramble for resources has often led to the development of research techniques that use the ample labor and natural environment available to scientists (Cueto 1989). Proyecto Huemul stands apart as one of the few instances in Latin America of state-sponsored scientific research that was able to import the advanced scientific equipment necessary to carry out its research objectives. Perón authorized the purchase of the equipment and instruments that formed the foundation of the continued development of Argentine physics long after Richter's dismissal. The crown jewel was the synchro-cyclotron obtained from the Netherlands, which historian Thomas Glick identifies as "the central element in training a distinguished generation of Argentine nuclear scientists who early on produced internationally recognized results in the field of radioisotopes" (Glick 1994, 502).

Finally, international networks play a critical role to understanding the success of science in Latin America. In this case, the debate over Perón's claims of success took place principally among scientists in the United States and Europe, in international scientific journals, and the *United Nations World*, one of the few journals at the time to link science and international relations on a global platform. Scientists articulated significant doubts based largely upon the fact that Richter seemed to have no verifiable connections to the international scientific community. No one stepped forward to confirm Richter's biography, the German university in Prague where Richter claimed to have first undertaken atomic research had burned during the war, and Richter's purported mentor had died.[5] In contrast, the members of the scientific commissions that ultimately ended Richter's tenure were themselves deeply embedded in national and international scientific networks.

Looking at Proyecto Huemul through this framework reveals the ways in which, by design, the project of atomic research in Argentina had many of the necessary—though perhaps not in and of themselves sufficient—characteristics of successful research in Latin America. Setting aside the question of its technical failure, the project was positioned for multiple avenues of success due to its links to public utility, Perón's nationalist vision of a "New Argentina," and the use of cutting-edge technology. The historical evidence indicates, however, that even a technically successful Proyecto Huemul would nonetheless have failed to accomplish all of the economic and political goals Perón assigned to atomic research. For example, apart from the goal of supporting the energy demands of growing industrialization and urbanization, Perón also claimed that the government would sell atomic energy in "one-and-a-half liter bottles for industrial and household use" (Clarín 1955). The world is still waiting for bottled atomic energy, but fully understanding the promise and political power of statements such as this one requires an examination of the interaction between technology, politics, and the state and the need for populist leaders like Perón to have stable technological solutions—even if their stability was restricted to cultural and social rather than scientific domains—sooner rather than later.

Politicizing Technology

The terminology surrounding Proyecto Huemul reveals how, within the context of a populist regime, time-sensitive and objective-oriented technological projects outbid open-ended and knowledge-oriented scientific institutions for Perón's interest and the Peronist state's financial support. This language demonstrates that Perón cast Proyecto Huemul as a technological rather than a scientific project. Historically speaking, up until the beginnings of the Industrial Revolution in eighteenth-century Europe, science and technology existed largely independently of one another within the traditions of moral philosophy and craftsmanship. As science emerged as a discipline and vocation, technology gradually shifted from its traditional craft base to that of science. It was only in the latter half of the twentieth century that society firmly wedded technology with science and differentiated it from craftwork.

Even though the history of science and technology over the past century arguably suggests a close if not indistinguishable relationship between the two terms, there are still identifiable differences. For example, Gregory Derry (1999) argues that we can define science as a way of "understanding the world" through both knowledge and theory; technology, in contrast, is a way of "controlling the world" using tools that science has sometimes greatly influenced and sometimes not (133). James Feibleman (1972) similarly identifies differences along the lines of the principles and practices identified with science and technology: the scientist uses experiments to test hypotheses and develops tools from theory, but the technologist uses trial and error, is aided by human intuition, and develops tools for solving real-world problems.

The perception of Proyecto Huemul as technology gained visibility primarily in its name: *project* implied a relatively stable undertaking with defined goals and a finite timetable. The project took place at the "pilot plant" on Huemul Island, which suggested that Argentina was developing and manufacturing a product through technological means. Such a formulation contrasted with the creation of an "institute" for open-ended "research" into atomic physics, which is in fact what Gaviola had originally proposed to the Peronist government in 1946 and what later became of the facilities and equipment in Bariloche (Mariscotti 1985). Following Derry, the pilot plant was a way for Argentina to "control" and not simply "understand" the world. Aided by human intuition and tackling real-world problems, Richter was a valuable technologist: someone who created stable technologies—tools, medicines, power plants—that would make Argentina more productive.

This argument becomes more persuasive when we look at the roles that Richter and Perón played in the project's narrative. Richter was the classic "inventor" of a technology who possessed a "secret" that enabled him to create something wholly new and empowering. Months after the announcement of the project's success, Perón described his initial encounter with Richter in August 1948 as a conversation in which the scientist explained his project to Perón "so well that I now have a thorough understanding of nuclear fusion" (Mundo Atómico 1951). Note that this meeting took place within a week of Richter's arrival in Argentina when he could speak only a few words and phrases in broken Spanish, and Perón's German was reported to be equally inadequate (Mariscotti 1985, 96–97). Richter possessed the atomic secret, and although he was able somehow to communicate this to Perón (or, clearly, Perón exaggerated this point), he was unwilling to make his secret publicly available.

In another example, Perón's statement on "bottling atomic energy" suggests the power of a stable, populist message that such a technologically conceived project could communicate. According to Perón, bottled atomic energy would serve to "provide light, cook food, and heat the iron" (Clarín 1955). Despite the "industrial use" that Perón claimed was an important component of the practical application of atomic energy, these specific examples of its use were important to his domestic constituents. In his analysis of the relationship between science and the Argentine military, Eduardo Ortiz (1996) suggests that the predilection for "very concrete project[s] with a clear end result, hopefully achievable with a finite amount of money and within a definite period of time" was "consistent with the objective-oriented 'flowchart' approach to reality which was then prevalent in the planning circles of the Army" (176). While we cannot overlook Perón's military background and the military government of which he was a critical part immediately before his election as president, Perón's populist appeal transformed the technological needs of the military into stable applications of knowledge suitable as solutions for social and national problems.

Indeed, Proyecto Huemul is not the only example of a technological project that received priority over scientific research in Peronist Argentina. For example, Kurt Tank's management of the Instituto Aerotécnico in Córdoba and his focus on the production of a fighter jet were as much a part of the Peronist public relations machine as Richter's atomic research. Tank's prototypes were much more visibly successful than the fusion experiments on Huemul Island. On February 8, 1951, Tank successfully flew the *Pulqui II* before a large audience at an air show in Buenos Aires (New York Times 1951a, 4). One could also argue that Perón's promotion of public health was undertaken in part at the expense of basic research in biology. Ramón Carrillo's Ministry of Public Health pursued populist projects like hospitals, rural health clinics, free physical exams, and vaccination programs while researchers such as Bernardo Houssay and Luis Leloir, whom Peronists had thrown out of the state universities, continued to engage in Nobel Prize-winning scientific research made possible only through the help of private funding from domestic and international sources.[6]

Clearly, technologically conceived projects with short-term results meant more to Perón because of the immediate political gains such projects could achieve. In this way, we can view Perón's use of Proyecto Huemul and similar projects as examples of a *political technology*, which Manuel Tironi and Javiera Barandiaran define elsewhere in this volume as "a set of embodied practices that produce knowledge, are adaptable to new political demands, and seek to transform the state." If Tironi and Barandiaran's concept is sufficiently malleable, then we might consider Perón's support for science and technology as the deployment of a populist political technology characterized by an emphasis on technology over science, the promise of a better future for the nation and its citizens through technology, the importance of the individual leader (e.g., Richter and Perón), and the practice of the press conference complete with audacious claims shrouded in mystery.

Conclusions

Most attempts at determining the meaning of Proyecto Huemul have centered on explanations of its failure. It was easy to blame Richter, and most people did. In 1955, the CNI had every reason to blame Perón directly for the project's failure but chose instead to emphasize Richter's managerial inadequacies and scientific self-delusion. Most scientists and government bureaucrats from the 1950s to the present day subscribe to the "mountebank and swindler" interpretation of Richter's role in the state-sponsored scientific project. Mariscotti, who in the process of researching his book personally interviewed Richter, felt after speaking with the scientist that "strictly speaking, [Richter] knew little of physics" but was "seductive" in explaining his "atomic secret" (cited in Yappert 2003).

However, the importance of this case lies in the question of why Perón undertook the project in the first place. Standard models of the unidirectional diffusion of Western science ("imported magic") and its active or passive reception by peripheral nations only go so far. The question of research in Latin America more generally is equally problematic, and understanding the relationship between Latin America and the centers of scientific and technological development has long presented difficulties to politicians, scientists, and historians. Common approaches include the narrative of the diffusion of knowledge outward from the centers of Western civilization and investigation into whether the periphery "actively" or "passively" receives knowledge (Glick 1994). A model of diffusion and reception assumes a deficiency in the scientific research that occurs outside of the center; equality or commonality between science in core and periphery is not possible (Basalla 1967; Basalla 1991).

As Cueto (1997) has demonstrated, the stabilization of knowledge entities such as atomic energy is rarely that simple, and understanding the history of such entities requires careful investigation of the local context. Despite the technical failure of Argentina's atomic research program, it is important to understand the opportunity presented to Perón by the conjuncture of Argentina's economic strengths and weaknesses, the state of atomic research worldwide immediately following World War II, Argentina's connections with international scientific networks, and Perón's relationship with Nazi Germany. These factors suggest ways in which scientific ideas and technologies move that differ from the standard models, in that Argentina was positioned to participate meaningfully and equally in the global development of atomic energy—not simply to receive the technology of atomic energy fully formed.

If, in the long run, Perón failed to deliver a bottle of atomic energy to every Argentine home, he did nevertheless succeed in achieving a number of short-term political goals. Therefore, understanding that Perón translated Proyecto Huemul into a *politically* stable technical undertaking reveals that the meaning of atomic research in Peronist Argentina lies beyond debates of fact or fiction, success or failure. Populist political movements like Peronism had much to gain from funding such technological projects because technology offered the hope of practical benefits to a broad section of people in a timely manner. Electric lighting, indoor plumbing, and steady jobs meant more to working class Argentines than advances in scientific knowledge. The promise of a bottle of atomic energy in everyone's home made Perón's otherwise abstract investment in the development of Argentina's industrial and energy sectors a concrete populist technological project that appealed to people throughout the nation and, as a political technology, promoted continued support of his regime.

As a consequence, putting aside Richter's failure to achieve a viable source of atomic energy, its pursuit by Peronist Argentina speaks to our understanding of technology's links to the state, modernity, and nation building and to the importance of understanding the processes surrounding science and technology rather than their outcomes. Perón's vision for a "New Argentina" was that of an industrialized, technologically equipped, and modern nation powered by cheap energy. From this perspective, the translation of atomic energy to the particular context of Argentina was not unlike what took place in other societies. Just as France turned to atomic research, power, and weapons in part to recapture a sense of prominence and importance on the world stage (Hecht 2009), Peronist Argentina made a not so inconceivable attempt as well.

Indeed, the Argentine case was very similar to that of physics in cold war Mexico, which Gisela Mateos and Edna Suárez-Díaz examine in this volume. In both instances, the pursuit of atomic energy accentuated the sense of modernity linked to nationalism. However, while a narrow interpretation of both cases would lead to a judgment that they failed, Mateos and Suárez-Díaz rightly note that the path of Mexican physics reflected positive political choices. To be sure, the case of Proyecto Huemul included a more mixed set of political entanglements than the Mexican case; and Richter's failure effectively masked the very contingencies that would have made possible a different stabilization of the form of knowledge that became "atomic energy"—one authored by Argentine physicists connected to an international network of scientists, technically capable, and supported by significant government funding. Still, although the specificities of Perón's use of Proyecto Huemul as a political technology further obscured the missed opportunity of state support for atomic research, this case exemplifies how the Latin American experience helps us to understand more broadly the relationship between science, technology, politics, and the state.

Notes

1. This narrative draws from several sources, cited throughout. There is little debate as to the facts of the case other than the motivations of the main players in the drama and the scientific question of what Richter did or did not hope to achieve.

2. The figure of 62.5 million pesos comes from the *Libro Negro de la Segunda Tiranía* (Comisión Nacional de Investigaciones 1958). US dollar amounts were calculated from contemporary currency exchange data and inflation figures from the US Bureau of Labor Statistics.

3. Kenneth Lipartito's (2004) analysis of the failure of AT&T's Picturephone to achieve commercial success inspired this essay's approach to explaining the failure of nuclear research in Argentina.

4. The Comisión Nacional de Investigaciones published its findings in the *Libro Negro de la Segunda Tiranía* (Black Book of the Second Tyranny). Its title equates Perón with a popular interpretation of Juan Manuel de Rosas, who ruled Argentina from 1835 to 1852 (the "first tyranny"), as strictly authoritarian and antidemocratic.

5. Mariscotti outlined his research methods, which included tracking down Richter's former classmates, in advance of the publication of his book (Sanchez 1983).

6. Houssay won the 1947 Nobel Prize in medicine for his discovery of the role the hormone of the anterior pituitary lobe plays in the metabolism of sugar. His Instituto de Biología y Medicina Experimental received support from the Sauberan Foundation in Argentina and the Rockefeller Foundation in New York. Leloir won the 1970 Nobel Prize in chemistry for his discovery of sugar nucleotides and their role in the biosynthesis of carbohydrates. Leloir's Instituto de Investigaciones Bioquímicas de la Fundación Campomar received funding from Argentine textile industrialist Jaime Campomar, the Rockefeller Foundation, and a later grant from the US National Institutes of Health.

References

Balseiro, José Antonio. 1988. *Informe referente a la inspección realizada en la isla Huemul en Septiembre de 1952*. Buenos Aires: Comisión Nacional de Energía Atómica.

Barbero, Maria Inés, and Marcel Rougier. 2002. La producción historiográfica respecto de las relaciones internacionales de la Argentina del período 1930–1955: Temas, problemas, y enfoques recientes. In *Sobre Nazis y Nazismo en la cultura Argentina*, ed. Ignacio Klich, 129–156. College Park, MD: Hispamérica.

Basalla, George. 1967. The spread of Western science. *Science* 156 (May 5): 611–622.

Basalla, George. 1991. The spread of Western science revisited. In *Mundialización de la ciencia y cultura nacional: Actas del Congreso Internacional "Ciencia, descubrimiento y mundo colonial,"* ed. A. Lafuente, A. Elena, and M. L. Ortega, 599–603. Madrid: Ediciones Doce Calles.

Batteau, Allen. 2010. *Technology and Culture*. Long Grove, IL: Waveland Press.

Bracker, Milton. 1951. Latin republics loath to raise *Prensa* issue. *New York Times*, March 25, 122.

Carassai, Sebastián. 2010. The formation of a post-Peronist generation: Intellectuals and politics in Argentina through the lens of *Contorno* (1953–1959). *Americas* 67:219–251.

Cardenas, Enrique, Jose Ocampo, and Rosemary Thorp, eds. 2000. *Industrialization and the State in Latin America: The Postwar Years*. Vol. 3 of *An Economic History of Twentieth-Century Latin America*. New York: Palgrave.

Clarín. 1951. El professor Richter explicó ampliamente el descubrimiento atómico. *Clarín*, March 26.

Clarín. 1955. Venta embotellada de energía. *Clarín*, October 7.

Comisión Nacional de Investigaciones. 1958. *Libro negro de la segunda tiranía*. Buenos Aires: Comisión Nacional de Investigaciones.

Cueto, Marcos. 1989. Andean biology in Peru: Scientific styles on the periphery. *Isis* 80:640–658.

Cueto, Marcos. 1997. Science under adversity: Latin American medical research and American private philanthropy. *Minerva* 35:233–245.

Derry, Gregory. 1999. *What Science Is and How It Works*. Princeton: Princeton University Press.

Feibleman, James. 1972. Pure science, applied science, and technology: An attempt at definitions. In *Philosophy and Technology: Readings in the Philosophical Problems of Technology*, ed. Carl Mitcham and Robert Mackey, 33–41. New York: Free Press.

Gerchunoff, Pablo, and Damian Antúnez. 2000. De la bonanza Peronista a la crisis de desarrollo. In *Los años Peronistas (1943–1955)*, ed. Juan Carlos Torre. Vol. 8 of *Nueva historia Argentina*. Buenos Aires: Editorial Sudamericana.

Glick, Thomas. 1994. Science and society in twentieth-century Latin America. In *Latin America since 1930: Economy, Society and Politics*, ed. Leslie Bethell, 463–535. Vol. 6 of *The Cambridge History of Latin America*. Cambridge: Cambridge University Press.

Goñi, Uki. 2002. *The Real Odessa: How Perón Brought the Nazi War Criminals to Argentina*. London: Granta.

Hecht, Gabrielle. 2009. *The Radiance of France: Nuclear Power and National Identity after World War II*. Cambridge, MA: MIT Press.

Hess, David. 1995. *Science and Technology in a Multicultural World: The Cultural Politics of Facts and Artifacts*. New York: Columbia University Press.

Lipartito, Kenneth. 2004. Picturephone and the information age: The social meaning of failure. *Technology and Culture* 44 (January):50–81.

Luca de Tena, Torcuato, Luis Calvo, and Esteban Peicovich, eds. 1976. *Yo, Juan Domingo Perón*. Barcelona: Editorial Planeta.

Mariscotti, Mario. 1985. *El secreto atómico de Huemul: Crónica del origen de la energía atómica en la Argentina*. Buenos Aires: Sudamerica/Planeta.

Martinez, Tomas Eloy. 1984. Perón and the Nazi war criminals. Working Paper no. 144. Washington, DC: Latin American Program, Wilson Center.

Mundo Atómico. 1951. Perón speaking to reporters, 29 June 1951. *Mundo Atómico* 2 (5).

National Archives and Records Administration. 1957. Personal History Statements—Dr. Ronald W. Richter, February 20, 1957, Attachment no. 2, Separate Sheet no. 3, page 8. Ronald Richter, Box 134, 190/A/24/6. Foreign Scientists Files. Joint Intelligence Objectives Agency, Record Group 330. National Archives and Records Administration, Washington, DC.

New York Times. 1951a. Perón commends ex-Nazi: Acknowledges presence of air expert in Argentina. *New York Times*, February 9.

New York Times. 1951b. Text of the report by Perón on "controlled" atomic energy. *New York Times*, March 25, 8.

Ortiz, Eduardo. 1996. Army and science in Argentina: 1850–1950. In *National Military Establishments and the Advancement of Science and Technology: Studies in Twentieth-century History*, ed. Paul Forman and Jose Sanchez-Ron, 153–184. Dordrecht: Kluwer.

Rock, David. 1987. *Argentina, 1516–1987: From Spanish Colonization to Alfonsín*. Berkeley: University of California Press.

Sanchez, Edgard. 1983. The glib "atomic scientist" who fooled gullible Perón. *Dimensión: The Post*, November 6, D1–D2.

Thirring, Hans. 1951. Is Perón's A-bomb a swindle? *United Nations World*, May, 1–2.

Vessuri, Hebe. 1997. Bitter harvest: The growth of a scientific community in Argentina. In *Scientific Communities in the Developing World*, ed. Jacques Gaillard, V. V. Krishna, and Roland Waast, 307–335. London: Sage.

Warren, Virginia Lee. 1951. Perón is scornful of atomic skeptics. *New York Times*, March 26, 1.

Yappert, Susana. 2003. Entrevista: Mario Mariscotti, físico. *Rio Negro On Line*. December 1. Available at http://www.rionegro.com.ar/arch200312/01/o01s19.php.

14 Peaceful Atoms in Mexico

Gisela Mateos and Edna Suárez-Díaz

This chapter provides an interconnected history of the promotion and peaceful uses of atomic energy in Mexico during the first decades of the cold war. This was a rich period in the development of Mexican science and its institutions, which exhibited both a strong nationalistic sentiment and an intensified international character. In large part this is because, as Greg Grandin has said, the cold war's "transcendental force" relied on the "politicization and internationalization of everyday life" (Grandin 2010, 4). National contests over politics, labor, the control of natural resources, and the direction of science policies did not escape the global conflict. This tension between the local and the global constitutes the overarching theme of our chapter.

In the global realm, the period between 1949 and the mid-1980s was punctuated by the creation of international agencies involved in the promotion of science—in particular, the promotion of the peaceful uses of atomic energy after 1953—and by the rise of American hegemony in scientific and technological matters. But despite the impetus of postwar America and its obvious influence in shaping world science and technology, scientific knowledge did not flow unidirectionally and unaffected through the southern border. Local contexts opposed, modified, and cross-fertilized international and global trends, thus providing evidence in support of the more recent view that American hegemony was co-constructed, instead of being a one-sided product of American preeminence after World War II (Jasanoff 2005; Krige 2008).

The Mexican case is particularly helpful to illuminate such complex interchanges and dynamics, if only because the cold war largely intersects with what is called the postrevolutionary period in Mexican history.[1] In this chapter we focus on the years between 1945 and 1970, marked by a growing economy (the so-called Mexican miracle) largely supported by the state monopoly of leading industries and relative political stability (Katz 2004). The construction of a new social regime, including a new sense of nationhood and modernity, the creation and recreation of new institutions, and the consolidation of new political and intellectual elites, are some of the consequences of the Mexican revolution (1910–1917) and its aftermath. In this chapter, we focus on the building of Mexican nuclearity to illustrate the modernizing nationalism of the Mexican postrevolutionary state. As defined by Gabrielle Hecht, *nuclearity* is "the degree to which a nation, a program, a policy, a technology, or even a material count[s] as 'nuclear.' ... [It is] a spectrum, not an on-off condition. Both nuclearity and its implications emerged in substantive ways from the dynamics between cold war and postcolonial visions of the world" (Hecht 2006, 26–27).

To speak of how Mexican science was shaped during this period requires us to acknowledge the often-contradictory role of international and national politics, and the intense circulation of people, materials, instruments, and knowledge between Mexico and other countries, in particular the United States. This focus on circulation refuses the idea that knowledge (science and technology) is imported from big centers of production and passively consumed at the periphery. Instead, we argue that local contexts are equally productive of reality. In this chapter, we point to a complex interaction of local and global institutions, discourses, bureaucracies, and material resources.

We start this chapter with a brief revision of the literature on cold war science, including a critical incorporation of some recent trends in the field. We introduce two requirements for a social history of science in Latin America and Mexico: a symmetrical treatment of global and local historical contexts, and the need to write interconnected transnational narratives to account for the co-construction of US scientific and technological hegemony after World War II. We then focus on the formation of a nuclear physics community and the early development of Mexican nuclearity.

This chapter also traces the formation of a body of nuclear experts and the development of scientific research in several fields (radiochemistry, nuclear physics, radiobiology). Creating this body of expertise involved the international circulation of students and researchers and produced a national scientific elite whose role was not limited to the scientific sphere. Scientists such as Nabor Carrillo and Manuel Sandoval-Vallarta were instrumental in the construction of the nationalistic institutions and ideology that characterized the new Mexican state. At the local level they also acted as experts and as political agents, deriving their authority from the extensive international networks to which they belonged.

The chapter, moreover, charts Mexico's troubled path toward nuclearity, and the role of nationalism in the decision to build and, eventually, halt the nuclear energy project. Finally, the concluding remarks discuss the prominent role of Mexico in signing the first atomic weapons nonproliferation treaty in the context of cold war goals and anxieties. It raises a set of specific questions for further understanding global histories during this period.

The Social History of Science in Latin America Today

In a recent paper (Mateos and Suárez 2011), we reviewed at length the state of historical research on cold war science, emphasizing the relative scarcity of such studies in

Latin America. For a number of reasons, one being the recent availability of archival sources, the cold war has become a hot area of study for historians of science. Because physicists played such a prominent role during and after World War II, it is no surprise that historians of physics started focusing on the particularities of this field in different contexts (e.g., Wang 1999; Kaiser 2002; Gordin 2010). This historiography led to national histories of nuclear programs and high energy physics, and a fruitful analysis of scientific practices and instruments in a number of scientific fields, enabling detailed accounts of the intimate relationship between science, industry, and the military (Forman 1987; Galison and Hevly 1992; Seidel 1986).

Two results of these studies are crucial to our present essay. First, the exponential growth in military and state funding had a significant impact on particular scientific agendas related to the cold war (Forman 1987). The impact of cold war priorities on particular fields of research has been the focus of historical studies covering quantum electronics (Forman 1987), physics (Kevles 1995; Seidel 2001), molecular biology (Kay 2000; De Chadarevian 2003), biochemistry (Creager and Santesmases 2006; Creager 2009), ecology (Bocking 1995), and oceanography (Hamblin 2005).

Second, the study of international and transnational collaborations during this period gives us a more nuanced sense of the co-construction of American hegemony (Krige 2008). Some of this literature has incorporated crucial analytical categories from the field of postcolonial studies, including the idea that symmetrical accounts illuminate the different ways in which local contexts shape attitudes, practices, and goals at the hegemonic centers. In cold war studies, this symmetrical perspective is illustrated by research on the development of the atomic programs in India (Abraham 1998), Pakistan (McMahon 1994), Iran (Leslie and Kargon 2006), South Africa (Hecht 2006), and China (Wang 2010). Moreover, there is a growing trend toward *transnational* histories of science and the dissolution of national, disciplinary, and institutional histories. Circulation, standardization, international collaboration, and the evolving American global hegemony constitute central areas of interest and the lens through which the history of the period is analyzed from a more symmetrical perspective.

Compared to the developments discussed above, scholars of the history of cold war science in Latin America and Mexico have been rather slow to respond to the debates taking place in the international community of historians of science. This is somewhat unexpected, given that critiques of science and technology have abounded in Latin American social sciences since the late 1950s and especially after the 1970s. Such critical accounts arose as a reaction to theories of development which prescribed that economic and political progress would arise from the transfer of science and technology from North to South (Basalla 1967). Dependency theories reflected the growing self-consciousness of Latin American, Asian, and African countries that resulted from postcolonial realities and, equally interesting from our perspective, the cold war context (Sosa-Alvarez 2000) and the correlated construction of the third world. The idea of the third world originally had a political connotation as an alternative path to the two-sided confrontation. Only later, in the 1970s and 1980s, was it translated into economic terms.

The critical analysis of dependency theories recognized the peripheries' historical agency, but its view of science and technology still does not permit the accommodation of recent findings in the field of STS. Dependence theories reinforced a history of science focused on national histories of disciplines and institutions, as opposed to the current emphasis on transnational histories. And while they acknowledged asymmetries and differences, the tension between the local and the global was assumed to be a result of the asymmetrical structure of the world, without a detailed analysis of the role of minor partners in the constitution of the larger partners. Exceptions include Cueto (2006 and 2007), Harwood (2009), Soto Laveaga (2009), and Birn and Necochea López (2011).

The study of this period has only recently been challenged and moved beyond discussions of development and dependency theories by a new generation of historians of cold war Latin America. The work of Daniela Spenser and her colleagues (including Seth Fein, Eric Zolov, and Steven Bachelor, among others) has opened new venues and themes for historians working on the period (Joseph and Spenser 2008). This historical work focuses on grassroots movements, but also on the ambivalence of Latin American countries toward the United States, and how the bipolar confrontation was related to the dirty wars taking place in many of these countries. Besides introducing new themes, this new generation of historians emphasizes the transnational character of the period in radical ways. They do not see the history of cold war Latin America as a marginal episode. Instead, they see the history of the confrontation between the two superpowers as a series of cultural, economic, and political struggles that took place in arenas outside of the United States and the Soviet Union. Thus, there is an interesting dialectic. These new historical studies underscore the ambivalence and autonomy of national projects within the region, but they also incorporate them as essential arenas in the broader conflict.

Mexicans resisted aligning themselves politically with their powerful northern neighbor and participating in the bipolar contest for global power (Bachelor 2008). More important to Mexican nuclearity and cold war-related science was the complex position of the country's foreign policy toward weapons development and testing. While Mexicans did not engage in mass protests against weapons and weapons testing in the 1950s and 1960s, the government's nationalistic nonaligned discourse and its foreign policy regarding independence from the United States compensated for Mexico's rightist policies at home and its increasingly close economic relations with its northern neighbor (Fein 2008, 197).

A symmetrical account of science and technology during this period recognizes that Latin American countries had a very different geo-sociopolitical experience of the cold war than their northern counterparts. As De Greiff and Nieto (2006) have observed, the cold war period was, indeed, a very "hot war" time for many Latin American countries, punctuated by several coups d'état, the impact of the dirty wars on leftist movements, and social agitation in many countries from the 1960s through the 1980s. In all these events, the more or less open intervention of the United States and international communist hysteria played crucial roles. While the dirty war in Mexico was less intense than those in other Latin American countries, such as Chile and Argentina, the cold war did impact Mexican national politics, by minimizing leftist opposition between the 1940s and 1950s, and repressing rural, syndical, and student movements in the 1960s and 1970s (Katz 2004). Cold war rivalries also played a role in the scientific arena. Mexican public universities tolerated Marxist ideology, and several Mexican academic programs formed collaborations with the Soviet bloc and China during this period with some success. Thus, a symmetrical historical account requires us to see the history of Latin American and Mexican science from the standpoint of social upheaval and the alliance of local elites not only with their American counterparts but also with other powers. Mexico was politically and economically allied with the United States, but it also acted as a wall of resistance against US interventionist policies in Latin America.

In the decades after World War II, the alliance between the US and Mexican governments solidified. This, in turn, nurtured the academic exchanges between the two countries that had begun in the 1920s (Sosa-Alvarez 2000). As Miller (2006) and others have shown, scientific collaboration and internationalism were effective tools for US foreign affairs. Mexican science students traveled to the United States to get doctoral degrees in engineering, physics, and the life sciences at prestigious institutions including Harvard, MIT, Michigan, Columbia, and Chicago. These academic exchanges with US institutions helped establish long-term binational collaborations. Mexican students also attended research institutions in France and the United Kingdom.

Modernization and Mexican Nuclearity

Two interconnected projects took place in Mexico after Hiroshima and Nagasaki: the development of a nascent project in nuclear energy, and the establishment of studies in nuclear physics. The political and scientific actors were for the most part the same. MIT physicist Manuel Sandoval-Vallarta, Harvard civil engineer Nabor Carrillo, and physicist Carlos Graef played the most active roles in the promotion and institutionalization of nuclear research and the nuclear energy project.

Sandoval-Vallarta and Carrillo illustrate how interconnected Mexican scientific elites were with their American counterparts.[2] Sandoval-Vallarta, for instance, worked with Vannevar Bush at MIT starting in the 1930s. In 1934 he became associate professor in the MIT physics department until his resignation in 1946. He had returned to Mexico in 1943 and became an advisor and expert on atomic matters for the Mexican government soon after Hiroshima.[3] Meanwhile, Carrillo kept his Harvard connections as an invited professor on soil mechanics during the 1940s and 1950s, while becoming the leading promoter of nuclear energy in Mexico (Domínguez Martínez 2000). His position as president of the national university (UNAM) from 1953 to 1961 further allowed him to act as the main promoter of Mexican nuclear capabilities. Mexico was one of the ten members of the 1946 UN Nuclear Energy Commission (which was subsequently disbanded in 1952). Because there were few Mexican physicists, Sandoval-Vallarta and Carrillo served as the sole government experts attending the UN meetings, and Sandoval-Vallarta acted as president of the commission in December 1946.

Both Sandoval-Vallarta and Carrillo mediated between Mexico's international and national nuclear postwar concerns. On August 22, 1945, in a quick reaction to the atomic bombings, the Mexican government announced that uranium, thorium, actinium, and other radioactive element reserves were national property (Rojas 1989; Vélez Ocón 1997). In 1946 another presidential decree established that radioactive materials could be exploited only by the state, following the constitutional principle that mineral resources belong to the Mexican nation. In 1948, the Alemán Law (submitted by President Miguel Alemán) was passed, which put radioactive reserves under the administration of the Comisión de Fomento Minero (Commission on Mine Development). These decisions marked the beginning of the state's ownership of nuclear energy production, and nuclear energy formed part of the Mexican government's nationalization of all energy sources to form state monopolies.[4]

However, as claimed before, Mexican nuclear knowledge depended on the transnational circulation of Mexican experts. Shortly after the passage of the Alemán Law, a Mexican commission was invited to watch the Bikini Atoll test. The government sent Carrillo and Colonel Juan Loyo González as experts to witness the atomic explosions. In the following decades, representatives from the Mexican Army acted as observers of nuclear research at different facilities, although they never had control over the instruments and their participation in decision making was limited.

In accordance with international developments—namely, the announcement of the Atoms for Peace initiative by President Dwight Eisenhower in 1953—local commissions for the promotion of nuclear energy were created around the world. After the 1955 Geneva Conference on Peaceful Uses of Atomic Energy, the Mexican government created the Comisión Nacional de Energía Nuclear (National Commission on Nuclear Energy; CNEN) in December 1956, starting operations in January 1957. Attorney José

María Ortiz Tirado was appointed as its first director, and the advisory board included Sandoval-Vallarta and Carrillo.

Since its beginnings, the CNEN was operationally connected with other international agencies created in the same years. In 1957 the International Atomic Energy Agency (IAEA) was established by the United Nations. Simultaneously, the Inter-American Nuclear Energy Commission (IANEC) was created with support of the American States Organization, under the leadership of the United States. IANEC, which promoted the peaceful uses of atomic energy in Latin American countries and held a series of meetings starting at Brookhaven National Laboratory in 1957, followed by meetings in Buenos Aires (1959), Río de Janeiro (1961), Mexico City (1962), and Santiago de Chile (1964), that focused on nuclear energy production and the application of radioisotopes in medicine and agriculture.

Despite the international character of nuclear policies, national contexts continued to shape local priorities. The most distinctive feature of the Mexican nuclear program was its civil, nonmilitary character. According to nuclear physicist and former Institute of Nuclear Research director Carlos Vélez Ocón (1997), President Adolfo Ruiz Cortines (1952–1958), an outspoken pacifist, decided that CNEN should take the name "nuclear" and not "atomic" because the latter was suggestive of war and armament. In this, the Mexican project was different from other nuclear projects in Latin America, such as those of Brazil (Hurtado de Mendoza and Vara 2007) and Argentina (Ribeiro de Andrade and Muniz 2006; Hagood, this volume), where atomic energy was considered a national security resource requiring military participation. On the other hand, the Mexican project shared with other Latin American countries, like Chile and Argentina, the discourse of modernization through nuclearity (Hagood, this volume; Tironi and Barandiarán, this volume), especially during the presidency of Miguel Alemán (1946–1952).

Alemán viewed science and technology as instrumental to the modernization of the country. When he decided to construct a new campus for the National University (UNAM) in southern Mexico City, the first stage of construction included new buildings for scientific institutes and schools, which were started in 1950 and inaugurated in 1954. The decision to buy a Van de Graaff accelerator as part of the new Institute of Physics installation at UNAM became the symbol of Mexico's scientific modernization. Indeed, studies of nuclear physics began around this instrument (Mateos, Minor, and Sánchez 2012).[5] Adriana Minor (2011) has shown how the Van de Graaff accelerator acquired different meanings that were generated by the convergence of diverse interests, as reflected in the government propaganda in the Mexican press between 1950 and 1953. Equally important, the Mexican nuclear physics community grew and consolidated in the following decades, enabling the circulation of people, practices, and substances, as well as the establishment of a solid tradition in instrument building.

While at the local level a "nationalistic modernization" was taking place, events at the international level also had an effect on Mexico's nuclearity. As a result of Eisenhower's Atoms for Peace initiative, three processes affecting this area took place: the promotion of different aspects of atomic energy via exhibitions, conferences, and international agencies; the open distribution to friendly countries of radioisotopes for research and applications; and the circulation of students and researchers to academic institutions in the United States, such as Oak Ridge National Laboratory.

The first exhibition on radioisotopes arrived in Mexico in 1961, sponsored by the IAEA. This exhibition took place in an itinerant bus, in which different instruments showed the public how radioisotopes were used in research, industry, and agriculture. Another exhibition soon followed, the Atoms for Peace program "Atoms in Action," held from April 2 to May 2, 1962. This exhibition was located at a military campus in Mexico City known as Campo Marte. Two of the main instruments in this exhibition were a research nuclear reactor, fabricated by the Lockheed Company, and a gamma irradiator. Through the publicity on the uses of radioisotopes and nuclear materials, nuclear energy was validated as harmless and promoted as a source of enormous benefit for people's future. This promotion of nuclear energy in Mexico was made possible by international agencies like the IAEA and, importantly at the Pan-American level, by the IANEC.

Radioisotopes were also crucial political instruments in promoting the peaceful uses of nuclear energy, becoming de facto transnational ambassadors for nuclearity. Because of their applications in cancer treatment and food technology, radioisotopes were portrayed as one of the positive uses of atomic energy (Creager and Santesmases 2006; Creager 2009). In 1947 the US Atomic Energy Commission started a program to distribute radioisotopes to allied countries in Europe and South America. Radioisotopes were shipped from Oak Ridge National Laboratory, but Mexico was not included until 1949. Given the importance of radioisotopes as a fundamental tool in medicine and biology, the Mexican physicists were urged to learn how to produce and use them.

By the early 1950s, a new generation of Mexican physicists and chemists was being educated. In 1954 Augusto Moreno, the director of the radiochemistry section at the Institute of Physics at UNAM, attended the first Oak Ridge course on radioisotopes. In the following years, many other Mexican physicists and chemists attended these courses, including Ariel Tejera. After coming back to Mexico, these scientists were responsible for the first courses on radioisotopes, beginning in 1958 at the UNAM campus and sponsored by CNEN. They also wrote a Spanish manual to guide students in the uses and applications of radioisotope technologies. Another example is the research by Fernando Alba, Virgilio Beltrán, Ariel Tejera, Augusto Moreno, and Tomás Brody, together with Silvia Bulboulian and Adelaida Palacios, who participated in the

UN scientific committee for the study of atomic radiation effects, measuring the atomic fallout of Strontium 90 and Cesium 137 in Mexico City.

The Tensions of a Mexican Nuclear Project

During the 1960s, Mexican students and researchers traveled back and forth to the United States, encouraging the formation of different research groups at UNAM and CNEN. Scientific research in Mexico covered theoretical and experimental physics and chemistry, radiochemistry, and animal genetics. The application of radioisotopes extended to medicine, the electric industry, and agriculture. However, the CNEN research laboratories were distributed throughout Mexico City.[6] This situation prompted Carrillo to promote the construction of a nuclear center that would bring together national nuclear research. Unfortunately, after Miguel Alemán, the next two Mexican presidents, Adolfo Ruiz Cortines (1952–1958) and Adolfo López Mateos (1958–1964), gave weak support to CNEN. López Mateos, in particular, is considered the "great denuclearizor of Latin America." When Carrillo asked López Mateos for his support of the nuclear center, he is said to have answered: "we are not a rich country to spend our resources on expensive toys" (Anonymous 1996, 24). However, Carrillo was not discouraged. He spoke with various state ministers and managed in 1964 to convince López Mateos to support the creation of nuclear research facilities in the town of Salazar, a few miles northwest of Mexico City.

The nuclear program and the nuclear physics community were reinforced by the construction of the Salazar center, which concentrated the human and material resources of Mexican nuclearity. Numerous physicists, chemists, biologists, and engineers returned to Mexico after getting their PhDs in different institutions, mainly in the United States, and were incorporated at Salazar.[7] The first laboratories (including the radioisotope group) moved to Salazar in 1968, even though the center was not officially inaugurated until 1970. A small group of scientists from the Salazar center, the UNAM, the Instituto Politécnico Nacional (National Polytechnic Institute), and the Comisión Federal de Electricidad (Federal Electricity Commission; CFE) formed an informal "Reactor Group," with the purpose of promoting the uses of nuclear reactors in Mexico. The group included Juan Eibenschutz, Augusto Moreno, Marcos Moshinsky, Ariel Tejera, Mario Vázquez Reyna, and Bruno de Vecchi, among others. These meetings produced a five-year plan for the development of reactors in Mexico. As a result, CNEN authorities bought a Tandem accelerator for research, set in use at Salazar in 1968. Also in 1968, a subcritical reactor TRIGA MARK III fabricated at the General Atomics Company in the United States was acquired and put to use through an agreement with the IAEA.

During the same years, nuclear experts and some sectors in the Mexican government encouraged the construction of nuclear energy plants (Rojas 1989). The same experts moved between academic research and the energy sector in Mexico. One of their first projects was a double-purpose plant for water desalinization and energy production, which was set to be built on the border between Mexico and the United States. The plant touched Baja California and the Sonora states in Mexico and Arizona and California on the US side. However, it was never built because the United States distrusted the building of a binational nuclear project, and because of the high cost involved (Rojas 1989, 113). Also, since 1966 the CFE had asked the CNEN to evaluate whether a nuclear energy plant with 600-megawatt potency would be adequate. As a result, in 1969, the Programa Nacional de Reactores de Potencia (National Reactors Program) was created by the CNEN; two international committees sent by the IAEA traveled to Mexico from September 1 to 10, 1969, and July 27 to August 1, 1970, in order to evaluate the viability of the Mexican project.

Related to these events, in 1972 the CNEN was transformed into the Instituto Nacional de Energía Nuclear (National Institute of Nuclear Energy; INEN) whose new aims included the use of nuclear reactors for the production of electric energy in Mexico and the construction of the first nuclear plant. The debates surrounding the acquisition of the plant's nuclear reactor reflected the tensions between the global and local context. As part of its foreign policy toward the United States, Mexico wanted to avoid compromises on nuclear tasks. Those who wanted a national Mexican nuclear industry that was not dependent on the United States contested the purchase of a reactor from the United States. Thus, discussions revolved around which type of reactor Mexico should buy: an enriched-uranium reactor that would tighten Mexican dependency on the US fuel supply, or a natural uranium reactor with heavy water, which some of the INEN scientists were convinced Mexico could produce. Finally, the decision was taken to build the only nuclear energy plant in Mexico. The Laguna Verde plant was built in the southern state of Veracruz in the Gulf of Mexico using a General Electric boiling water reactor (BWR) with enriched uranium. It started operation in 1990 and was run by the state-owned electric company CFE (Rojas 1989; Azuela and Talancón 1999).

However, the nuclear energy program was profoundly affected by several local and global developments during this period. The international oil crisis of the early 1970s and the rising costs, due to security issues, of the nuclear industry slowed down the construction of nuclear plants around the globe (Rojas 1989). In 1972 and 1973, moreover, the Mexican economy entered a transitional crisis that increased its foreign debt through the early 1980s. In the meantime, the social agitation of students and labor unions that took place between 1966 and the mid-1970s spread to the nuclear industry. A devastating strike of INEN's workers in 1976, combined with the unfavorable international context and the discovery of the low quality of Mexican uranium reserves, put an end to the government's plans to extend the nuclear project beyond the Laguna Verde plant—effectively slowing down research on nuclear science in

Mexico. Thus, the entanglement of national and international contexts was responsible not only for the uses and promotion of nuclear energy, but also for its constraint and eventual end.

Conclusion

We have offered an interconnected history of the nuclear project during the cold war and the postrevolutionary regimes in Mexico. Collaboration, the circulation of materials, people, instruments, and techniques, and the intervention of international and national agencies were crucial for this entanglement of national and international contexts, and of academic and governmental spheres in Mexico.

Mexican nuclearity was in large part defined by the state and the nationalistic character of the postrevolutionary regimes, which ruled over techno-political decisions. The Mexican local context provided the conditions for a civil nuclear energy project and the creation of a community of nuclear physicists, radiochemists, and radiobiologists at UNAM and CNEN-INEN. In turn, these engineers and scientists provided the postrevolutionary state with material and discursive power. They were not only political actors in the construction of the new Mexican state and its institutions, but also creators of new scientific realities in Mexico. Mexican scientists were instrumental in adapting new technologies and practices and applying them to new contexts and problems, at times struggling in the midst of local resistances and/or reconfiguring those technologies and practices to suit divergent goals.

A symmetrical perspective allows us to see the local resistances to world trends, such as the military uses of atomic energy. In the context of Mexico's postrevolutionary regimes, nonmilitarization was a positive political option requiring political decisions and actions. Though Mexico was not a member of the nonaligned movement, which began in 1961, it has remained an observer country, supporting the principles of the movement. The pacifist position of Mexico was reflected in its active promotion of the Treaty on the Non-Proliferation of Nuclear Weapons, also known as the Tlatelolco Treaty (Tratado para la Prohibición de Armas Nucleares en America Latina y el Caribe), which was signed in Mexico City in 1967 by 33 countries (Cuba not included). This treaty had profound consequences for the development of Mexican nuclear capabilities because it constrained nuclear research and the production of nuclear energy.

What is the pending agenda for an interconnected history of cold war science in Mexico? To answer this question, a few historiographic issues should be considered. First, scholars in this field need to include international archival sources. Given the interconnection of global and local science in this period, this is a crucial requirement. A few historians have done so, but they are still a minority (Cueto 2007; Harwood 2009; Soto Laveaga 2009; Birn and Necochea López (2011). If STS and its constituent fields, such as the history of science and technology, are to go beyond historiographies grounded in dependency theories, studies in this area need to emphasize the nonisolationist character of national projects. Second, the cold war period needs to be problematized as a historical category (Heyck and Kaiser 2010). The long decades between the end of the 1940s and the late 1980s include changing priorities and deep transformations in civil society, the global economy, and the geopolitical order, to name a few areas of change (Kuznick and Gilbert 2010; Whitfield 1996). The interruption of the exponential growth in science funding at the end of the 1960s and the beginning of the 1970s is an example of the transformations punctuating the cold war years. Third, a symmetrical account requires us to investigate how the power struggles between the Soviet and American blocs played out in Latin American countries (Morley 2008). Research needs to be done on the influence of Soviet and Chinese science in Latin American settings, in particular in public universities (Katz 2004), where leftist ideologies survived and were tolerated. In the case of Mexico, there are signs of the Soviet Union's presence in the scientific community, for example in the Spanish translation of Soviet scientific textbooks, the number of student fellowships for Mexican scientists to study in the Soviet Union, and the failed attempt for an exhibition on the peaceful uses of atomic energy in 1958.

Finally, the processes by which postwar American hegemony was built need to be carefully analyzed as a co-construction. The circulation of students and scientists, data, materials, and scientific practices gives shape to shared practices and knowledge. Circulation requires standardization and, as such, can be a channel for hegemonic views, consisting of measures, methods, goals, and institutional bureaucracies. But circulation also means restricting what is allowed to travel across borders: radioisotopes and fallout measurements traveled relatively easily, whereas enriched materials like uranium and reactors did not. Moreover, Mexico's nationalistic regimes and scientists created obstacles that slowed the expansion of the US nuclear industry and militaristic applications for nuclear science (by means of the Tlatelolco Treaty).

Also, it needs to be emphasized that nuclear energy was not only involved in the production of US hegemony; it was also involved in the production of regional hegemonies within Latin America. Mexico played a leading role in nuclear science during the cold war years, as did Brazil and Argentina. Mexico's geopolitical importance derived from its strong commitment to the nonproliferation of nuclear weapons worldwide, and its leadership role in pacifist diplomacy during the 1970s. In 1982, Mexican UN diplomat Alfonso García Robles received the Nobel Peace Prize because of his involvement in the drafting and passing of the Tlatelolco Treaty at the United Nations. With this treaty, Latin America became the first nuclear weapon-free zone in the world.

To understand how this happened, we need transnational histories because they provide more nuanced understandings of how knowledge hegemonies are co-constructed. Mexican nuclearity during the cold war period is an excellent case in point.

Acknowledgments

This research has been supported by research grants provided by UNAM-PAPIIT project number IN303111 and CONACYT 152879. We want to thank Eden Medina, Ivan da Costa, and Christina Holmes, the editors and organizers of the workshop "STS in Latin America: Beyond Imported Magic," which took place August 2012 at Indiana University, Bloomington, as well as the workshop participants for their valuable comments on a previous version of this chapter.

Notes

1. The postrevolution is a common periodization in histories of twentieth-century Mexico. According to Mexican historians Josefina Zoraida Vazquez and Lorenzo Meyer (2006), it starts after 1920 and has its peak during the Cardenist regime. This moment is characterized by agrarian reform, the nationalization of the oil industry, and the corporatization of workers and peasants within the official party. This period extends until the arrival of neoliberal governments in the mid-1980s. However, there are ample differences between the postrevolutionary regimes of the 1930s and 1940s, and the industrialization period of the 1950s and 1960s.

2. This is in sharp contrast to Domínguez Martínez's (2000) view that nuclear physics in Mexico was the result of an "exogenous dynamics" (96). Such views are related to the "magic bullet" perspective.

3. The cause of Sandoval-Vallarta's resignation from MIT and his return to Mexico are still unclear. He claimed that in the context of World War II and the military projects at MIT he was required to adopt US nationality, but he refused to give up his Mexican nationality. However, Adriana Minor García is writing a PhD dissertation at UNAM on Sandoval-Vallarta's biography and has found documents that suggest otherwise.

4. The oil industry was nationalized by President Lázaro Cárdenas in March 1938, creating the state-owned company PEMEX. The electric industry was nationalized in September 1960 by President Adolfo López Mateos, who created the Comisión Federal de Electricidad (CFE).

5. In 1950, UNAM bought a 2 MeV Van de Graaff generator that cost one million pesos at the time (around US $125,000). The purchase resulted from early negotiations between Manuel Sandoval-Vallarta, Carlos Graef, and Nabor Carrillo and the American company High Voltage Engineering Corp. These negotiations later continued between the American company and UNAM.

6. In 1954, the Companía de Luz y Fuerza Motriz de Mexico imported radioisotopes, Iridium 192, and Cobalt 60 from Canada to analyze the quality of the junctures in the pressure tubes.

7. For example in 1955, eight Mexican students were sent to study nuclear engineering at the University of Michigan: Bruno de Vecchi Appendini, Carlos Vélez Ocón, Arnulfo Morales Amado, Vinicio Serment Cabrero, Antonio Magaña Plaza, Miguel Angel Barberena Vega, Luis Gálvez Cruz, and Roberto Treviño Arizpe (Vélez Ocón 1997).

References

Abraham, Itty. 1998. *The Making of the Indian Atomic Bomb: Science, Secrecy and the Postcolonial World*. London: Zed Books.

Anonymous. 1996. *40 años de usos pacíficos de la energía nuclear en México*. Mexico: ININ.

Azuela, Luz Fernanda, and José Luis Talancón. 1999. *La historia de la energía nuclear en México, 1945–1995*. Mexico City: CEPE, IIS, IG, and Plaza y Valdés.

Bachelor, Steven J. 2008. Miracle on ice: Industrial workers and the promise of Americanization in cold war Mexico. In Joseph and Spenser 2008, 253–272.

Basalla, George. 1967. The spread of Western science. *Science* 156:611–622.

Birn, Anne-Emmanuelle, and Raúl Necochea López. 2011. Footprints on the future: Looking forward to the history of health and medicine in Latin America in the twenty-first century. *Hispanic American Historical Review* 91:503–527.

Bocking, Stephen. 1995. Ecosystems, ecologists and the atom: Environmental research at Oak Ridge National Laboratory. *Journal of the History of Biology* 28 (1):1–47.

Creager, Angela. 2009. Radioisotopes as political instruments, 1946–1953. *Dynamis* (Granada, Spain) 29:219–240.

Creager, Angela, and Maria Jesús Santesmases. 2006. Radiobiology in the atomic age: Changing research practices and policies in comparative perspective. *Journal of the History of Biology* 39:637–647.

Cueto, Marcos. 2006. Excellence in twentieth-century biomedical science. In *Science in Latin America*, ed. Juan José Saldaña, 231–240. Austin: University of Texas Press.

Cueto, Marcos. 2007. *Cold War, Deadly Fevers: Malaria Eradication in Mexico, 1955–1975*. Baltimore: Johns Hopkins University Press.

De Chadarevian, Soraya. 2003. Mice and the reactor: The "genetics experiment" in 1950s Britain. *Journal of the History of Biology* 39:707–735.

De Greiff, Alexis, and Mauricio Nieto. 2006. What we still do not know about South-North technoscientific exchange: North-centrism, scientific diffusion, and the social studies of science. In *The Historiography of Contemporary Science, Technology, and Medicine*, ed. Ronald E. Doel and Thomas Söderquist, 239–259. New York: Routledge.

Domínguez Martínez, Raúl. 2000. *Historia de la física nuclear en México, 1933–1963*. Mexico City: CESU-UNAM/Plaza y Valdés.

Fein, Seth. 2008. Producing the cold war in Mexico: The public limits of covert communications. In Joseph and Spenser 2008, 171–213.

Forman, Paul. 1987. Behind quantum electronics: National security as basis for physical research in the United States, 1940–1960. *Historical Studies in the Physical and Biological Sciences* 18: 149–229.

Galison, Peter, and Bruce Hevly, eds. 1992. *Big Science*. Stanford: Stanford University Press.

Gordin, Michael D. 2010. *Red Cloud at Dawn: Truman, Stalin and the End of Atomic Hegemony*. New York: Picador.

Grandin, Greg. 2010. Living in revolutionary time: Coming to terms with the violence of Latin America's long cold war. In *A Century of Revolution: Insurgent and Counterinsurgent Violence during Latin America's Long Cold War*, ed. Greg Grandin and Gilbert M. Joseph, 1–41. Durham: Duke University Press.

Hamblin, Jacob Darwin. 2005. *Oceanographers and the Cold War: Disciples of Marine Science*. Seattle: University of Washington Press.

Harwood, J. 2009. Peasant friendly plant breeding and early years of the green revolution in Mexico. *Agricultural History* 83:384–410.

Hecht, Gabrielle. 2006. Negotiating global nuclearities: Apartheid, decolonization, and the cold war in the making of the IAEA. *Osiris* 21:25–48.

Heyck, Hunter, and David Kaiser. 2010. Focus: New Perspectives on Science and Cold War: Introduction. *Isis* 101:362–366.

Hurtado de Mendoza, Diego, and Ana María Vara. 2007. Winding roads to big science: Experimental physics in Argentina and Brazil. *Science, Technology and Society* 12:27–48.

Jasanoff, Sheila. 2005. *Designs on Nature: Science and Democracy in Europe and the United States*. Princeton: Princeton University Press.

Joseph, Gilbert M. 2008. What we know and should know: Bringing Latin America more meaningfully into cold war studies. In Joseph and Spenser 2008, 3–45.

Joseph, Gilbert M., and Daniela Spenser, eds. 2008. *In from the Cold: Latin America's New Encounter with the Cold War*. Durham: Duke University Press.

Kaiser, David. 2002. Cold war requisitions, scientific manpower, and the production of American physicists after World War II. *Historical Studies in the Physical Sciences* 33:131–159.

Katz, Friederich. 2004. La guerra fría en América Latina. In *Espejos de la guerra fría: México, América Central y el Caribe*, 11–28. Mexico City: SRE/Ciesas.

Kay, Lily. 2000. *Who Wrote the Book of Life? A History of the Genetic Code*. Palo Alto: Stanford University Press.

Kevles, Daniel J. 1995. *The Physicists: The History of a Scientific Community in Modern America*. Cambridge, MA: Harvard University Press.

Krige, John. 2008. The peaceful atom as political weapon: Euratom and American foreign policy in the late 1950s. *Historical Studies in the Natural Sciences* 38:5–44.

Kuznick, Peter J., and James Gilbert, eds. 2010. *Rethinking Cold War Culture*. Washington, DC: Smithsonian Books.

Leslie, Stuart W., and Robert Kargon. 2006. Exporting MIT: Science, technology, and nation-building in India and Iran. *Osiris* 21: 110–130.

Mateos, Gisela, A. Minor, and V. Sánchez. 2012. Una modernidad anunciada: Historia del Van de Graaff de la ciudad universitaria. *Historia Mexicana* 61 (1):415–442.

Mateos, Gisela, and Edna Suárez. 2011. Mexican science during the cold war: An agenda for physics and the life sciences. *Ludus Vitalis* 20 (37):47–69.

McMahon, Robert J. 1994. *The Cold War in the Periphery: The United States, India, and Pakistan*. New York: Columbia University Press.

Miller, Clark A. 2006. An effective instrument of peace: Scientific cooperation as an instrument of U.S. foreign policy, 1938–1950. *Osiris* 21:133–160.

Minor, Adriana. 2011. Instrumentos científicos en movimiento. Historia del acelerador Van de Graaff del Instituto de Física de la UNAM (1950–1983). Master's thesis, Universidad Nacional Autónoma de México.

Morley, Jefferson. 2008. *Our Man in Mexico: Winston Scott and the Hidden History of the CIA*. Lawrence: University Press of Kansas Press.

Ribeiro de Andrade, Ana M., and R. P. A. Muniz. 2006. The quest for the Brazilian synchrocyclotron. *Historical Studies in the Physical and Biological Sciences* 36 (2):311–327.

Rojas, José Antonio. 1989. Desarrollo nuclear de México. Mexico City: Universidad Nacional Autónoma de México.

Seidel, Robert W. 1986. A home for big science: The Atomic Energy Commission's laboratory system. *Historical Studies in the Physical and Biological Sciences* 16: 135–175.

Seidel, Robert W. 2001. The national laboratories of the Atomic Energy Commission in the early cold war. *Historical Studies in the Physical Sciences* 32 (1):145–162.

Sosa-Alvarez, I. 2000. América Latina, retos y resultados en la investigación. Paper presented at the 19th International Congress of Historical Sciences, Oslo, Norway.

Soto Laveaga, Gabriela. 2009. *Jungle Laboratories: Mexican Peasants, National Projects, and the Making of the Pill*. Durham: Duke University Press.

Vázquez, Josefina Zoraida, and Lorenzo Meyer. 2006. *México frente a Estados Unidos. Un ensayo histórico, 1776–2000*. Mexico City: Fondo de Cultura Económica.

Vélez Ocón, Carlos. 1997. *Cincuenta años de energía nuclear en México 1945–1995*. Mexico City: Universidad Nacional Autónoma de México.

Wang, Jessica. 1999. *American Science in an Age of Anxiety: Scientists, Anticommunism, and the Cold War*. Chapel Hill: University of North Carolina Press.

Wang, Zuoye. 2010. Transnational Science during the cold war: The Case of Chinese/American Scientists. *Isis* 101 (2):367–377.

Whitfield, Stephen J. 1996. *The Culture of the Cold War*. Baltimore: Johns Hopkins University Press.

Zolov, Eric. 2008. ¡Cuba sí, Yanquis no! The sacking of the Instituto Cultural México Norteamericano in Morelia, Michoacán, 1961. In Joseph and Spenser 2008, 214–252.

15 Neoliberalism as Political Technology: Expertise, Energy, and Democracy in Chile

Manuel Tironi and Javiera Barandiarán

> Chile's unique policy path can largely be attributed to a uniquely powerful and ideologically coherent team of free-market technocrats, with a long-term vision for the Chilean economy.
> Fourcade-Gourinchas and Babb 2002, 545–546

Neoliberalism has had a profound impact on contemporary Chile. Neoliberal policies redefined sectors and institutions in industry (Ffrench-Davis 1980), labor (Foxley 1983), health (Ossandón 2009), the city (Portes and Roberts 2005; Sabatini 2000), and the environment (Liverman and Vilas 2006), from the 1970s through today. Many say that nowhere else has neoliberal restructuring been more extended and aggressive (Klein 2008; Lave, Mirowski, and Randalls 2010). In addition, the link between neoliberalism as a set of policies and as an epistemological framework related to the Chicago School of Economics (Van Horn and Mirowski 2009) is embodied in Chile by the infamous Chicago Boys—a group of Chicago-trained economists, endorsed by the military regime, who overhauled the Chilean economy in the late 1970s and early 1980s.

Although an abundant literature exists on neoliberalism in Chile, we identify two accounts still missing from this history. First, neoliberalism has been understood more as an epochal and abstract force than as situated practices. More detailed analyses of how neoliberalism unfolded in specific sites and through specific controversies are needed to interrogate the material and knowledge practices that enact neoliberalism. Second, while a robust literature has focused on the arrival of neoliberal ideas and the implementation of neoliberal policies in the 1970s, little has been said about how neoliberal ideology adapted to the post-dictatorship settings of the 1990s and 2000s.

To tackle these gaps, we examine neoliberalism as a *political technology*. Neoliberalism as *technology* means it is applied knowledge about how to define, order, and calculate the world. Neoliberalism as a *political* technology draws attention to how this applied knowledge is used pragmatically and purposefully to transform the state and society. Because Chile's neoliberal experiment was pursued aggressively, it generated tensions and conflicts about how economic beliefs shape government's technical decision-making practices that are more evident to the researcher than those observed in similar experiences of neoliberal restructuring around the world. Furthermore, Chile's experience sheds light on the role of scientific and technical expertise in government across authoritarian and democratic regimes, adding to the growing STS literature on science, the state, and democracy (e.g., Ezrahi 1990; Jasanoff 2004). We argue that neoliberalism is active, malleable, and productive. In Chile, it operated as a set of purposeful practices to change the role of experts and the state; yet these practices were pragmatic enough to adapt to the demands of Chile's post-Pinochet return to democracy.

We analyze two cases that illustrate crucial moments in neoliberalism's Chilean trajectory. The first case examines how energy policies were neoliberalized in the late 1970s. In 1979, the National Energy Commission, led by the Chicago Boys, canceled Chile's growing nuclear energy program. Using new techniques and definitions, the National Energy Commission imposed a "pure" economic evaluative framing that displaced traditional energy planning principles: the role of the state was minimized, political considerations were erased, and the predominant engineering culture changed to one that privileged the expertise of economists. The second case examines how neoliberalism was deployed thirty years later in the highly divisive controversy over whether to approve HidroAysén, a project to build five mega-hydroelectric dams in Patagonia. We argue that, as a political technology, neoliberalism was first performed through economists and later was scripted in the practice of science at large. The successful inscription of neoliberal logic into democratic institutions reflects its capacity to adapt to social and political challenges, including the arrival of democracy.

Neoliberalism: From a Thought Collective to a Political Technology

The definition of *neoliberalism* is as contested as the subject itself. Some observers define neoliberalism as an extension of neoclassical economics (Harvey 2005). Others argue it is better understood as the active promotion of market-based solutions to a broad range of issues (Lave, Mirowski, and Randalls 2010). Neoliberalism has also been defined as an ideological movement that disempowers the state (McCluskey 2003) or as a US-led "global empire" (Hardt and Negri 2000). As Mirowski (2009) observes, it is not unusual to see neoliberalism represented as monetarism, Thatcherism, Reaganism, or Howardism. Meanwhile, Latin American scholars have usually linked neoliberalism with the augmentation of poverty and inequality since the late 1970s, thus equating neoliberalism with the rise of privatization processes and consumerism (Moulian 2002).

Beyond these differences, there is a tendency to understand neoliberalism as an abstract, ideological, macrosociological force. For example, Plehwe asserts,

"Neoliberalism must be approached primarily as a historical 'thought collective' of increasingly global proportions" (2009, 2; see also Plehwe, Walpen, and Neunhöffer 2007). Plehwe uses the term *thought collective* to refer to "a set of shared values and principled beliefs" that allows "community members to effectively communicate across disciplines and audiences in the pursuit of hegemonic strategies" (2009, 35). Whether it is considered an antistate political thought, a free-market ideology, or a *homo economicus*–based imaginary, neoliberalism has thus been framed as a cultural entity—a ubiquitous, collective, and, according to critics, compulsory cosmology.

Scholars have, moreover, examined the rollout of neoliberalism in several Latin American countries, including Argentina (Grimson and Kessler 2005; Teubal 2004), Chile (Foxley 1983; Ffrench-Davis 1980; Gárate 2012), Brazil (Amann and Baer 2002), and Bolivia (Assies 2003; Spronk and Webber 2007). However, such studies have tended to focus on the political economy of national neoliberal transformations without accounting for the micro dynamics at work.

Applying tools and insights from STS, we argue for a micro, situated, and practice-based analysis of the processes of neoliberalization. Our analysis does this by comparing the controversies surrounding the development of two large energy projects, nuclear development in the 1970s and hydropower in the 2000s. The comparison shows how neoliberal assumptions were practically deployed to harness specific ideas about energy development, the environment, and the common good. We claim that neoliberalism needs to be understood as a set of embodied practices that produce knowledge, are adaptable to new political demands, and seek to transform the state. This is an argument not for a new definition of neoliberalism, but for a new analytical tool with which to study it, one that we call *political technology*.

Lakoff and Collier define political technology as "a systematic relation of knowledge and intervention applied to a problem of collective life" (2010, 244). The term *technology* forces the analysis of politics to transcend ideologies and emphasizes "the techniques and practices that give a concrete form to this new political rationality" (Foucault 2001, 410). For Foucault this new political rationality was the eighteenth-century liberal state, but it can also be extended to twentieth-century neoliberalism (Rose 2004).

The prefix "political" points to the productive and intervention-oriented nature of these techniques. As a political technology, neoliberalism is a program to transform the state, the common good, the role of politics, and the modes of decision making in a predefined direction (Mitchell 2005; Mirowski and Plehwe 2009; Goldman 2005). Though closely related to constitutive co-production (Jasanoff 2004), a political technology is prescriptive and interventionist.[1] "It defines and regulates targets of intervention according to a normative rationality" (Lakoff and Collier 2010, 262).[2] In the case of neoliberalism, this normative rationality contains a clear vision of the desired society—one constituted by the market itself. Neoliberalism thus produces its own world that realizes and confirms the parameters of a neoliberal epistemology (Callon 1998; Callon, Millo, and Muniesa 2007; MacKenzie 2008; Pinch and Swedberg 2008).

Understood as a political technology, that is, as applied knowledge related to government intervention, neoliberalism can also be viewed as a form of embodied scientific expertise. Neoliberal ideas cannot be abstracted from the fate and fortunes of economics as an academic discipline (Mitchell 2005). Neoliberalism promoted an increasingly mathematical and scientific approach to economics that both displaced Keynesianism and became the undisputed expert knowledge (Dezalay and Garth 2002; Van Horn and Mirowski 2009). The development of this approach was aided by think tanks (Mitchell 2009), universities (Fischer 2009), and several national and international institutions (Centeno and Silva 1998; Dezalay and Garth 2002; Woods 2006). Indeed, its capacity to produce and percolate through technocratic elites around the world is part of neoliberalism's success. Chile's Chicago Boys are exemplary. They displaced Keynesian economists, engineers, and lawyers to become government experts with considerable power. They were pro-science, internationalist, averse to politics, educated in the United States (particularly in the University of Chicago School of Economics), right-wing conservatives, and heavily present in government offices from the 1970s onward (Dezalay and Garth 2002; Markoff and Montecinos 1993; Silva 1991; Valdés 1995).

While Chile's neoliberal experts triumphed over discordant values, meanings, and practices, these triumphs also produced moments of controversy. As Sarewitz (2004) shows, when competing values and interests are at stake, actors mobilize science and the institutions at their disposal to persuade others to join their position (Sarewitz 2004). Controversies expose these practices and the political work that goes into separating "scientific facts" from "values" (Gieryn 1999) and "expert" from "lay" (Callon, Lascoumes, and Barthe 2009). This raises questions such as, How is the role of the state delineated? Who carries authoritative knowledge? Who are the incumbents in a controversy? And how are notions like "representation," "common good," or "risk" organized (Callon, Lascoumes, and Barthe 2009; Jasanoff 2004)?

We examine neoliberalism as a political technology by examining how economic practices were deployed to end the development of nuclear power in Chile in the 1970s, shortly after Pinochet came to power and implemented a neoliberal restructuring of the economy, in comparison with the practices involved in the promotion of hydroelectric power in the 2000s, when neoliberalism and democracy were fairly consolidated (tables 15.1 and 15.2). Our analysis draws from the central finding in STS scholarship that technologies are malleable and adaptable (Bijker and Pinch 1984; MacKenzie and Wajcman 1985) and shaped by users (Akrich 1992; Oudshoorn and Pinch 2003; Wilkie and Michael 2009). As a political technology, neoliberalism must therefore be understood as a set of techniques that transform users and contexts, just as it is transformed by them.

Table 15.1
Timeline of important events in Chilean energy politics

1973	Military government begins (under General Pinochet).
1975	The National Nuclear Energy Commission, together with state-owned companies Endesa and Chilectra, develops the Nucleoelectric Energy Plan. Endesa also proposes to develop HidroAysén.
1977	The National Energy Commission is created.
1979	The National Energy Commission cancels the nuclear power project.
1982–1983	Markets for electricity (DFL1) and water are created.
1987	Chilectra is privatized.
1989	Endesa is privatized. Pinochet loses a national referendum and will step down.
1990–1994	Transition to democracy: a number of legislative and constitutional changes are adopted, including the approval of the Environmental Impact Assessment process in 1994.
2005	Endesa, now a private, multinational company, again proposes HidroAysén.
2011	HidroAysén's Environmental Impact Assessment is approved. Legally, the project could now be built.
2012	HidroAysén is suspended: Colbún, a partner of Endesa in this venture, says "conditions don't exist to move ahead." This is the most recent information available at the time of writing.

Table 15.2
Summary of the three "purifications" under dictatorship and the three "transitions" to democracy

Dimensions	Case #1 (1970s) Nuclear *Processes of purification*	Case #2 (2000s) HidroAysén *Processes of transition*
1. Social goods	Energy redefined to its economic essentials. Political goals (national security, environment) stripped from energy planning.	Environment included in energy planning, but fragmented through bureaucratic controls.
2. The state	Distorts proper social and economic performance and is best left out of decision making. State-owned companies are corrupt.	Protects the environment and should be at the center of decision making, but it is weak and centralized.
3. Experts	Economists are the ultimate experts. Engineers are unaccountable and incompetent.	The expertise of natural scientists, environmental engineers, and social scientists is included, but delegitimized.

We use the comparison of these two energy projects to describe how neoliberal ideas have been purposefully and practically applied in the energy and environmental sectors, and how these ideas and associated practices have changed alongside Chile's democratization process. The first section of this paper examines how Pinochet's economic team, organized under the National Energy Commission, made electricity fit into a neoliberal world. We argue that the Energy Commission's economists were able to end the military's nuclear energy program through three technical "purifications": (1) social goods were redefined as economic concerns; (2) the state was eliminated from decision-making for distorting a neoclassical definition of optimal firm performance; and (3) engineers were marginalized, to the benefit of economists.[3] In the second section, we move forward thirty years to examine how the environmental impact assessment policies that were used to evaluate projects like HidroAysén challenge the neoliberal premises previously held by the Energy Commission. In response to democratic demands of the time, environmental impact assessment policies implemented three "transitions": (1) social goods were redefined as environmental concerns; (2) the state was put at the center of decision making; and (3) natural scientists became the required experts. Each of these transitions challenged the neoliberal world created in the 1970s, but the reach of the challenge was undermined by practices we detail that left important aspects of neoliberalism intact.

Neoliberalism in the Making: The Chilean Nuclear Energy Plan

Just two years after the military took power, the National Nuclear Energy Commission (Comisión Chilena de Energía Nuclear) submitted to the government a national plan to develop nuclear energy (Plan de Energía Nucleoeléctrica). Together with Endesa, the state-owned electricity generating company, and Chilectra, the state-owned electricity distribution company, the engineers at the Nuclear Commission proposed a detailed technical and economic project to introduce the first commercial nuclear plant by 1990. At the heart of the nuclear program, and its expectations for technological and industrial development, lay a vision of engineers as the standard-bearers of Chile's technological grandeur, embodied in prestigious state institutions like Endesa and Chilectra. As in Mexico or Argentina (see Hagood, this volume; Mateos and Suárez-Díaz, this volume), nuclear power in Chile was also seen as an agent of modernization.

By the mid-1970s, the nuclear energy plan was one of the most important technological programs in Chile and it seemed irreversible. Fascination with nuclear technology was strong across Latin America, and Chile was competing with Argentina—which was also positioning itself as a nuclear power—for geopolitical authority. For years the Chilean government had trained several dozen army engineers in nuclear operations and engineering, signed several assistance and research agreements, and created networks of institutional and technical allies, including with the United States and the United Kingdom. Above all, Endesa and Chilectra, national symbols of Chile's technological capabilities, had been enrolled in the nuclear project as part of an epic narrative of technological and industrial development. A former military engineer who worked at the Nuclear Commission summarized: "for twenty years, [the army] put a lot of dough into nuclear energy, and not only in steel and construction, but in production, training of people, preparation" (interview by MT, 2012). By 1979, the National Nuclear Energy Plan had evaluated different sites for nuclear power plants, resolved legal and financial considerations, and produced a computer model to design the optimal energy matrix. Nuclear power seemed not only necessary, but also inevitable. By the early 1980s, however, the nuclear power plan was dead.

The nuclear energy plan was submitted for evaluation to the National Energy Commission (Comisión Nacional de Energía), created in 1979 to produce an "efficient market" for electricity generation and distribution. Created following orders from Pinochet's new economic team, the new Energy Commission was meant to correct deviances in the energy sector, where inefficient and co-opted state companies (e.g., Endesa and Chilectra) had a monopoly. The economists worried that electricity operations were guided not by economic criteria but by political incentives, leading to a distorted price system (Rudnick, O'Ryan, and Bravo 2001). As a former Energy Commission officer explained, creating an electricity market required addressing "the issue of [market] prices, then establishing a framework in which the private sector could somehow enter [into the market], and then decentralizing it [to break the monopoly]" (interview by MT, 2012). The new Energy Commission, led by Chicago-trained economists, thus set out to reframe energy production through new economic expertise.

Nuclear energy was the first significant application of the Chicago Boys' reasoning to a large-scale technical project. The economists' primary objective was to *isolate* the economic evaluation of energy projects, and the nuclear energy plan in particular. The Energy Commission saw its mission as secluding decision making from contaminating factors, so the final decision would result from a narrow evaluation that reflected economic reasoning in its strictest, purest form. The economist-experts at the Energy Commission set out to purify the decision on nuclear power in the following three ways, and so came into conflict with the engineers at the Nuclear Energy Commission.

First Purification: Eliminating Political Elements from Economic Decisions
The Energy Commission's experts first set out to demarcate economic elements from noneconomic ones, and to eliminate the latter. Only after disentangling economics and politics would energy prices reflect "real" economic values. The economic experts did not believe that a political project like the National Nuclear Energy Plan could produce "real market prices."

In contrast, the engineers at the Nuclear Energy Commission regarded nuclear energy as a geopolitical issue, as reflected in the way they introduced the project in their 1975 report:

> Today's energy crisis is far from solved, and continues to produce profound changes in economic and energy structures around the world; it is altering the force equilibrium in large political-economic influence zones and between States, and is effectively generating new forms of power that could lead to unexpected international confrontations. Economic development and the survival of nations has [sic] a very important relation with their capacity to generate energy. Here lies the importance of this vital world problem. (CCHEN/Endesa 1975, 3)

For nuclear engineers, energy was above all a political issue of cold war politics, including a balance of power, strategy, and economic planning. They also identified some benefits: they praised nuclear energy for its environmental benefits and cleanliness compared to other sources (CCHEN/Endesa 1975, 9). Moreover, nuclear energy had to be assessed as part of Chile's modernization. It could trigger industrial and technological development to an extent never before seen in Chile:

> The benefits derived from [the nuclear] project will reach important aspects such as: formation and training of human resources required by a nuclear development of this magnitude, and the following: enhancement of scientific-technical level and infrastructure, preparation of the industry and improvement of quality standards. (CCHEN/Endesa 1975, 10)

In response to such political arguments, the Energy Commission recast nuclear energy in economic terms. The Commission did so by redefining "social benefit": "The basic objective of electricity planning is to determine the generation, transmission and distribution infrastructure that would serve demand, while securing the *maximum benefit for the community*" (CNE 1979, 22, emphasis added). Critically, they defined "maximum benefit for the community" in purely economic terms, arguing that if two energy programs seek maximum social benefit, their costs must be equal. Assuming inelastic demand, then demand should also be equal for both programs. These assumptions about costs and demand led them to equate maximizing social benefit with *minimizing total actual costs*: "social benefit" was purified to mean "the least expensive project."

The Energy Commission, furthermore, framed those elements most dear to nuclear engineers—modernization, industrialization, and technoscientific development—as technically unviable. For example, the Commission criticized the notion of *technology transfer*, an important element in the Nuclear Plan's argument. They stated that the nuclear engineers held exaggerated expectations for national technological improvement. Nuclear development was too complex for Chile whose "national participation in the construction of a first [nuclear] plant will be, in the best of cases, limited to the execution of public works, part of the assembly, and the monitoring of the project." In addition, the Commission felt technology transfer would impose severe opportunity costs on the state, because "a program of several plants would have to be planned ... generating important overcosts." In that case, they speculated, it is "worthwhile asking whether there aren't other activities in the country that justify this allocation of resources" (CNE 1979, 69). Too few plants would not catalyze development, making technology transfer an irrational strategy in the Energy Commission's view.

Second Purification: Eliminating the State from Energy Management
With political elements isolated, the economists moved to eradicate an even more ubiquitous entity from energy production: the state. Following a long tradition of electric engineering in Chile (Ibáñez 1983), the Nuclear Energy Plan put the state at the center of its project. The Nuclear Plan states: "a country's energy is a national asset and source of power. It constitutes one of the fundamental infrastructures for National Security. Hence its use, conservation, and development are a fundamental State concern" (CCHEN/Endesa 1975, 3). Endesa personified this "state concern" and national ethos. As a public servant from the National Nuclear Energy Commission recalled: "we thought it was appropriate that state enterprises were the ones in charge of [nuclear power], for example Endesa, that had prestige as [the national leaders in] energy production" (interview by MT, 2012). The Nuclear Energy Plan highlights the importance of Endesa as the Nuclear Energy Commission's main technical partner in several ways. The Plan cites an Endesa report to justify the need for a nuclear power plant to be operative by 1986 (CCHEN/Endesa 1975, 5). It emphasizes that Endesa's mathematical models lie behind all the feasibility studies (CCHEN/Endesa 1975, 7). And the report trusts Endesa with the responsibility of the nuclear power plants' operations (CCHEN/Endesa 1975, 11). In brief, for the National Nuclear Energy Plan, the participation of the national company was necessary to make nuclear power a state project, and therefore relevant, feasible, and irrevocable.

The Energy Commission, formed of Chicago-trained economists, had a radically different perspective. Far from seeing the state as an enabling element, they viewed it as a source of distortion that had to be eliminated from evaluating electricity projects. From the economists' perspective, state-owned firms were corrupt, particularly Endesa. The economists implemented new methods for evaluating energy projects, rejecting previous methods "because that [evaluative] function had been run by [state] companies too autonomously" (interview by MT, 2012). The Energy Commission assumed that agents in charge of managing a firm in which they have no investment will only maximize their own personal gains, not "social" ones. An important former Energy Commission officer recalled telling a room full of Endesa engineers, "Let's be clear, you have never felt that this company [Endesa] belongs to the state, nor to Chile. You feel this firm belongs to *you*, with the difference being that you have *never* put a dime in it" (interview by MT, 2012). The economists saw Endesa as the epitome of these deviant incentives and argued this point to justify Endesa's privatization. Another Energy Commission official went further, saying these deviances were so evident that even Pinochet would have understood the conflict and would have said: "'I [Pinochet] think that [the privatization of Endesa] should be done, you know why? ... Not for economic reasons, because I don't understand economics, but I have realized that these things create corruption and power'" (interview by MT, 2012).

Third Purification: Eliminating Engineering Expertise

The first two purifications, rendering the social economic and excluding the state, were not enough. The economists also felt it was necessary to destabilize the Nuclear Energy Commission's entire collective cosmology. This meant attacking its engineering culture.

Engineers were integral to the Chilean state's modernization goals. They were called on to help construct a "progressive and modern nation ... which integrates material development with social well-being" (Ibáñez 1983, 58). The best engineers worked at Endesa and Chilectra, the icons of "Chilean technology" and proud bearers of the state's national development strategy. An officer from the National Nuclear Energy Commission reminisced that the nuclear program was sustained by this unique mix of technical prowess and nationalism, characteristic of Endesa's engineering culture:

I also think that engineering, not so much from the point of view of knowledge but from that of the attitude and the concept of engineering that we had in the 1970s, as best expressed by Endesa, would have transformed the nuclear [program] into a veritable school. (Interview by MT, 2012)

Chilean engineers, like the nineteenth-century French engineers described by Hecht, "did not so much derive legitimacy from their technological achievements as the other way around. That is, their position within the state conferred legitimacy on their technologies" (Hecht 2009, 26). The mix of technical expertise with a "progressive and nationalizing mission" (Ibáñez 1983, 58) made engineers—trained in state universities and as public servants in state-owned companies—the only certified authority in electricity matters. The engineers greeted the economists' arrival on the electricity scene with extreme skepticism. A military officer occupying a high-rank position within the Nuclear Energy Commission recalled their reaction when the economists' Energy Commission came to evaluate nuclear energy:

We had an unpleasant time. ... [The Energy Commission's director] was mad with me for a while, because he was a systems analyst and looking forward to applying his [economic] rationale, but he himself told me when we met, "I don't know a thing about energy, nothing," and I responded, "How can it be possible to name someone that doesn't know a thing [about energy] as executive director [of the Energy Commission]?" (Interview by MT, 2012)

The new Energy Commission made every effort to delegitimize engineering culture. Following their assumption that economic agents cannot be rational if they have not invested in the firm they manage, the Energy Commission stigmatized engineers as individuals likely to fall into technological fantasies. Free from economic rationality, they thought, engineering creates technological monstrosities. According to a founder of the Energy Commission:

> While [nuclear energy] was run by Endesa, it was basically a rather technical project and something typical of technicians and engineers. ... They have always liked to build new and different things, especially if they are not framed within a rational economic system. ... The temptation of technicians is to build something different to see what happens. Whether or not it was worthwhile, or if somebody else paid for it, that's another story. (Interview by MT, 2012)

The economists embedded their delegitimating efforts in Chile's long-standing class divisions and the different status enjoyed until then by university graduates and graduates of the Military Polytechnic Academy. As graduates of the Military Polytechnic, engineers were now cast as incompetent. "I think that the main explanation for the nuclear program's disorder ... was where it was anchored, in the Chilean Commission of Nuclear Energy which, to be honest, was managed by extremely incompetent military polytechnicians," explained a former officer from the Energy Commission. The economists saw military polytechnic engineers as ill prepared for duties outside of combat-related activities because they studied engineering for fewer years than their civil peers from the Universidad Católica or the Universidad de Chile. A former Energy Commission officer explained:

> Actually, if you think about it, military polytechnic schools come from the French, and [military polytechnic engineers] are really sappers, guys that what they really should do is study how to rapidly assemble a bridge, how to pass a river. But in the [1970s] they generated a kind of school, a military polytechnic school, which in four years supposedly made you an engineer.[4] But in those four years these guys also have to follow the military career, so it is far less than four years. (Interview by MT, 2012)

Spain, moreover, was the international center of reference for military engineers. While Chilean nuclear engineers had for decades attended the doctoral program in nuclear engineering at Madrid's Polytechnic University, the economists regarded US institutions as the only valid source of knowledge. Non-US doctorates were considered inferior or, in the words of a former Energy Commission employee with a US PhD from a prestigious university, "indecent."[5] After the cancellation of the nuclear program, young engineers were sent to the United States to get their master's degrees. He explained: "once you had a couple of guys that understood the [nuclear] issue, the discussion was over. If they had sent an intelligent officer to study in a decent university, he would have realized that [nuclear energy] was not just a question of following the Spaniards" (interview by MT, 2012). He went on to attribute the same argument to Pinochet, himself a military man but by then presumably fully imbued with the new economic rationale. Pinochet stopped the nuclear power plan, asking: "'wait a minute, who among you has followed a course of study, who has a doctorate, who knows what you're talking about?'" The nuclear engineers might have responded, "'well, we had courses in Spain.' 'Don't tell me that, it is not the same. When you can really show me that you have degrees, we'll talk about what can and cannot be done'" (interview by MT, 2012).

The economists at the Energy Commission put into practice a political program to neoliberalize energy. Far from an abstract ideology, neoliberalism operated like a political technology: economists in government used neoliberal economic techniques to change evaluating practices and notions of social benefit, development, and good government. Optimization and elasticity theory were mobilized to eliminate political factors from energy planning. The assumption of the self-interested, rational agent was imposed to recast state-owned firms as inefficient and corrupt. And educational excellence was redefined to marginalize engineers and military polytechnicians by appealing to new US-centered and elitist sentiments. Hence, neoliberal principles—small states, free markets, instrumental rationality (table 15.2)—were not abstractions but technical practices of measuring, evaluating, and planning.

Neoliberalism Remakes Its World in (Chilean) Democracy: Environmental Impact Assessments

In 1979 the Energy Commission canceled the nuclear energy plan as a result of the economists' efforts. This was the first step in a series of neoliberal laws and policies that reshuffled Chile's energy and environmental sectors (table 15.1). In 1982 a new electricity law was passed (DFL 1), and water rights were created to set up new water and electricity markets. In 1987 and 1989 Chilectra and Endesa, respectively, were privatized. Chile's electricity and water laws are among the most market-enabling energy frameworks in the world (Bauer 1998; Budds 2004; Prieto and Bauer 2012) and are also a driving force behind the carbonization of Chile's energy sources and other environmental damages (Mundaca 2013).

In 1990 Chile transitioned to democracy and, to respond to local and global demands for greater environmental protections, adopted environmental impact assessments (EIAs). EIAs use expert and public opinion to evaluate and improve the environmental impacts of projects such as electrical or industrial plants before the government approves them for construction (Owens and Cowell, 2002). Today EIAs are in worldwide use (Pope et al. 2010). Although they began in the United States in 1969 as an environmental victory, they are increasingly seen as a neoliberal tool (Tecklin, Bauer, and Prieto 2011). Treating neoliberalism as a political technology, however, draws attention to EIAs as a site of transition and adaptation. This section compares economic practices and assumptions in energy politics in the 1970s and 2000s, and asks how neoliberal principles transitioned from dictatorship to democracy and gained legitimacy in the process. It also traces how EIAs transitioned from an environmentalist to a neoliberal evaluative tool.

EIAs matter because they are a "single window," administratively and politically, for the Chilean government to approve large investment projects—over 800 since 1993.[6] Increasingly, they are also a site of controversy. By 2010, the frequency of conflicts, together with pressure from the Organization for Economic Cooperation and Development (OECD), led legislators to create an autonomous EIA Agency that administers EIAs. Legislators drew on their experiences of EIAs gone wrong to justify these reforms. This included their previous experiences with energy projects at Los Robles, San Pedro, Castilla, Guacolda, Campiche, and La Higuera; threatened rivers like the Copiapó, Caren, and Choapa; and mines and industry at Los Pelambres, Tocopilla, Chañaral, Chuquicamata, Ventanas, Puchuncaví, Antofagasta, La Calera, Pascua Lama, and Valdivia.[7] Legislators also sought to isolate the technical aspects from the political aspects of the EIA evaluations, and saw the assessments as technocratic tools that used scientific knowledge as the "solution" to the "problem" of power (Cashmore and Richardson 2013).

HidroAysén was the first big test case for the reformed EIA Agency. The HidroAysén project's EIA, which was the largest EIA project to date, was approved by the government in May 2011. HidroAysén consists of five mega-dams on the Baker and Pascua rivers in Aysén, a region in southern Chile, as well as a 2,000 kilometer transmission line to central Chile. Endesa, then a state enterprise, originally conceived of HidroAysén in the 1970s, but the project was refloated in 2006 to respond to energy shortages following Argentina's interruption of gas exports to Chile. Endesa, now a private and foreign company, owns the project together with minority partner Colbún, a Chilean company.[8] HidroAysén divided the country and spurred massive social protests in May 2011 and February 2012.[9] Supporters argued that Chile needs the energy to grow, while opponents denounced the extension of a neoliberal model for inequitable and unsustainable growth that would also damage one of the world's last unindustrialized regions. To opponents, privatized Endesa represents corruption and power, just as the state-owned Endesa did earlier to the Energy Commission's economists, but for opposite reasons.

In contrast to the three "purifications" Chilean economists used to change evaluative practices in the 1970s, in the 2000s policymakers introduced the EIA with three "transitions" that implemented the new environmentalist and democratic policy while preserving neoliberal practices and assumptions. Such transitions of practice help explain the durability of neoliberal principles and show how policies, like EIAs, can mutate from environmentalist victories to neoliberal tools. Specifically, the EIA posed a challenge to the Energy Commission's brand of neoliberalism by

(1) expanding concerns beyond economics; (2) reintroducing the state into decision making; and (3) requiring the participation of environmental science experts. Each of these elements was introduced in ways that enabled neoliberal continuity so that the EIA, as a decision-making tool, divided (and continues to divide) policy makers and civil society actors. On one side are those who believe economic criteria alone should guide state policy. On the other are those who wish to expand the state's responsibilities to consider broader questions about sustainable and equitable development. To some extent these disputes are over the technical expertise that makes an EIA credible. However, these debates over what constitutes a good EIA are also about negotiating the neoliberal world built by economist-experts during Pinochet's regime and making it compatible with the new democracy.

First Transition: Bringing the Environment In

Through the EIA, the state evaluates new projects to improve their environmental performance. The Chilean EIA requires projects that may have significant environmental impacts, as defined by law, to submit a study of existing environmental conditions, identify which impacts the project expects to generate, and detail measures taken to minimize, avoid, or compensate for those impacts. Thus, the EIA introduces environmental criteria, including compliance with applicable environmental quality standards and emissions regulations, into the state's decision to authorize a project for construction. The EIA system puts noneconomic considerations—the environment—at the center of what Chicago Boy economists would consider purely economic investment decisions.

Environmental concerns, however, are also obscured as government evaluators face administrative obstacles to an adequate evaluation. First, the EIA Agency can exclude observations made by government agencies such as public works or forestry by controlling the ambiguous distinction between sectoral and environmental issues. Relying on a narrow reading of legal permits, the EIA Agency can argue that it alone has jurisdiction over environmental concerns, while government agencies have jurisdiction only over their sectoral permits. As a regional EIA Agency director explained:

> The observations made by government agencies are not binding for us [at the EIA Agency]. … The observations must be justified and within the agency's areas of competence … and expressed as a question, otherwise we have the obligation to exclude it from the next report. … People often do not understand this. We evaluate the environmental fraction of land use change. (Interview by JB, March 2011)

Thus, while government agencies contribute information to the evaluation process, they must express this information strictly in terms the EIA Agency considers "environmentally relevant." Otherwise, the EIA Agency can exclude the information, saying it is "not binding on them." In addition, the EIA Agency has discretion to define what counts as "the environmental fraction" of a permit. In contrast, evaluators at government agencies, though very familiar with the applicable regulations, were not able to clearly define "sectoral" and "environmental" issues without resorting to examples: permits emitted by agencies are "sectoral" because they represent a sector-specific concern, such as forest management. As the above quotation shows, there is a power struggle between the EIA Agency and government agencies with sectoral expertise in forests, soils, geology, etc., that shapes the distribution of authority in legal instruments like permits. Though the EIA Agency ultimately evaluates EIAs, the sectoral agencies have the legal power to enforce permits.[10] Evaluators found the process exhausting: one said he only wanted "to do his job well, technically, ignoring the politics." This meant giving up on the thousands of observations his agency originally made about HidroAysén and focusing on the two permits the agency is responsible for.

Second, the EIA Agency prioritizes adding clarity to the administrative aspects of the EIA process but not to environmental knowledge itself. Again, the EIA Agency director said:

What we aspire to do as the EIA Agency is to make environmental evaluation more transparent. ... We think that with clear rules, concise and precise instruments, we can improve the evaluation of projects, so they can be evaluated to a higher standard and even faster, because many investors depend on this to be able to execute their project. (Interview by JB, March 2011)

For evaluators, in contrast, shortening the time for evaluation is an attack on quality. They had thirty days to evaluate HidroAysén's EIA, which occupied "a cubic meter of information."[11] Subsequent revisions were even quicker: fifteen days, although HidroAysén added new data and maps. On average, the EIA Agency evaluates projects in just eight months, and only nineteen projects have been in evaluation for more than three (but less than four) years since the agency began.

State agencies' authority to protect a strong version of the environment was smothered under artificial distinctions between sectoral and environmental fractions and a strict adherence to rules and regulations. As a result, the introduction of environmental concerns into state decision making eroded the dominance of economic criteria, but only in ways that fragmented and rendered the environment technical through administrative practices, deadlines, and definitions. Interestingly, the technification of decision making has been seen as one of the main features of the (neo)liberal state (Ezrahi 1990). Thus, by introducing the environment in a fragmented and technified form, the EIAs did not limit the reach of neoliberal principles but reapplied them in a different way.

Table 15.3
EIAs by result, 1993–2011

	Number	Percentage
Approved	594	68
Rejected	47	5
Withdrawn	133	15
Not admitted	52	6
In evaluation	54	6
Total	880	100

Source: Database of EIA projects, July 2011. The EIA was voluntary until 1997. Only 712 EIAs of the total had both entry and exit dates.

Second Transition: Bringing the State Back In, but Keeping It centralized

Thirty-six government agencies evaluated HidroAysén, mobilizing staff from Santiago to Aysén. Citizens, communities, and NGOs offered 11,000 observations through public meetings and in writing. After all this material was evaluated, HidroAysén's EIA went up for a vote in a committee of regional representatives of the central government. As with all EIAs, the committee then had the choice to vote to approve, approve with conditions, or reject the project (table 15.3).

In general terms, then, the EIA puts decision making under state control, and particularly under *regional* state control, restoring the role of regional and local politics in decision making—a role that was neglected by the Energy Commission's economists. Indeed, Pinochet's economist-experts saw the evaluation of energy projects as a primarily technical and state-led exercise, which was constrained by Pinochet's harsh political repression, and so the Energy Commission in the 1970s and 1980s did not consider any participation of local actors whatsoever.

But local political voices are incorporated in ways that are also undermined. In 2010, Congress made the regional committees that vote on EIAs more "technical" by eliminating local politicians. Prior to this reform, EIAs were approved by a committee formed by three groups of people: (1) SEREMIs (Regional Ministerial Secretaries), who are individuals named by ministers of the executive government to represent the executive government in each region; (2) regional governors named by the president of the republic to lead the region; and (3) indirectly elected regional councilors. After the reform, regional councilors who lived in the region and had local ties were excluded to make the committee more "technical."

Municipalities are particularly marginal to the EIA process. In the case of HidroAysén, small, resource-strapped municipalities found it difficult to make technical evaluations for the EIA. Although municipalities are not required to make observations for the EIA, this is the only formal mechanism available to them to improve the projects that come to their area. Forced to rely on outside help, these municipalities face multiple pitfalls that result from Chile's market for expertise. For example, scientists produced for the Tortel Municipality (population 507) a model of the dams' impact on the Tortel coastal ecosystem. The municipality was initially glad to receive such a boost to the observations they had provided to HidroAysén's EIA. But the model later caused the municipality to face a crisis of legitimacy. Since the scientists who made the model were "pro-environment," the municipality itself was soon similarly regarded, therefore damaging their political communication with the central and regional government. With few possible external collaborators, and even fewer internal resources, municipalities in the region preferred to focus on the issues they thought they knew about, such as the local price of firewood or the needs of the families that would need to relocate.

The EIA's relation to local participation is thus ambivalent. Challenging the dismissal of the state, as had been promoted by economists in the 1970s, the EIA reintroduced local and regional state voices into public decision making. The challenge, however, was half-hearted. Technical and central state voices—as opposed to political and local ones—were introduced into decision making. To look at it one way, Hayek's vision of a small group of experts centrally managing the fate of the state (Centeno and Silva 1998) has not been challenged, but rather reworked to accommodate some—but arguably not enough—of the local actors in the decision-making process.[12]

Third Transition: Broadening Experts' Access, but Undermining Their Credibility
New experts from the natural sciences, environmental engineering, archaeology, and other disciplines are important participants in the EIA. HidroAysén hired eight of Chile's most prestigious universities to elaborate the baselines for the project's EIA. Baselines are a description of the state of nature in an area before construction. HidroAysén afforded unprecedented opportunities: a vertebrate biologist said they had funding for 30 scientists to do fieldwork in unexplored areas around the Baker and Pascua rivers, and to study lichens, forests, and coastal ecosystems like never before. For HidroAysén, the investment in university science was worth the added transparency. While the Energy Commission's economists had made every effort to delegitimize experts outside of economics, the EIA's legitimacy is based on a diverse knowledge platform.

Unfortunately, the new experts participate in conditions that undermine their credibility. A scientist who collects information on existing conditions for EIAs, but does not evaluate impacts for EIAs, explained:

The thing with the EIA system is that the consulting firms are in a vicious circle because the company pays you to do a study to evaluate the company's project's environmental impacts. The company is judge and jury in its own cause. [We are reluctant to be involved in EIAs] because of an ethical issue. We don't want to be involved with drying out a river where there are otters, and you tell the company and they say, "This can't go in the report because they won't let me build my thing." I know they go and remove the otters and hand in their report. ... The company-consultant relationship is toxic. (Interview by JB, November 2010)

Many scientists share feelings like these, including scientists who worked on HidroAysén's EIA. For many, the EIA breeds dishonesty and bad data. Researchers are under pressure to produce results to get the project approved: "They don't hire you to raise questions about the project." Interesting science is not produced: baselines are a "random pile of data" and produce "data, not information" (interviews by JB, March and June 2011). Scientists face a type of Faustian bargain: work on baselines for funding and access to hard-to-obtain data, in exchange for losing control over the results. Scientists lose control because the company owns the data; in fact, EIA scientists must receive permission from the company to publish their results. The baseline reports, moreover, are hard to obtain and the indicators are only partially available.[13] Control is also lost because scientists do not interpret their own data, which lends itself to what scientists call "cut and paste." An example is the disappearance of the otter in the quote above; consulting companies edit data to get the EIA approved.

Just as engineering experience was excluded from assessments of nuclear energy in the 1970s, so too are government-funded scientific groups today. The Center for Research on Patagonian Ecosystems (CIEP) was founded in 2005 to study Aysén's rich and varied ecosystems. CIEP was the first permanent scientific presence in Aysén, and it is funded by central and regional governments as well as by the private sector.[14] It seemed like a perfect collaboration: Endesa would have local ecology and biology experts, and the scientists would strengthen their local expertise. Endesa initially hired CIEP to do HidroAysén's baselines, but the center pulled out when it was accused of having a conflict of interest. According to a senior CIEP scientist:

The regional governor chairs the CIEP. Therefore, strictly speaking, the CIEP could not participate in HidroAysén's EIA because the governor also chairs the committee that votes on the project. So it couldn't be seen as a technical, scientific group. (Interview by JB, March 2011)

CIEP's status as a research center of local experts was too tenuous, and HidroAysén, far from bringing prestige, risked undermining the nascent scientific group. The social costs of participating in the EIA were too high for CIEP, despite the financial and access incentives. The only biologists and ecologists with a permanent presence in Aysén are thus excluded from participating in local fieldwork. In summary, Chilean scientists who participate in the EIA must take steps to protect their scientific legitimacy. This includes narrowing the scope of "science" to baselines. As an analyst at Endesa said, "to prepare baselines is a more pure scientific study, like taking a photo. ... To evaluate environmental impacts is different" (interview by JB, May 2011). Because the EIA is science on sale, it is seen as leading to untrustworthy data and does not bring prestige, as the CIEP experience illustrates.

Chile's return to democracy challenged the previous insistence of economists that the government should rely on US-trained economists as public experts. The introduction of EIAs, for example, required the participation of natural and environmental scientists. EIAs did not, however, challenge the use of markets to organize and distribute expertise, thus preserving a neoliberal logic that undermined the credibility of the scientists. As a test case of the reformed EIA, HidroAysén shows how, even after ambitious reforms, decision making and evaluative practices reflected neoliberal models.

Conclusion

Comparing how neoliberal economic principles shaped evaluative practices in the 1970s and 2000s helps explain the durability of neoliberal practices from dictatorship to democracy. This durability has to do with the flexibility of neoliberalism to respond to new challenges, such as demands for greater participation from the state and environmental experts in energy decisions. We argue that, like regulatory experiments whose results are always being tested and adapted (Lezaun and Millo 2006), neoliberalism operates in Chilean energy and environmental politics not as a totalizing force but as a political technology. It acts as an ensemble of thoughts, techniques, and knowledges that are applied to predetermined political goals by shaping the methods for knowing the world and the communities of experts responsible for producing that knowledge. The success of neoliberalism lies in the fact that, like many other technologies, it "isn't too rigorously bounded, ... doesn't impose itself but tries to serve, ... is adaptable, flexible and responsive" (de Laet and Mol 2000, 225). As a set of practices, neoliberalism is active, malleable, and productive.

Chile's current electricity market reflects the active efforts to create a neoliberal state described in this chapter. Energy production could have taken multiple routes in 1975, but it seemed likely to include a nuclear option. In just a few years, however, economists transformed how the state made energy decisions, not by appealing to the authorities' values, thoughts, or beliefs, but through new procedures to evaluate projects. Economists used optimization and elasticity theory to recalibrate "social goods" as "costs"; they promoted a *homo economicus* to recast state organizations as corrupt and distorting; and they marginalized engineers as incompetent and fantastical by imposing a new epistemic culture. To these economists an evaluative tool like the environmental impact assessment was an anathema. The EIA threatened to reverse their 1970s practices by introducing the environment, new local and regional state voices, and natural science experts into decision making. Yet neoliberal practices—such as privileging administrative efficiency, a small state, and the market—directed the potentially disruptive EIA toward the status quo. As a result, the EIA itself is a site of controversy between those who advocate for a more neoliberal state—guided by economic criteria and market mechanisms—and those who call for a state that can promote certain social goods, like sustainable development.

The tensions, contrasts, and ambiguities between these two moments in the unfolding of neoliberalism in Chile between the 1970s and 2000s are best illustrated by the changing nature of Endesa. By mediating between the state and the electricity market, Endesa has always played a crucial role in justifying neoliberalism. But its position has changed vis-à-vis the mutations of neoliberalism in the country. By construing Endesa as an example of state corruption, Energy Commission technocrats were able to justify the unfolding of neoliberal measures into the energy sector. Twenty years later, the environmental risks produced by large corporations like Endesa justified the need for EIAs. But these EIAs did not tame neoliberalism, they just reformatted it in a new, democratic shape. Our point is not to evaluate Endesa, but to stress to what extent it epitomizes the adaptable nature of neoliberalism and its capacity to promote, hamper, and modulate the world differently, according to the requirements of changing contexts.

To examine neoliberalism as a political technology provides rich insights into how the state is transformed, how economic epistemologies shape the ways scientific and technical experts participate in government decision making, and how neoliberalism is practiced through common evaluative tools. Three decades ago many Latin American countries adopted neoliberal policies, often during military dictatorships. Since the 1980s many countries in the region, including Chile, Argentina, Brazil, Uruguay, Peru, Paraguay, and others, democratized and, more recently, took a much commented "turn to the left." Yet critiques and complaints against neoliberalism persist. For example, scholars of the region regularly use neoliberalism to explain the failures of policies like the EIA in Chile, a policy that would have been labeled as environmentalist a few decades ago. This occurs because neoliberal ideas are inscribed in technologies, practices, forms of expertise, and evaluation tools that are both flexible and active—they intervene and transform the world. To analyze neoliberalism as a political technology helps us identify these knowledge-practices that give neoliberalism a tangible meaning, providing new insights into current social conflicts and political debates in the region.

Notes

1. Co-production is an idiom that seeks to "explain why the products of science and technology acquire such deep holds on people's normative instincts and cognitive faculties" (Jasanoff 2004, 38). Neoliberalism in Chile followed the same strategies that Jasanoff identifies—making identities, institutions, discourses, and representations—but what needs to be explained, in the case of neoliberalism, is the relative stability of certain normative instincts and cognitive faculties. This chapter argues that neoliberalism's "technological flexibility" is part of the answer.

2. Other examples of political technologies include, for example, preparedness for natural disasters, the precautionary principle, or "family values." See Lakoff and Collier (2010) and, in the same volume, Diprose (2010).

3. Latour (1993) uses the term *purification* to describe the separation of nature and society. This term adequately describes the first three movements of neoliberalization in the 1970s. We prefer the term *transition* to describe the second set of movements, under the transition to democracy, because these represent a change from the previous state and do not fully restore the earlier situation. By using different terms for the 1970s and 2000s, we highlight the process of change in neoliberal practices and the importance of the transition from dictatorship to democracy (itself operative at different macro and micro scales).

4. The engineering curriculum in a traditional university (like Universidad Católica and Universidad de Chile) takes six years.

5. In Chilean Spanish, *decent* and *indecent* are commonly used as antonyms. *Indecent* is commonly used to refer to something cheap or second-rate.

6. "Single window" refers to a common policy of setting up one-stop shops for all permitting and administrative work. These were introduced in many Latin American countries to increase efficiency and reduce bureaucracy.

7. See Legislative History 20.417, available online from Chile's Congressional Library. Legislative Histories are transcripts of legislative hearings for each law.

8. Endesa was privatized in 1989. Today the majority ownership is held by Enel, an Italian company.

9. The massive protests in February 2012 concerned several issues related to natural resource management. For an overview, see McAllister (2013).

10. This has been changing since the new Superintendencia de Medio Ambiente (Environment Enforcement Agency) came into operation during 2013.

11. The EIA is also submitted electronically. HidroAysén submitted PDFs that are not searchable and the resolution made the maps unviewable (this was later corrected).

12. Friedrich Hayek (1899–1992), Austrian economist and philosopher, is often identified as the key figure of neoliberal thought.

13. No one reported being denied permission to publish by HidroAysén. The unedited baseline reports are available only in a few offices in Aysén; the edited ones can be downloaded from the EIA Agency. HidroAysén has made weekly, but not daily, hydrology data available.

14. CIEP was created by a government program to kick-start science with funding for groups (as opposed to individuals) outside of Santiago.

References

Akrich, Madeleine. 1992. The description of technical objects. In *Shaping Technology / Building Society: Studies in Sociotechnical Change*, ed. John Law. Cambridge, MA: MIT Press.

Amann, Edmund, and Werner Baer. 2002. Neoliberalism and its consequences in Brazil. *Journal of Latin American Studies* 34:945–959.

Assies, Willem. 2003. David versus Goliath in Cochabamba: Water rights, neoliberalism, and the revival of social protest in Bolivia. *Latin American Perspectives* 30 (3):14–36.

Bauer, Carl. 1998. *Against the Current: Privatization, Water Markets, and the State in Chile*. Boston: Kluwer.

Bijker, Wieber, and Trevor Pinch. 1984. The social construction of facts and artifacts: Or, how the sociology of science and the sociology of technology might benefit each other. *Social Studies of Science* 14 (3):399–441.

Budds, Jessica. 2004. Power, nature and neoliberalism: The political ecology of water in Chile. *Singapore Journal of Tropical Geography* 25 (3):322–342.

Callon, Michel, ed. 1998. *The Laws of the Markets*. London: Blackwell.

Callon, Michel, Pierre Lascoumes, and Yannick Barthe. 2009. *Acting in an Uncertain World: An Essay on Technical Democracy*. Cambridge, MA: MIT Press.

Callon, Michel, Yuval Millo, and Fabian Muniesa, eds. 2007. *Market Devices*. London: Wiley-Blackwell.

Cashmore, Matthew, and Tim Richardson. 2013. Power and environmental assessment: Introduction to the special issue. *Environmental Impact Assessment Review* 39 (C):1–4.

CCHEN (Comisión Chilena de la Energía Nuclear)/Endesa. 1975. *Plan de Energía Nucleoeléctrica*. Santiago: CCHEN.

Centeno, Miguel, and Patricio Silva, eds. 1998. *The Politics of Expertise in Latin America*. London: Macmillan.

CNE (Comisión Nacional de Energía). 1979. *Factibilidad económica de una central nuclear en el sistema interconectado*. Santiago: CNE.

de Laet, Marianne, and Annemarie Mol. 2000. The Zimbabwe bush pump: Mechanics of a fluid technology. *Social Studies of Science* 30 (2):225–263.

Dezalay, Yves, and Bryant G. Garth. 2002. *The Internationalization of Palace Wars: Lawyers, Economists, and the Contest to Transform Latin American States*. Chicago: University of Chicago Press.

Diprose, Rosalyn. 2010. The political technology of RU486: Time for the body and democracy. In *Political Matter: Technoscience, Democracy, and Public Life*, ed. Bruce Braun and Sarah J. Whatmore, 211–242. Minneapolis: University of Minnesota Press.

Ezrahi, Yaron. 1990. *The Descent of Icarus: Science and the Transformation of Contemporary Democracy*. Cambridge, MA: Harvard University Press.

Ffrench-Davis, Ricardo. 1980. Liberalización de importaciones: La experiencia chilena en 1973–79. *Coleccion Estudios CIEPLAN* 4:39–78.

Fischer, Karin. 2009. The influence of neoliberals in Chile before, during, and after Pinochet. In Mirowski and Plehwe 2009, 305–346.

Foucault, Michel. 2001. The political technology of individuals. In *The Essential Foucault*, vol. 3, ed. James Faubion, 403–417. New York: New Press.

Fourcade-Gourinchas, Marion, and Sarah Babb. 2002. The rebirth of the liberal creed: Paths to neoliberalism in four countries. *American Journal of Sociology* 108 (3):545–546.

Foxley, Alejandro. 1983. *Latin American Experiments in Neoconservative Economics*. Berkeley: University of California Press.

Gárate, Manuel. 2012. *La revolución capitalista de Chile (1973–2003)*. Santiago: Ediciones Universidad Alberto Hurtado.

Gieryn, Thomas. 1999. *Cultural Boundaries of Science: Credibility on the Line*. Chicago: University of Chicago Press.

Goldman, Michael. 2005. *Imperial Nature*. New Haven: Yale University Press.

Grimson, Alejandro, and Gabriel Kessler. 2005. *Argentina and the Southern Cone: Neoliberalism and National Imaginations*. New York: Routledge.

Hardt, Michael, and Antonio Negri. 2000. *Empire*. Cambridge, MA: Harvard University Press.

Harvey, David. 2005. *A Brief History of Neoliberalism*. Oxford: Oxford University Press.

Hecht, Gabrielle. 2009. *The Radiance of France: Nuclear Power and National Identity after World War II*. Cambridge, MA: MIT Press.

Ibáñez, Adolfo. 1983. Los ingenieros, el estado y la política en Chile. Del Ministerio de Fomento a la Corporación de Fomento, 1927–1939. *Historia* (Wiesbaden, Germany) 18:45–102.

Jasanoff, Sheila, ed. 2004. *States of Knowledge: The Co-Production of Science and Social Order*. London: Routledge.

Klein, Naomi. 2008. *The Shock Doctrine: The Rise of Disaster Capitalism*. New York: Picador.

Lakoff, A., and S. J. Collier. 2010. Infrastructure and event: The political technology of preparedness. In *Political Matter: Technoscience, Democracy, and Public Life*, ed. Bruce Braun and Sarah J. Whatmore, 243–266. Minneapolis: University of Minnesota Press.

Latour, Bruno. 1993. *We Have Never Been Modern*. Cambridge, MA: Harvard University Press.

Lave, Rebecca, Philip Mirowski, and Samuel Randalls. 2010. Introduction: STS and neoliberal science. *Social Studies of Science* 40 (5) (September 28): 659–675.

Lezaun, Javier, and Yuval Millo. 2006. Regulatory experiments: Genetically modified crops and financial derivatives on trial. *Science and Public Policy* 33 (3):179–190.

Liverman, Diana M., and Silvina Vilas. 2006. Neoliberalism and the environment in Latin America. *Annual Review of Environment and Resources* 31:327–363.

MacKenzie, Donald. 2008. *Do Economists Make Markets? On the Performativity of Economics*. Princeton: Princeton University Press.

MacKenzie, Donald, and Judy Wajcman, eds. 1985. *The Social Shaping of Technology*. Milton Keynes: Open University Press.

Markoff, John, and Verónica Montecinos. 1993. The ubiquitous rise of economists. *Journal of Public Policy* 13:37–68.

McAllister, Carlota. 2013. Troublemakers: Rural Chileans fight to keep dams out of Patagonia. *Boston Review* (January 1). http://www.bostonreview.net/world/troublemakers.

McCluskey, Martha. 2003. Efficiency and social citizenship: Challenging the neoliberal attack on the welfare state. *Indiana Law Journal* 78:783–878.

Mirowski, Philip. 2009. Postface: Defining neoliberalism. In Mirowski and Plehwe 2009, 417–456.

Mirowski, Philip, and Dieter Plehwe, eds. 2009. *The Road from Mont Pèlerin: The Making of the Neoliberal Thought Collective*. Cambridge, MA: Harvard University Press.

Mitchell, Timothy. 2005. The work of economics: How a discipline makes its world. *European Journal of Sociology* 46 (2):297–320.

Mitchell, Timothy. 2009. How neoliberalism makes its world: The urban property rights project in Peru. In Mirowski and Plehwe 2009, 386–416.

Moulian, Tomás. 2002. *Chile actual. Anatomía de un mito*. Santiago: LOM Ediciones.

Mundaca, L. 2013. Climate change and energy policy in Chile: Up in smoke? *Energy Policy* 52:235–248.

Ossandón, José. 2009. The enactment of private health insurance in Chile. PhD thesis, Goldsmiths College, University of London.

Oudshoorn, Nelly, and Trevor Pinch, eds. 2003. *How Users Matter: The Co-Construction of Users and Technologies*. Cambridge, MA: MIT Press.

Owens, Susan, and Richard Cowell. 2002. *Land and Limits: Interpreting Sustainability in the Planning Process*. London: Routledge.

Pinch, Trevor, and Richard Swedberg, eds. 2008. *Living in a Material World: Economic Sociology Meets Science and Technology Studies*. Cambridge, MA: MIT Press.

Plehwe, Dieter. 2009. Introduction. In Mirowski and Plehwe 2009, 1–44.

Plehwe, Dieter, Dieter Walpen, and Gisela Neunhöffer, eds. 2007. *Neoliberal Hegemony: A Global Critique*. London: Routledge.

Pope, Jenny, Alan Bond, Angus Morrison-Saunders, and François Retief. 2010. Advancing the theory and practice of impact assessment: Setting the research agenda. *Environmental Impact Assessment Review* 41:1–9.

Portes, Alejandro, and Bryan Roberts. 2005. The free-market city: Latin American urbanization in the years of the neoliberal experiment. *Studies in Comparative International Development* 40 (1):43–82.

Prieto, Manuel, and Carl Bauer. 2012. Hydroelectric power generation in Chile: An institutional critique of the neutrality of market mechanisms. *Water International* 37 (2):131–146.

Rose, Nikolas. 2004. *Powers of Freedom: Reframing Political Thought.* Cambridge: Cambridge University Press.

Rudnick, H., R. O'Ryan, and R. Bravo. 2001. Liberalization of the Chilean electricity system and its effects on environmental performance. Mimeo.

Sabatini, Francisco. 2000. Reforma de los mercados de suelo en Santiago, Chile: Efectos sobre los precios de la tierra y la segregación residencial. *EURE: Revista Latinoamericana de Estudios Urbano Regionales* 26 (77):49–80.

Sarewitz, Daniel. 2004. How science makes environmental controversies worse. *Environmental Science and Policy* 7:385–403.

Silva, Patricio. 1991. Technocrats and politics in Chile: From the Chicago Boys to the CIEPLAN Monks. *Journal of Latin American Studies* 23 (2):385.

Spronk, Susan, and Jeffery R. Webber. 2007. Struggles against accumulation by dispossession in Bolivia: The political economy of natural resource contention. *Latin American Perspectives* 34 (2):31–47.

Tecklin, David, Carl Bauer, and Manuel Prieto. 2011. Making environmental law for the market: The emergence, character and implications of Chile's environmental regime. *Environmental Politics* 20:879–898.

Teubal, Miguel. 2004. Rise and collapse of neoliberalism in Argentina: The role of economic groups. *Journal of Developing Societies* 20 (3–4):173–188.

Valdés, Juan Gabriel. 1995. *Pinochet's Economists: The Chicago School in Chile.* Cambridge: Cambridge University Press.

Van Horn, Rob, and Philip Mirowski. 2009. The rise of the Chicago School of Economics and the birth of neoliberalism. In Mirowski and Plehwe 2009, 139–180.

Wilkie, Alex, and Mike Michael. 2009. Expectation and mobilisation: Enacting future users. *Science, Technology and Human Values* 34 (4): 502–522.

Woods, Ngaire. 2006. *The Globalizers: The IMF, the World Bank and Their Borrowers.* Ithaca: Cornell University Press.

16 Creole Interferences: A Conflict over Biodiversity and Ownership in the South of Brazil

Ana Delgado and Israel Rodríguez-Giralt

Back to Place: On How Creole Seeds became Public Entities

In Latin American history, the term *creole* (*crioulo* in Portuguese, *criollo* in Spanish) stands for "person native to a locality" and has been used to designate the descendants of those who emigrated from Europe.[1] It refers to those who were born *here*. Thus, *creole* evokes colonial stories of displacement, delocalizations, and relocalizations, stories of the coexistence of different worlds that occurred as encounters in the "contact zone" (see Cukierman in this volume). Although the history, experience, and meaning of the creole in Brazil has been much discussed and even questioned (Ribeiro 2000), less has been written about how Brazilian law has been a pioneer in the acknowledgment of creole seeds (*semente crioula*). As defined by the new Brazilian Act on Seeds, creole seeds are those seeds that have been developed, adapted, or produced in situ by local farmers and by landless and indigenous people over the years. In this context, *creole* thus refers to a set of relations and practices that are anchored in a certain space (here) as well as referring back to a certain time (tradition). Unlike commercial seeds, creole seeds are presented as embodying diversity and being old. Yet, in a sense, they are always new entities, as they are endlessly adapting to local environmental conditions, always in a temporary state, waiting for their next realization.

In this chapter, we explore how the once neglected local seeds have been called back into being as legal and public entities under the name of "creole" and how they have coexisted with an already present legal entity, the commercial seed. As we shall see, this coexistence developed as a conflictive situation in which creole seeds, despite being diverse and temporary entities, had to be stabilized somehow. Attempts at integrating the creole seeds into national legal, scientific, and bureaucratic systems resulted in a number of interferences, *in-between* zones in which new ways of identifying, classifying, and registering the seeds had to be produced and new forms of ownership recognized. This chapter tells the story of those encounters and generations. It does so by evoking the idiom of displacement, conviviality, and disruption that is, in a way, the idiom of creole. This story of coexistence connects to other partial episodes in the Brazilian agrarian history, a history of conflict. We shall introduce here three such episodes to provide some insights on how creole seeds became public entities in Brazil.

The first episode in our story is about the earlier oblivion and displacement of local seeds—in many ways, this was a global story that took place in many locations. One of the ways in which modernity arrived in Brazil was through the "imported magic" of agrarian development and new technologies that would revolutionize agriculture. Part of this "magic turn" of agrarian development relied on the promise of social progress and wealth, and part relied on a trick of oblivion and displacement. While local practices were labeled as old and backward, agrarian experts from the national public system promoted standard methods for large-scale agriculture that were to turn Brazil into a modern nation. As the "local" seeds, knowledges, and markets were displaced, standardized varieties of crops were promoted as public entities. Expert breeders developed a number of methods to turn seeds into stable scientific objects. That done, those seeds could be inscribed in national registries of risk and properties, thus becoming public entities of the modern Brazilian state. As standardized objects, commercial seeds could be identified and tracked at any time, as well as linked to a certain owner. A number of intermediary practices and bureaucracies have been mobilized to ensure that the seeds will behave as expected, despite being dispersed across long distances. Commercial seeds are key elements within a configuration that articulates itself within large scales, standard objects, remote political decision making, and the promotion of a type of innovation that comes from elsewhere (as it is a sort of imported innovation). In this chapter, we argue that the recognition of local creole seeds as public entities disrupted such topological configurations and generated a different one.[2]

The second episode of our story begins with how those forgotten entities, local seeds, were brought back into existence as new public entities under the name of creole. As they "came back," they brought about a new topological configuration in which the local, national, and global levels intersect in new ways and the old (tradition) became a source of innovation. In Brazil, although creole seeds are acknowledged as valuable entities to be conserved and protected because they are old and local, this notion of the value of the old and the local is, somewhat ironically, quite new and global (Torgerson 1999). Hosting the Rio Summit in 1992, Brazil was an early signer of the Convention on Biological Diversity (CBD) and later became one of the seventeen acknowledged "megadiverse countries." Biodiversity conservation became an important element in the construction of the new global Brazilian identity. The CBD acknowledges the sovereignty of the nation state over the biodiversity situated within its national borders (see UNEP 1992). Other international treaties also acknowledge local farmers' rights to seeds,[3] acknowledging their role as preservers of in situ biodiversity.[4] These international treaties acknowledge an ecological order in which local knowledge and seeds constantly change, adapting to dynamic ecosystems. Innovation (the production of the new) appears in this case as diversification and ongoing changes that occur in situ. That bond to the place is what entitles farmers' ownership of seeds. As producers and preservers of in situ biodiversity, local communities are bestowed with the right to "benefit sharing" by the CBD. Thus, the CBD and other global treaties make visible those entities that had fallen into oblivion during modernization.[5] Yet existing forms of intellectual property are made to regulate entities that are stable, can be reproduced, and can be traced to individual invention, while local seeds are always changing and, as the CBD states, are the product of successive generations rather than of individuals. Hence, how is local biodiversity to be accommodated within existing national regulations? The CBD does not say, but rather transfers the problem to the national level. To deal with this problem, in 2003 the Brazilian state passed the New Act on Seeds (Federal Law No. 10,711 of August 5, 2003).[6] The new Brazilian law on seeds is an attempt to integrate difference: it regulates both creole and commercial seeds, assuming that, despite their disparities, they can coexist under the same legal framework. As such, the law reveals a conflicting tension between diversification and standardization.[7]

The third episode is about ways in which creole seeds became an issue of public concern within the struggles of the social movements. Unequal access to land can be traced back to colonial times and has been one of the most prominent political battlegrounds in the constitution of the modern Brazilian state. The struggle for the land, *a luta pela terra* as it is known in Portuguese, has been a conflict between small and large-scale farming and thus a conflict of scales and spaces as much as one of exclusions and dispossessions. In a globalized bioeconomy, this conflict has translated into new scales and locations, with seeds emerging at the center of political dispute. As the issue of biodiversity went global with the CBD and with the consolidation of global environmentalism in the 1990s (Hajer 2000), the Via Campesina (International Peasant Movement) became more involved in global political arenas. They not only appealed to the notion of "the peasant," thus bringing a forgotten figure back to the political scene, but they did so by emphasizing a new role for peasants as preservers of biodiversity. In these global arenas, the emergence of local biodiversity as a political claim came together with fights against genetically modified organisms (GMO), which in many ways were two sides of the same struggle (Kinchy 2012). The social movements of La Via Campesina have led actions and protests against GM crop monocultures and monopolies around the globe. Perhaps especially in postcolonial contexts, such actions are reminiscent of previous struggles, stories of dependency now articulated against globalizing agroindustry pressures. La Via Campesina's actions are attempts at breaking such global dependencies by gaining control over the local. In their affiliation with La Via Campesina, some sectors of Brazilian rural movements also joined those initiatives. Among other things, they occupied the experimental fields of Syngenta (a leading GM seeds multinational) in the south of Brazil, transferring the practice of occupying the land that had been so central in Brazilian rural activism to the dispute over seeds. Seeds thus have become an entity embodying the political struggle and identity of the environmentalist sector of rural movements in Brazil.[8] In such a climate of dispute, the Brazilian rural movements and civic organizations exerted pressure on the Brazilian state to legally acknowledge creole seeds (Santilli 2011), which eventually resulted in the enactment of the new law on seeds in 2003. Furthermore, together with Via Campesina, they also promoted a number of courses and technical schools for farmers to learn about forgotten local biodiversity. Since then, young members of the rural movements have studied agroecology in the Latin American School of Agroecology of Via Campesina[9] in the south of Brazil and in other technical schools.[10] Some of those young farmers became "agroecology technicians," bringing together the emergence of the creole seeds with a certain expertification of rural movements (Delgado 2010).[11]

In their mission to protect local farmers and biodiversity, these activist technicians have traveled across the Brazilian countryside, searching for forgotten knowledge and seeds, locating them where they still exist, identifying, and collecting them. These have again been moments of encounter and translation in which local practices were assessed and selected through ecological criteria of sustainability. As ecology is a relatively new science, what is really local and what is global, and what is new and what is old in those encounters, are still matters of debate. The aim of this chapter, however, is not to establish such neat distinctions between those scales, but rather to look at the contact zones, the *in-betweenness*, where such scales "interfere," generating new social orders (or topologies). It is in these overlapping zones between the global, the national, and the local, between innovation and tradition, that creole seeds have emerged as a public entity.

The conflicting coexistence of creole and commercial seeds is a coexistence of differing topological orders. Attempts at integrating creole seeds into state bureaucracies and expert practices have triggered a cascade of what others have termed *interferences* that this chapter sketches in a number of partial stories.[12] Each of those stories takes us back to the unruly nature of creole seeds and emphasizes that attempts at capturing them as standard entities have constantly failed. In the next section, we describe how attempts to turn creole seeds into public entities relied on a tension between practices that produce diversification and practices that produce standardization—a tension that has been only partially and precariously resolved, generating new orders. Concretely, the interfering coexistence of creole and commercial seeds has generated a number of impermanent and locally bounded practices for identifying, registering, and owning the seeds. The chapter ends by introducing the notion of *transient standardization* to illustrate the generative character of such an interfering coexistence of creole and commercial seeds. Transient standardization, we conclude, adds interesting nuances to existing understandings of coexistence. *Coexistence* may refer to the ideals of symbiosis or synthesis, as it presupposes that different, even antagonistic, actors can be neutralized or integrated (for a discussion see Lezaun 2011). *Transient standardization*, on the other hand, names a set of indefinite and precarious interventions to posit a sort of conflicting and never fully integrated cohabitation of hostile actors, life forms, and ways of living.

Selecting Seeds: Making Difference, Making Identity

Appealing to place and tradition, the Brazilian Act on Seeds acknowledged the category of creole seed in 2003. The new act turned an absence into a presence and, in so doing, connected different scales—the global, the national, and the local—in a new way. There are several means through which these connections were drawn in the making of creole seeds.

There is one practice that is central to the making of any seed: selection. This is common to all ways of practicing agriculture: some seeds are always selected and stored for the next harvest. But whereas large-scale agricultural selection practices are oriented to national and global markets, in small-scale agriculture, the practice of selection has a local scope. In large-scale agriculture, seeds are selected by professional breeders on the basis of scientific methods and a set of standard criteria. Selection is targeted to the production of genetically identical individuals. Thus, the production of plant varieties is performed by "making identity" (a variety being a kind of identical individuals). The result is a set of standard seed varieties. Commercial seeds are made to be stable objects: being genetically identical, they are expected to work in predictable ways. By making standard objects, producers ensure that these identical seeds will behave in the same way even if they are sown in different places (or such is the scientific expectation). As we will see, only standardized seeds can circulate as commodities in national and global markets.

In the making of creole seeds, the practice of selection works in an inverse way: here, selection produces diversity rather than identity. Local farmers have traditionally selected seeds by following a number of criteria that vary from one year to the next and from one community to the other. Hence, local knowledge produces not standard but rather unruly entities. Random natural selection also plays a role here, as some seeds will fail while others will become adaptive, resulting in an unpredictable transformation of the seed in each harvest.[13] Combining variable criteria and random natural section results in variable seed varieties.[14] While commercial varieties of seeds are genetically identical, creole varieties of seeds are genetically diverse and so to a certain extent are made to behave unevenly. As we will see in the following sections, in order to identify and track the creole seeds, they have to be situated within a certain community and place. Because of the local seeds' ecology, they are bound to the place from which they come. Creole seeds cannot travel far because they always belong to a certain place and community (and when they do travel, they become something else). They are not identical objects, but are always changing; their identity is their lack of identity, or rather, they have a sort of transient identity anchored to a place (they are in situ). This bond to a place and the local scale makes them central to the struggles for farmers' rights.

By introducing the category of creole seeds (together with local and traditional seeds), the new law on seeds aims to articulate this new and indeterminate entity. It defines creole, local, and traditional seeds as those seeds that have been "developed, adapted or produced by the local farmers, landless people and indigenous people and that have well defined phenotypic features that are recognized by the corresponding communities and that following the criteria of the Ministry of Agriculture, take into consideration sociocultural and environmental descriptors, and that are characterized as substantially different from commercial seeds" (Federal Law No. 10,711 2003). As such, even though the law is an attempt to define the quality and identity of creole seeds, it actually acknowledges a high degree of indeterminacy to the extent that creole seeds are acknowledged to be the result of using variable sociocultural criteria within selection practices. In spite of that variability, though, the law also adds that creole seeds need to be recognized by both the local communities and the authorities and bureaucracies of the Ministry of Agriculture. In practice, this dual recognition has turned out to be no easy task. In what follows, we describe how a number of bureaucratic and technical devices were put in place in an attempt to define and capture creole seeds. We also describe how these devices have systematically failed. Creole seeds are evanescent entities that escape existing ways of ordering, and as such, they generate a new way of ordering that has been revealed in a number of interferences.

Interference 1: Ensuring

Built on different practices of selection—one that produces identity and another that produces diversity—creole and commercial seeds are ascribed different practices for the production of safety (and risk). Contrary to the production of commercial seeds, local traditional communities diversify their seeds, ensuring that they will behave differently, thereby producing a seed stock that is supposed to be resilient to changing local environmental conditions: if one variety fails, another will succeed. Arguing for this ecological advantage of creole seeds, the Brazilian rural movements of Via Campesina and the National Agroecology Association (ANA) have launched a number of projects to promote farmers' adoption of creole seeds. They link diversification to security not only in an ecological sense, but in the sense of ensuring farmers' economic

A Conflict over Biodiversity and Ownership in the South of Brazil

autonomy (autosubsistence), liberating them from dependency on distant markets (food sovereignty). In the system of large-scale agriculture, safety is performed in a different way: through the mediation of the state and distant markets. Being standardized objects, commercial seeds can be inscribed in national registries. As we will see, inscription in those registries turns the seeds into objects of national systems of rural credits and insurance as well as into a commodity. If a disaster occurs, the state will cover the farmers' losses. For the activists of La Via Campesina, on the other hand, the main source of risk is large-scale agriculture based on monocultures. They see diversification as the safest option for farmers, a view the government would only partially acknowledge by the legal category of creole seed. If creole seeds were to be integrated in the national bureaucracies that regulate agriculture, ANA claimed, farmers growing them should also benefit from the national system of insurance. Yet attempts to include the creole seeds in the system of national insurance triggered a number of further problems.

La Via Campesina's campaign "Seeds: Heritage of the People for the Good of Humanity" was initiated in 2003 to promote food security through biodiversity conservation. Some sectors of the Brazilian movements and ANA joined the campaign (Bionatur/MST 2005), promoting the adoption of local seeds by farmers. In 2006, ANA reported that farmers who had attempted to adopt creole seeds had ended up in a drastic situation (Londres 2006): in 2004–2005, a severe drought had caused the crop to fail. The ANA report highlights the seriousness of the situation: most farmers depend financially on their yearly harvest. While farmers growing commercial seeds would be covered by state national insurance, farmers growing creole seeds were not protected in the event of adverse weather conditions that would ruin their crops. That type of rural insurance[15] and the state system of providing rural credit have been two crucial elements in promoting the modernization of the Brazilian countryside, operating as key incentives for peasants to adopt the new modern technologies, including commercial seeds. But the insurance did not include creole seeds, thus discouraging the adoption of agroecological practices. That was the main concern of ANA and social movements: that exclusion from the insurance would stop farmers from adopting and preserving creole seeds (Londres 2006). This exclusion revealed a line of friction: a collision between a state system of bureaucracies and the conservationist project of the social movements. It also elicited a number of reactions and negotiations between social movements and the state on how to include creole seeds within the national insurance system, as discussed below.

Interference 2: Registering

The main hindrance to including creole seeds in the national insurance system was the variable character of these seeds, which resists standardization. To become part of the national insurance plan, creole seeds would have to be registered in the RNC (the national registry for the protection of seeds)[16] and in the National Mapping of Risky Climatic Zones,[17] as is regularly done for commercial seeds. To register the seeds, first a number of protocols and inscription practices have to be followed.[18] These practices ensure the identity and quality (and eventually the property) of the seeds. Once the quality of the seeds has been established, another set of practices to establish matches between seeds and regions is performed. The Mapping of Risky Climatic Zones ensures that certain seeds (identified with scientific names) are suitable for cultivation within certain climatic regions. Both the Registry and the Mapping are inscription devices that allow the state to track where seeds can go and to whom they belong. The standardization of the commercial seed provides stability for national commercial networks, as everybody (buyers, sellers, insurers) expects that the seeds will behave in predictable ways. By applying a number of scientific procedures and breeding methods, seeds are made to have certain stable features (both genotypic and phenotypic). Genetic vigor, size, and productivity are usually prioritized as criteria of selection for commercial seeds. The stable features of these standardized seeds are translated into a list of descriptors. By encoding seeds into a list of descriptors, it is possible to delocalize them, as the list will allow for the identification of the seeds by actors in other places (e.g., agrobusiness experimental fields). These scripts are used to inscribe commercial seeds in the national registries and the Mapping of Risk Climatic Zones, rendering them eligible for national insurance.

Commercial seeds thus emerge as an object in an intricate interplay among state bureaucracies, scientific practices, and markets.[19] Creole seeds, however, escape and upset those relations because they are made to change and be diverse. They are selected by using variable sociocultural criteria such as good taste and softness, and they have coevolved in a continuous adjustment to local environmental conditions. While large-scale farming relies on a standardized set of practices, small-scale agriculture has traditionally been performed through a set of practices that is heterogeneous, diffuse, and orally transmitted (Richards 1986). It depends on a type of communication that is versatile and negotiated in situ. A local agricultural practice is thus adapted and reinvented every time it is transmitted. Handed down through oral practices, local traditional knowledge rarely encodes material seeds into inscriptions (i.e., a stable list of descriptors).

As we have seen, for ANA and the rural movements, the diversity and variability of the seeds and the local knowledge underlying them is what makes them valuable for ecological and political reasons. In 2004–2005, however, farmers producing creole seeds were at risk of losing their harvest to unpredicted environmental conditions, prompting a crisis that revealed the farmers' dependence on the national insurance. And because of this dependency, farmers would tend to adopt creole seeds only if the seeds were enrolled into national bureaucracies. But since creole seeds resisted standardization and escaped normal inscription practices, the question of how to register them became crucial for both state representatives and representatives of social movements. This problem triggered a number of actions and frictions. In 2006, ANA reported that "the inscription form of the National Registry assumes a high level of genetic homogeneity that does not exist in creole varieties" (Londres 2006, 17), thus identifying the official registering practices rather than the variability of the seeds as the source of the problem: registering practices excluded farmers from the insurance. In response to the drought, ANA and the Ministry of Agriculture agreed on a provisional solution to include farmers who had been planting creole seeds in 2004–2005 into the national insurance program. This temporary measure, which was called "Regional Validation,"[20] produced an impermanent way of enrolling seeds: seeds should be registered according to the places in which they were produced and the organizations (social movements and communities) that had produced them. Furthermore, these descriptions would be based on variable sociocultural and environmental criteria. This registry would necessarily be impermanent, as it would have to be updated for every harvest. In this way, a variable registry was negotiated that would adapt to the changing character of creole seeds. The registry was an attempt to include creole seeds within the national system of agriculture without completely neglecting their variable character. However, as we shall see, this type of alternative registering practice led to further ruptures in relation to the quality and identity of the creole seeds.

Interference 3: Accounting

In large-scale agriculture systems, private enterprises and state expert agencies assume responsibility for maintaining the quality and identity of seeds over time, so that the seeds can be relied upon and therefore commercialized. These actors ensure that the physical features of seeds correspond with the list of descriptors in the registry from one year to the next. For the seed to have a clear identity (i.e., being clearly different from any other seed), those features have to be kept stable. In this system, if the variety is not kept "pure," it cannot be registered for commercialization. If a variety goes for a set number of years without being maintained and used, its property rights may expire and the variety falls into the public domain and disappears from the market. This is one of the ways in which biodiversity loss may occur. In small-scale agriculture, practices for keeping seeds work in an inverse way, since it is not the genetic identity of seed that is to be maintained, but the diversity. Therefore, attempts at integrating creole seeds within existing state bureaucracies raised a difficult question: Who would take responsibility for maintaining the quality of an ever-changing entity such as the creole seed? That was a problem that the social movements and ANA had to face with the implementation of impermanent practices of registration such as regional validation. How could the representatives of these social movements give an account of all those varieties of creole seeds, distributed in different communities, always changing, and constantly being exchanged? Action had to be taken to cope with this problem. At that time, community-based seed banks became a popular practice promoted by the activists and technicians of social movements, particularly in the southern region of Brazil. Local seed banks are a way to bring together otherwise dispersed seeds so they can be localized, described, and cataloged. As noted earlier, such catalogs need to be continually updated due to the variable character of creole seeds. Figure 16.1 shows a record system proposed by rural movements for cataloging and inscribing seeds at a local level. This inscription form includes information not only about the physical features of the seed but also about the location, the historical background of the seeds, and the community that has preserved it over the years.

Making community-based catalogs of creole seeds is one possible way to enable social movements to identify creole seeds and hence to render their quality accountable to state authorities. In this way, creole seeds could be registered, farmers' rights over local seeds would be acknowledged, and the seeds would be included in the national insurance program. Realizing that the representatives of social movements could account for the quality and identity of creole seeds, the Ministry of Agriculture launched a permanent registry of such organizations. This solution, however, triggered new problems. Different varieties of creole seeds could have the same name in different regions, and varieties with morphological similarities (or even the same genotype) could have different names in different places (Santilli 2011). This lack of one-to-one correspondence between seeds and local names rendered identification difficult. Another problem noted by ANA was that those farmers who were growing creole seeds but did not belong to any social movement or civic organization would be excluded from national insurance and their right over seeds would not be fully acknowledged.

Interference 4: Commercializing

In acknowledging creole seeds, the 2003 new act on seeds was mainly concerned with the quality and identity of those seeds. By recognizing creole seeds as the product of communities, however, the law conflicted with the existing Brazilian regulations on property rights. A particularly difficult conflict was with Law 9456/1997,[21] which regulates property rights on plant varieties in Brazil. Following international treaties and agreements,[22] the Brazilian system of property rights on plants is organized around individual property (i.e., individual people or enterprises being the holders of those rights). The law states that when individuals have obtained a seed variety that is new, homogeneous, and stable, they hold exclusive rights to the reproduction and trading

RECORD FOR THE POPULAR REGISTRY OF CREOLE SEEDS.
Record for rescued Creole varieties.

Diverse seeds

Cultivar_____Variety:_____
Name of the farmer:_____
Address _____
Telephone:_____ Email: _____
Belongs to: ( )Movement ( )Organization ( )Group. Who? _____

Seed color: _____ Flower color: _____
Type of grain: _____ Grain: ( )hard ( )soft
Plant height: _____ Features: _____
Diseases: _____

Cycle: _____ Month to sow: _____ Month to harvest:_____
Plant: ( ) single ( ) in collaboration with _____
Soil type and fertility: _____
Production by hectare: _____ Overall production: _____
Seed for: ( ) family consumption ( ) animal use ( ) market
What do you like most about this variety? _____
From whom did you get this seed? _____
Is there someone else seeding/sowing this variety in the community?
Who? _____
Do you cultivate other seeds, trees and/or native animals? Which? _____
Amount of seeds rescued: _____
Place: _____ Date: ___/_____/_____
Organization/Movement that rescued them: _____
Person responsible for the information: _____

Describe the history of the seed: origin, who was cultivating it, features, legends and knowledge of the harvest, etc. (use the back page)

Figure 16.1

A local seed record form for the registry and "rescue" of creole varieties. Issued by MPA (Movimiento de los Pequenos Agricultores)/Via Campesina. Translated from the original by the authors.

of that genetic material.[23] This property rights system depends on the possibility of creating a stable object (e.g., commercial seed) that can be identified and related to a legal person in a one-to-one relation. The possibility of establishing clear ownership depends on the possibility of standardizing seeds by using scientific methods and then inscribing those seeds. Creole seeds, however, have been acknowledged as products of community knowledge: a kind of knowledge that was variable, diffuse, and distributed within the local communities through generations. On this basis, social movements came to claim forms of intellectual property rights focused on communities, rather than individuals, as holders of rights.[24]

By the legal category known as "farmers' privilege,"[25] the Brazilian legislation ensures that local farmers can keep seeds from one year to the next and that they can exchange and even trade those seeds from individual to individual. However, it imposes limitations on the trading of seeds by social movements and local communities. Decree 5153/2004 establishes that "family farmers, agrarian reform settlers, and indigenous peoples, when acting as *individuals*, can distribute, exchange, and sell seeds among themselves; however, their organizations can distribute seeds only among farmers who are members of these specific organizations, and farmers' organizations cannot sell seeds (without being registered) in any circumstance" (Santilli 2011, 54). Such restrictions would hamper the agenda of social movements, as they prevented effective ownership by farmers of their seeds. Not only would the restrictions stop the free exchange of seeds, but they would also be contrary to culturally established rules for exchange (precisely the kind of exchange needed to make the diversification of seeds successful). Furthermore, they would prevent trading within the internal networks of social movements, which some considered key to protecting local agriculture from the dependencies and fluctuations of international and national markets. Legal restrictions on the rights of social movements to trade within internal networks activated a number of actions, such as the creation of Bionatur.[26] Among other things, Bionatur aimed at recovering varieties that had fallen into the public domain in order to bring them back into use.

Thinking Back: Transient Standardization

In trying to describe this conflict of seeds and ownership in Brazil, we eventually came to the conclusion that there was no way to tell the story of the conflict in a linear way. Our multiple attempts at doing so resulted in oversimplifying the conflict. For the conflict was messy, and actors repeatedly returned to the same point: the variability of the creole seed. The identity (or rather lack of identity) of the creole seed is the recursive element articulating this conflict. Central to the different practices and strategies for capturing the identity of creole seeds is the definition of where and to whom they belong. However, the creole seed escapes standardization and definition. As legal and public entities, creole seeds constantly emerge as disruptive and transitory entities. This diversity and variability is crucial for creole seeds, as it is what makes them valuable for ecological and political reasons. Their introduction as public and legal entities has thus sparked a conflicting coexistence with commercial seeds, in which two ways of doing agriculture and two differing modes of spacing and timing coexist. Commercial seeds enable large-scale agriculture, which is delocalized (or nationally localized) and organized around mobile and stable commodities. Creole seeds are an important factor in a more localized and strongly diversified order. Yet both configurations, or topologies (to use the term coined by Law and Mol 2001), must somehow coexist. This is what the 2003 act on seeds tries to consolidate by setting out a system that will accommodate these two different life forms and ways of living. Nonetheless, the continuous attempts at producing compatibility between the two have instead detonated further disruptions and interferences. It is no easy task to integrate creole seeds within Brazilian agriculture policies. By emphasizing the uneven and unstable character of seeds, social movements and civil organizations situated creole seeds as belonging to local traditional farmers; at the same time, they also developed a number of strategies intended to integrate local farmers within the state system of insurance. Ironically, integrating creole seeds within state institutions seems to be the only way to preserve them. If they are not included within those institutions, the seeds might appear as too risky for farmers. This reflects a tension between attempts at promoting diversification and the local, on the one hand, and the need to stay attached to state institutions and standardizing practices.

As we have shown, creole seeds constantly intersect and conflict with existing forms of ordering, producing new ones. By enacting the 2003 act on seeds, the state acknowledged the underdetermined character of these seeds. In a conflictive collaboration with social movements, the state has generated impermanent solutions as an attempt at integrating creole seeds as national entities. Expiring inscription devices such as creole seed records and variable registries that need to be periodically updated are attempts to capture the instability of creole seed so as to include them within state bureaucratic practices of inscription and registering. Such practices result from the multiple, and partially failed, attempts at harmonizing standardizing practices with practices that produce diversity. These expiring practices of inscription deploy what we have termed *transient standardization*, practices attuned to novel forms of common property and the goals of maintaining diversity. Transient standardization devices allow one to trace back how, where, and by whom creole seeds were made while they also remain open for transformation. As in open source culture, the expectation is that the community of users will improve the quality of the shared entity (generating more adaptive seeds, in this case). Yet even when they are ascribed to a community and a place, creole seeds become recursive entities, always in the making, awaiting their next realization. In this way, they not only embody tradition but are in a process of constant innovation.

Transient standardizations, therefore, result from attempts at conciliating a multiplicity of dimensions: innovation and tradition; the global, the local, and the national. Yet they also materialize the failure of the integration that the new act on seeds pursues. As a story of coexistence, the story of the creole seeds suggests that interferences can be generative, an important observation given that coexistence is often understood as a sort of biopolitical symbiosis. Legal and discursive frameworks, such as those in Europe and in relation to the coexistence of GM and conventional crops, may eventually imply a sort of neutralization or bypassing of antagonistic differences "by the erection of a ... complex system of governance meant to guarantee the indefinite cohabitation of hostile actors, moralities, life forms, and forms of life" (Lezaun 2011, 17). The coexistence of creole and commercial seeds in Brazil, however, emerges as an ever-unstable biopolitical configuration that works itself out on the basis of failed attempts at overcoming differences. In the introduction to this chapter, we called upon the idiom of creole to designate such a coexistence without integration. Naming local seeds as creole is far from a casual choice, but it explicitly evokes colonial and postcolonial stories of hybridism, conviviality and unresolved differences as well as entailing a claim for what belongs *here*. Creole seeds embody an attempt at gaining control over the local, opposing globalizing dependencies. As described in this chapter, such creoleness appears as a specific way of articulating the political. Creole seeds and commercial seeds emerge as ever-differing entities, their mutual externality creating an ambiguous, complex, but also generative pattern of interrelations. Far from having the effect of reducing, neutralizing, or integrating differences, the coexistence in this case takes the form of a set of indefinite, dynamic, and precarious interventions that guarantee a conflicting and never fully integrated cohabitation of hostile and yet deeply interdependent actors, life forms, and ways of living.

Acknowledgments

Author Ana Delgado was able to work on this chapter during her postdoctoral period, financed by the Research Council of Norway (Project No. 187969/O10). Israel Rodríguez-Giralt was financed by the Catalan Research Agency (BP-A 00107) during the writing of the chapter. We are also grateful to the Centre for the Studies of Science and Humanities, University of Bergen (Norway), for its economic support during the writing of this chapter. Thanks to Kjetil Rommetveit and Roger Strand for commenting on an early version.

Notes

1. The origin of the term *criollo* is disputed, but it appears to be derived from the Spanish verb *criar* ("to breed, rear," from the Latin *creare*, "to create") (Schnepel 2004). The word was created to refer to slaves born in the Americas rather than Africa and to elites born in the Americas as opposed to *peninsulares* born in Spain. It was introduced into Portuguese as *crioulo*, where it came to designate Brazilians of African descent.

2. Here we use the notion of topology to refer to social topology, a critical approach developed mainly by Law and Mol (2001) to raise awareness of the limitations of other spatial metaphors (mainly, the notion of the network itself) being used in the social sciences to understand the complex and not always coherent dynamic of the social (Mol and Law 1994). Although *topology* suggests a particular interest in spatiality, and in particular in the attributes of the spatial that secure continuity for objects as they are displaced through a space (Law 1999, 6), social topology also works as a critical tool to address social performativity in its broadest sense (Moreira 2004). Networks, regions, fluids, and fires, some of the topologies identified so far, name ways of connecting localities as well as events and actions in particular ways (Latour 1997; López and Domènech 2008; Moreira 2004).

3. International Treaty on Plant Genetic Resources for Food and Agriculture. TIRFAA-FAO, 2001. See Schmitt and Almeida (2011).

4. See Convention on Biological Diversity, article 8 on "Traditional Knowledge, Innovation and Practices" (UNEP 1992).

5. Some would claim that it does so by reinventing tradition. (See, for example, Ellen, Parkes, and Bicker 2000; Agrawal 2002.) Although that kind of discussion touches upon some issues presented in this chapter, it falls outside its scope.

6. Presidência da República, Casa Civil, Subchefia para Assuntos Jurídicos http://www.planalto.gov.br/ccivil_03/leis/2003/L10.711.htm.

7. This kind of tension is also articulated on the global scale, with the FAO and the CBD promoting practices of diversification, small-scale agriculture, and farmers' rights, while the Agreement on Trade-Related Aspects of Intellectual Property Rights (TRIPS) of the World Trade Organization (WTO) and the Cartagena Protocol legitimize individual property, standardization, and large-scale agriculture.

8. Our descriptions here are restricted to those sectors of the rural movements that have chosen to go green and that are affiliated with Via Campesina.

9. See Latin American School of Agroecology of Via Campesina, http://escolalatinoamericanadeagroecologia.blogspot.com.br/.

10. *Agroecology* has been defined as "the science of sustainable agriculture" (Altieri 1995); as used by social movements in Brazil, it also includes social sustainability.

11. This chapter builds on ethnographic work carried out by one of the authors, Ana Delgado, in 2004–2006 in the south of Brazil. In that work, Delgado focused on a project that was

performed within Via Campesina's campaign "Seeds: Heritage of the People for the Good of Humanity" in that region, particularly as it related to the Bionatur project, presented later in this chapter. In southern Brazil, the modernization of the countryside had been particularly intense. In this region, both monocultures of GM soybean and rural movements were more marked than in other parts of the country. The ethnographic work consisted mainly of participant observation, in-depth interviews, and a collection of documents and drawings produced by social movements and local farmers. For a more detailed description of this work, see Delgado (2010) and Delgado and Strand (2009). For the particular purpose of this paper, we build on documents produced by social movements and civic organizations; we primarily base our descriptions of the conflict for creole seeds on a report produced by ANA (National Association for Agroecology) in 2006.

12. We take the notion of interference from Moreira (2004). Going deeper into the social topological argument, he uses an operating room to illustrate the different topologies built upon each other to produce a particular event (surgery). In this account, he develops the notion of interference (Moreira 2006), borrowing it from Michel Serres's (1972) and Gabriel Tarde's (1999) work, to name those generative copresences and codependencies through which the social is reenacted and depicted in much more complex, partial, and noncoherent ways.

13. In every harvest, farmers select seeds in accordance with a number of phenotypic features. A set of seeds whose features do not appear as immediately interesting are also stored.

14. Different kinds of seeds are similar in their phenotype but vary in their genotype, the borders between kinds being always rather diffuse.

15. SEAF (Seguro da Agricultura Familiar).

16. In Brazil there are two main registries to ensure the quality and safe trading of seeds, the RNC (Registro National de Cultivares) and RENASEM (Registro Nacional de Sementes e Mudas); see Brazil's Ministry of Agriculture website, http://www.agricultura.gov.br/vegetal/mercado-interno/sementes-mudas.

17. Zoneamento Agrícola de Risco Climático do MAPA (Ministry of Agriculture). See http://www.agricultura.gov.br/politica-agricola/zoneamento-agricola.

18. *Inscription* is a classic term used in actor-network theory stories to explain how networks are built up, are managed, or extend beyond the *hic et nunc* (Latour 1987). The term refers to those operators—either a scientific diagram, an equation, or an object—that behave as an abstraction, as a contraction of a large number of different and distant elements (Latour and Woolgar 1979). As stable and mobile means, they are crucial to turn the world into a center of calculation (Latour 1987).

19. For an excellent discussion of how similar entanglements perform and mobilize certain political rationalities, see Tironi and Barandiarán in this volume.

20. *Validação regional* (see Londres 2006, 20).

21. Known in Brazil as "Lei de Cultivares" (Plant Variety Protection Law). Available at World Intellectual Property Organization (WIPO) website, http://www.wipo.int/wipolex/en/details.jsp?id=517.

22. These were mainly the TRIPS (Agreement on Trade Related Aspects of Intellectual Property Rights) and UPOV (International Union for the Protection of New Varieties of Plants).

23. See Law 9456/1997, WIPO website.

24. For a more detailed description on farmers' rights that have been negotiated in Brazil, see Schmitt and Almeida (2011).

25. "Privilégio do agricultor," in "Lei de Cultivares."

26. Bionatur is a social enterprise and commercial network launched by the MST (Movimento dos Trabalhadores Sem Terra, translated into English as Landless Rural Workers' Movement). It aims to enable exchange and trading within the Brazilian rural movements.

References

Agrawal, Arun. 2002. Indigenous knowledge and the politics of classification. *International Social Science Journal* 54 (173):287–297.

Altieri, M. 1995. *Agroecology: The Science of Sustainable Agriculture*. Boulder: Westview Press.

Bionatur/MST. 2005. *Sementes patrimônio dos povos a serviço da humanidade*. São Paulo: Pontocom.

Delgado, Ana. 2010. Activist trust: The diffusion of green expertise in a Brazilian landscape. *Public Understanding of Science* (Bristol, England) 19 (5):562–578.

Delgado, Ana, and Roger Strand. 2009. Looking north and south: Ideals and realities of inclusive environmental governance? *Geoforum* 41 (1):144–153.

Ellen, Roy, Peter Parkes, and Alan Bicker, eds. 2000. *Indigenous Environmental Knowledge and Its Transformations*. Amsterdam: Harwood Academic Press.

Hajer, Martin. 2000. *The Politics of Environmental Discourse: Ecological Modernization and the Policy Process*. Oxford: Oxford University Press.

Kinchy, A. 2012. *Seeds, Science and Struggle: The Global Politics of Transgenic Crops*. Cambridge, MA: MIT Press.

Latour, Bruno. 1987. *Science in Action: How to Follow Scientists and Engineers through Society*. Cambridge, MA: Harvard University Press.

Latour, Bruno. 1997. Trains of thought: Piaget, formalism, and the fifth dimension. *Common Knowledge* 6 (3):170–191.

Latour, Bruno, and Steve Woolgar. 1979. *Laboratory Life: The Social Construction of Scientific Facts*. Beverly Hills, CA: Sage Publications.

Law, John. 1999. After ANT: Complexity, naming and topology. In *Actor Network Theory and After*, ed. John Law and John Hassard, 1–15. Oxford: Blackwell.

Law, John, and Annemarie Mol. 2001. Situating technoscience: An inquiry into spatialities. *Environment and Planning D: Society and Space* 19:609–621.

Lezaun, Javier. 2011. Bees, beekeepers, and bureaucrats: Parasitism and the politics of transgenic life. *Environment and Planning D: Society and Space* 29 (4):738.

Londres, Flavia. 2006. A nova legislação de sementes e mudas no Brasil e seus impactos sobre a agricultura familiar. Accessed April 2012. Available at http://aspta.org.br/2011/05/estudo-sobre-a-nova-legislacao-de-sementes-e-mudas-no-brasil-e-seus-impactos-sobre-a-agricultura-familiar/.

López, Daniel, and Miquel Domènech. 2008. On inscriptions and ex-inscriptions: The production of immediacy in a home telecare service. *Environment and Planning D: Society and Space* 26 (4):663–675.

Mol, Annemarie, and John Law. 1994. Regions, networks and fluids: Anaemia and social topology. *Social Studies of Science* 24:641–671.

Moreira, Tiago. 2004. Surgical monads: A social topology of the operating room. *Environment and Planning D: Society and Space* 22 (1):53–69.

Moreira, Tiago. 2006. Heterogeneity and coordination of blood pressure in neurosurgery. *Social Studies of Science* 36 (1):69–97.

Ribeiro, Darcy. 2000. *O povo Brasileiro: A formação e o sentido do Brasil*. São Paulo: Companhia das Letras.

Richards, Paul. 1986. *Coping with Hunger: Hazard and Experiment in an African Rice Farming System*. London: Allen and Unwin.

Santilli, Juliana. 2011. *Agrobiodiversity and the Law: Regulating Genetic Resources, Food Security and Cultural Diversity*. New York: Earthscan.

Schmitt, Claudia J., and M. P. Almeida. 2011. Mecanismos de propiedades intellectual e dereitos dos agricultores: Percepções, conflitos e cercamentos. Paper presented at XV Congresso Brasileiro de Sociologia, July 26–29, 2011, Curitiba (PR).

Schnepel, Ellen M. 2004. *In Search of a National Identity: Creole and Politics in Guadeloupe*. Madison: University of Wisconsin Press.

Serres, Michel. 1972. *L'Interférence*. Paris: Editions de Minuit.

Tarde, Gabriel. 1999. *Monadologie et sociologie*. Paris: Institut Synthelabo/Les Empea cheurs de Penser en Rond, Le Plessis-Robinson.

Torgerson, Douglas. 1999. *The Promise of Green Politics: Environmentalism and the Public Sphere*. Durham: Duke University Press.

United Nations Environment Programme (UNEP). 1992. *Convention on Biological Diversity*. Nairobi: United Nations Environment Programme.

17 The Juridical Hospital: Patient-Citizen-Consumers Claiming the Right to Health in Brazilian Courts

João Biehl

Introduction

A retired bus driver, Edgar Lemos lives in a lower-middle-class neighborhood of Porto Alegre, the capital of the southern Brazilian state of Rio Grande do Sul. Dealing with significant motor difficulties, Edgar had to wait for more than a year for a specialized neurological appointment at a nearby public hospital. He was finally diagnosed with hereditary cerebral ataxia in November 2008. The neurologist prescribed the drug Somazina, which is not included on any governmental drug formulary.

Raised in a destitute family, Edgar had worked since the age of eight. He was proud of the gated brick and mortar house he had built himself on the top of a hill. Edgar's ataxia affected not only his mobility but also his sense of dignity and worth, as it made him dependent on the care of his wife and two adult daughters. Religion had become an important source of emotional sustenance and a complement to his pharmaceutical treatment. While Edgar felt that Somazina was helping to halt the degeneration of his motor abilities, he was also taking a variety of other drugs, from statins to antihypertensives and anxiolytics, to soothe additional symptoms.

During a conversation I had with him over his dining room table in August 2011, Edgar opened a box containing the five medicines that make up his regimen. As he held each one in turn, he said, "This one I don't judicialize, this one I don't judicialize. ... I only judicialize this medicine [Somazina] because I went into debt paying for it." A monthly supply of Somazina costs about US $200.

After paying for the drug out of pocket for several months, Edgar had to take out a bank loan. Unable to keep up with house expenses and his loan interest, he had "no other alternative but to judicialize." He learned about the Public Defender's office (Defensoria Pública) from other patients also waiting for specialists' referrals at the public health post and filed a lawsuit to compel the state to pay for his medication. The Porto Alegre district judge issued a court injunction on his behalf, and Edgar received the medicine for several months, but then "the delivery stopped." He filed a new claim and won another injunction for three additional months of treatment.

As state attorneys were appealing the judge's decision, Edgar nervously anticipated having to renew the lawsuit again.

I asked Edgar why he was not judicializing the other drugs he was taking. "I know that the state cannot give everything to everyone," he stated. "I have to do my part and pay for whatever I can."

Across Brazil, patients like Edgar are seeking, and sometimes achieving, access to health care through the courts, a phenomenon that has been termed the "judicialization of health" (Biehl et al. 2009; Ferraz 2009; Marques and Dallari 2007). Though patients are suing the government to provide everything from baby formula to complex surgeries, a large number of lawsuits are for access to prescribed drugs (Scheffer, Salazar, and Grou 2005).

In this chapter, I explore how right-to-health litigation has become (in the wake of a successful universal AIDS treatment policy) an alternative for many Brazilians seeking to access health care, now understood as access to pharmaceuticals that are either on governmental drug formularies or are available only through the market. Throughout, I show how the relations between individual bodies, political subjectivities, medical technologies, and state institutions are compellingly rearranged along this judicialized front. Poor people are not waiting for medical technologies to trickle down; they are leveraging public legal assistance and a receptive judiciary to hold the state accountable to its mandate and to their medical needs, now. The chapter's ethnographic vignettes thus pave a path toward a relatively unexplored frontier of science and technology studies: that zone where technology, medicine, and law intersect in unexpected and deeply personal ways and where our ideas about the social life of technology and our notions about how medicalization and biopolitics operate from the bottom up must be rethought.

From the Right to Health to the Right to Pharmaceuticals

The 1988 Brazilian Constitution declared health a "right of all persons and the duty of the State" (Constituição Federal do Brasil, 1988), and the creation of the country's Unified Health System (SUS) extended health coverage to all citizens. Judicialization stems from an expansive definition of the meaning of the right to health and also, in part, from the passage of a landmark law in 1996 establishing free universal access to antiretroviral (ARV) therapies for HIV-infected individuals (Biehl 2007b). Ministry of Health policies and a 2000 ruling by the Federal Supreme Court further advanced the right to medicines as part of the constitutional right to health (Supremo Tribunal Federal 2000).

SUS provides health services and medicines free of cost. As part of a broader process of decentralization and in an effort to improve the administration of SUS, the Ministry of Health divided responsibilities for pharmaceutical distribution among three levels of government. Federal, state, and municipal tiers of government are responsible for purchasing and distributing medicines according to specific drug formularies. The federal health ministry continues to finance high-cost medicines they call "specialized medicines," which are distributed by state health secretariats. Municipal governments are responsible for purchasing low-cost "essential medicines," which are dispensed at local public pharmacies. State governments finance and distribute "special medicines" that their state residents require but that do not appear on either of the other two formularies. In addition, the federal health ministry funds strategic programs for the control of certain infectious diseases such as HIV/AIDS, tuberculosis, and leprosy as well as rare disorders such as Gaucher's disease.

Despite these laws, policies, and judicial rulings, the experience of patients in realizing access to medicines has been uneven. Today, about 200 thousand Brazilians take ARV drugs paid for by the government. At the same time, many citizens go to local public pharmacies only to find that essential medicines are out of stock and that the newer medicines they seek are not included in official formularies. Decentralization delegated responsibility but did not ensure sustainable funding and technical capacity at local levels. Regional and municipal governments have not been able to adequately budget and administer the growing complexity of medical needs and technological and infrastructural demands within an already complex health system.

With a population of about 200 million people and an economy on the rise, Brazil has one of the fastest-growing pharmaceutical markets in the world, with an estimated total value of more than US $25 billion in 2012 according to a business association (SINDUSFARMA 2012). Public and private doctors increasingly prescribe and patients demand new medicines, some of uncertain benefit. Newer medicines, however, are often available only through private purchase. Unable to pay out of pocket (as in Edgar's case) or to find low-cost generics at public pharmacies, patients are increasingly suing the government to obtain what they need. People often use the expression *entrar na justiça*, "to enter the judiciary" or, literally, "to enter justice," to refer to their lawsuits.

Para-Infrastructures and Political Experimentation

For the past five years, I have been coordinating a multisited ethnographic study of right-to-health litigation in the southern Brazilian state of Rio Grande do Sul, which has the highest number of such lawsuits in the country.[1] Implementing collaborative evidence-making practices, our research team moved across domestic, clinical, judicial, and administrative domains to track the interconnection of sites and the interplay of scales that the judicialization of health calls on and calls into question. Some of the core queries that guided our investigation included the following: Is the judicial system an effective venue for implementing socioeconomic rights? Which social fields and practices of citizenship and governance are crystallized in the struggles over pharmaceutical access and administrative accountability? How is it possible to gauge the market's influence on medical demands and practices as well as on the public institutions of the world's sixth largest economy?

While examining the tense negotiations of the constitutional right to health in daily life, I often had a sense that social roles and political positions were out of place: the judiciary was a sort of pharmacy, the public defender was acting as a physician, the physician as an activist, the patient association as legal counsel, and the patient-citizen was becoming a consumer, among other translocations and displacements. I found Michel Foucault's (2008) tentative reflections on biopolitics and neoliberalism helpful as I tried to understand the form and reach of these novel medico-socio-legal realities, in particular his observations on the frugality of government in contexts where market exchange determines value. But these realities also contravened Foucault's reflections, as they underscored the importance of the juridical subject to late-liberal political economies.

In his 1978–1979 lectures at the Collège de France, Foucault argued that we can adequately analyze biopolitics only when we understand the economic reason within governmental reason, suggesting that the market shapes and even determines governmental logic. In Foucault's words, "The market constitutes a site of veridiction-falsification for governmental practice. Consequently, the market determines that good government is no longer simply government that functions according to justice" (2008, 32).

The ways and means of right-to-health litigation in Brazil reveal an intense experiential-political-economic field. Here the penetration of market principles in health care delivery is unexpectedly aligned with the juridical subject of rights. The rational choice-making economic subject (necessarily a consumer of technoscience) is also the subject of legal rights. The right to life is claimed in between the clinic, the court, and the marketplace. How are the interpenetrating domains of health, therapeutic markets, and the law emerging as implicit and explicit sites for claiming political rights and confronting political failures?

Jonathan Spencer has written about anthropology's difficulties in "drawing bounds round 'the political'" (2007, 29). While classic anthropology limited politics to formal and functional analyses (a "politics without values"), the anthropology of politics that emerged in the 1980s and 1990s as a necessary and invigorating corrective (as exemplified by subaltern studies) "deliberately exclud[ed] the state from the domain of authentic politics" (Spencer 2007, 23). In the intervening decades, the anthropology of politics has moved to include a consideration of the state and development (Ferguson 1994), of transnational politics and neoliberalism (Comaroff and Comaroff 2011; Ong 2006), and of the affective domains and subjective experiences of political life (Biehl, Good, and Kleinman 2007). And while much recent anthropological and STS scholarship has productively applied Foucault's concept of "biopolitics" to a variety of contexts (Tironi and Barandiarán, this volume; see also Fassin 2007; Nguyen 2010; Ong and Collier 2005; Rabinow and Rose 2006), we are only beginning to capture the fluidity and fragility of biopolitical processes and their entanglement with the market as a testing ground for techniques of governance and self-fashioning (Edmonds 2010).

Clearly, anthropologists and STS scholars have stayed attuned to politics—even as the substance of what is considered "political" has varied with disciplinary conversations—whether as activists concerned with the inequalities of the field or in their theoretical concerns with such issues as postcolonial disorders, structural violence, social suffering, and biopolitics (Comaroff and Comaroff 2011; Das 2007; Farmer 2003; Good et al. 2008; Scheper-Hughes 1992). Most compellingly, anthropologists have begun to examine the politics involved in the formation of "para-infrastructures" such as humanitarian interventions and therapeutic policies (Biehl and McKay 2012, 1210; Fassin and Pandolfi 2010; McKay 2012).

While Stephen Collier (2011) has explored how Soviet urban infrastructures reveal political and economic rationalities and negotiations over the form of the (post)social state, other scholars such as Nikhil Anand (2012) and Hannah Appel (2012) have shown how infrastructures (such as water networks and oil enclaves) form critical sites of engagement and negotiation for corporations and states and their subjects (or citizens) in everyday life. However, with the term *para-infrastructure*, I mean to call attention to, and account for, the interstitial domain of political experimentation that becomes visible in people's case-by-case attempts to "enter justice" in Brazil. There is no predetermined strategy of control in the judicial para-infrastructure. Norms are constantly in flux, and numerous parties—state and market institutions as well as experts, legal representatives, and citizens—can manipulate levers of access. While laying claims to life, facing off over and disputing responsibility, evidence, and costs-benefits, these various parties bide their time and become at once empirically present and permeable.

Although precarious, para-infrastructures such as the judicialization of health significantly inform the ways of living that people take up in the context of ailing or inadequate public institutions as well as the scope and reach of governance in real time. Attention to such "intermediary power formations," as I considered them elsewhere (Biehl 2007b, 94), and to the growing "judicialization of politics" (Comaroff and Comaroff 2006) presents new ethnographic quandaries. They compel us to engage and think through the ambiguous political subjectivities and social formations that crystallize amid the blurring of distinctions between populations, market segments, political movements and constituencies, and collective objects of intervention or disregard.

Moving across various scales of anthropological analysis, this chapter brings into view lives and living forged across exceedingly complex and often contradictory institutions. The experiences presented here of lawyers, patients and their families, doctors, advocates, policy makers, and judges do not and cannot perfectly cohere. I try to describe the entanglements of the judicialization of health without claiming that it is seamless. Instead, I urge readers to consider how this new political phenomenon compels sick persons, laws, experts, officials, and commodities to shuttle between the home, the hospital, public offices, and the courtroom, remaking those spaces and themselves.

Ethnographic realities can help us to refine, complicate, and even dislodge totalizing assumptions about neoliberal structural adjustments, market-driven societies, and technology transfer (see Marques, this volume). In the Brazilian judicialization of health, rather than seeing a top-down biopolitical model of governance in which population well-being is the object of knowledge and control, we see instead a struggle over the utility and purpose of government by multiple private and public stakeholders. At stake here are the ways in which government (qua drug regulator, purchaser, and distributor) facilitates a more direct relationship, in the form of technology access, between atomized and ambiguous political subjects of rights and interests and the biomedical market.

Surprisingly, the decentralization of state authority has created the space for a return of the juridical subject but in an altered form. Neither entirely controlled by nor fully accountable to the state or the market, those who inhabit this new political subject position negotiate the constraints and possibilities of a technological society using jurisprudence. They work through available legal mechanisms and instantiate new sociopolitical domains to engage and adjudicate their demands, making abstract human rights concrete. These various developments, in turn, end up consolidating the judiciary as a critical site of politics—and of political economy.

The Diseased Citizen and Justice in the Absence of Adequate Public Policy

"Welcome to the juridical hospital," said Paula Pinto de Souza, the lawyer in charge of right-to-health litigation at the Public Defender's office in Porto Alegre, during our first encounter in August 2009. Her office is where the poor get free legal assistance and where the majority of the lawsuits requesting medicines from the state originate. Souza did not mince words in describing what she thought the state's biopolitics had become: "When there are no defined public policies, or when they exist but are not executed, or when policies are not in touch with new maladies and medical advancements ... what do we have? We have a diseased citizen."

When people finally access public institutions, all their vulnerabilities are exposed and they have become quite sick. Souza continued, "We are beyond preventive medicine here and the concept of health as physical, mental, and social well-being is no more. When this infirm person comes to me, the cure is most likely no longer possible. Her right to health has been profoundly injured by public power."

Souza speaks of her work in the office as an attempt to ameliorate human suffering and to restore rights to the sufferer. For the public defender, this means indicting local politics: "The person comes here sick and wronged by the failure of public policies. This is the medicine that I practice here: to help people survive with dignity. Even if the medication might not bring them life, the claim is also for their dignity."

The judiciary, in her view, can acknowledge the person's medical emergency and call on "the state writ large (federal, regional, municipal) to take on its responsibility to provide the prescribed treatment." Souza is adamant that "it is not the role of the judiciary to make public policies." Yet without judicialization, she reasons, state politics would remain populist and only electorally focused, failing to uphold constitutionally mandated responsibilities: "The government lacks political will to make public policies work. There is no concern with the human being, but a lot of concern with publicity. Forget about infrastructure. When it is election time, *then* medicines get disbursed, drug formularies updated."

In the past five years, right-to-health litigation, particularly over access to medicines, has become a subject of contentious debate throughout Brazil and has attracted international attention (Azevedo 2007; Economist 2011). At a conference a few years ago on "Accessing Medicines via Courts," a senior official in the health secretariat of Rio Grande do Sul affirmed the state's commitment to addressing the issue of pharmaceutical dispensation by SUS "in a manner that is more comprehensive, more just, and that benefits a growing number of people." But instead of speaking of specific policies the government might champion, the public administrator highlighted the secretariat's "partnership" with the General Attorney's office in addressing "frontier issues in knowledge and technology" that are increasingly "at the center of public services."

In his comments, the public administrator reduced the complex reality of right-to-health litigation to instances of demand for select and largely ineffective medical technologies recently brought to the market: "We try to guarantee the availability of medicines. But it is extraordinarily perverse that we have to guarantee the most expensive medicines, which have no effect whatsoever. The laboratories use patients to increase profits."

The public administrator mentioned the lack of convincing data on the efficacy and safety of drugs and asked whether "a medical professional has the right to prescribe whatever he wants, independent of protocols and scientific proof." Several times, he emphasized "public disinformation," "the draining of public health funds," and the "inequality" that the demand for new medical technologies by a selective population has inaugurated. "We are talking about public money here."

This official's rendering reflects what has crystallized as the myth of judicialization: an apparently clear line between good and bad science, between need and interest, between unconscionable for-profit medicine and responsible public health officials. In the context of a process of protocol writing and resource allocation that is assumed to be perfect and fully certain, right-to-health litigation represents the unwelcome intrusion of special interests into the sphere of biomedical rationality and economic certainty. As the public administrator proudly announced, the state's Attorney General's office created its own taskforce of medical consultants to verify or disqualify claims for treatment access and efficacy.

With the judicialization of the right to health, courts have become battlefields of veridication-falsification and a politics of one-case-at-a-time medical rescues. But cases like Edgar's show that at a deeper level, right-to-health litigation also makes the judiciary a site for the emergence of a new kind of citizenship, and in which the state's complicity with the commodification of care—in collusion with the market—is exposed for public critique. The judicialization of health, while never fully free of market forces, represents a rare opportunity to make the state care, to respond to the need for palliation, and to attend to citizens' social suffering.

The Pharmaceuticalization and Judicialization of Health Care

While the justiciability of the right to health is of increasing interest internationally (Gauri and Brinks 2008; Yamin and Gloppen 2011), the volume of individual right-to-health lawsuits in Brazil stands out. In 2009, 5,536 cases appealing high court rulings related to the right to health reached the Superior Court of Justice, and about half of these cases were for access to medicines. In 2009, the federal health ministry spent US $47.8 million on court-attained drugs, a significant increase from the $20.4 million spent in 2008 and $4.2 million spent in 2007 and dwarfing the 2003 federal expenditure of $58,800 (Collucci 2009). There are currently more than 240,000 health-related lawsuits under review in state and federal courts in Brazil. Almost half of all lawsuits (about 113,000) have been filed in the state of Rio Grande do Sul (Zero Hora 2012).

HIV/AIDS activists were among the first to successfully equate the constitutional right to health with access to medicines, and the rights-based demand for treatment has now "migrated" to other diseases and groups. As I documented in the book *Will to Live: AIDS Therapies and the Politics of Survival* (Biehl 2007b), an incremental change in the concept of public health has also been taking place. In terms of both delivery and demand, public health is now understood less as prevention and primary care and more as access to medicines and community-outsourced care; that is, public health has become increasingly pharmaceuticalized and privatized.

Today, a variety of actors—industry advocates, public health and private-practice physicians, medical researchers, and patient associations—have vested interests in making high-technology medicine accessible to all. In the process, the country is becoming a profitable platform of global medicine. It is estimated that almost 50 percent of the adult population (about 60 million people) uses pharmaceuticals on a daily basis.

Government-purchased medicines make up a formidable market in Brazil (Gertner 2010). The health ministry spent more than US $2.5 billion on the acquisition of drugs in 2007, accounting for 10.7 percent of its total expenditures that year, twice the 2002 percentage (Vieira 2009). However, new drugs are often available only for private purchase. Furthermore, we know that drug prices in Brazil are, overall, 1.9 times higher than in Sweden and 13.1 times greater than the mean bulk-unit prices listed in the International Drug Price Indicator Guide.

Let me pause to unpack what I mean by describing the judicialization of the right to health as part of a broader pharmaceuticalization of care and of public health. First, the concept of "pharmaceuticalization" builds on and revises the related notion of "medicalization," understood as a modern form of social control that obscures the political, economic, and social determinants of health by approaching disease and treatment in exclusively biomedical terms (Conrad 2007). Scholars have traced the public health, policy, and treatment consequences of the medicalization of a range of complicated social problems, from hunger and malnutrition to substance abuse and depression. In particular, critics note that the phenomenon has led to an overemphasis on access to health care (especially medicines) in health policy at the expense of equally needed improvements in financial and food security, education, housing, and environmental conditions (Lantz, Lichtenstein, and Pollack 2007). Medicalization, it is argued, strains health care systems, national economies, and household finances alike. Sociologist Peter Conrad and colleagues (2010) have gone as far as to estimate that in 2005 the pervasive medicalization of social conditions cost the United States $77 billion—3.9 percent of total domestic spending on health care.

The concept of pharmaceuticalization, however, stands for something more complex than an increase in the quantity of medications that societies consume (Biehl 2007a). In the last decade, medical anthropologists have critiqued the medicalization paradigm for being overly deterministic (Lock 2003). While the culture of biomedicine is undeniably powerful, people do not simply become the diagnostic categories applied to them—they inhabit them to greater or lesser degrees, refuse them, or redefine and deploy them to unanticipated ends (Biehl 2005; Han 2012; Petryna, Lakoff, and Kleinman 2006). Likewise, both policy debates and patient struggles surrounding access to pharmaceuticals are part of broader transformations in public health (Biehl 2007b; Ecks 2008). Understanding pharmaceuticalization requires moving beyond the unidirectional construction of patient subjectivity by medical diagnostics and treatments to account for the entanglement of multiple social forces and markets, the chemical concreteness and circulation of pharmaceuticals and illnesses, and the role of patients' agency and desires.

I also want to highlight that contemporary processes of pharmaceuticalization have historical antecedents in international health policies and interventions. While health development programs once focused primarily on large-scale public health measures (e.g., sanitation, availability of clean water, hygiene), in recent decades, global health organizations have increasingly focused on access to pharmaceuticals as an indicator of health care development.

This trend is crystallized in the "Essential Medicines" list of the World Health Organization (WHO), first proposed in 1975 and then codified in a published list revised every two years (Greene 2010). According to historian Jeremy Greene, while the idea that public health should be rooted in essential medicines "has taken on somewhat of a moral universality ... and commonsensical status" (2011, 28), creating such a taxonomy of fundamental drugs has revealed ambiguities and raised difficult questions. Access to new medical technologies and treatment strategies is increasingly thought of as a human right, like shelter, education, and clean water—but how are "essential" medications selected? Can effective but new and experimental treatments be considered "essential"?

As the HIV/AIDS epidemic increased in severity in the early 1990s, the WHO did not identify any ARVs as essential medicines because they were very expensive and had only recently been developed. The disease, however, claimed a larger and larger portion of total deaths in developing countries, and activists forcefully challenged the absence of ARVs on the Essential Medicines list (Greene 2011, 23). While the WHO now considers some ARVs essential, the HIV/AIDS epidemic continues to provoke the difficult question of whether access to treatments that extend lives—but ultimately do not save them—should be considered a human right. These questions of which things are "truly indispensable" to health and living and who is legally and financially responsible for making these things available are central to how the people described in this essay both invoke and critique biopolitics: from Edgar's comment in the beginning of the chapter—"I know that the state cannot give everything to everyone"—to the state official's deployment of evidence-based medicine to both rationalize care delivery and authenticate misrecognition and disregard.

In Brazil, pharmaceuticals clearly have become key elements in the state's public health arsenal. As AIDS activism migrated into state institutions and the state played an increasingly activist role in the international politics of drug pricing, AIDS became, in many ways, the "country's disease." Yet, while new pharmaceutical markets have opened and ARVs have been made universally available (in the case of AIDS, the state is actually present through the dispensation of certain medicines that carry high political stakes), it is up to individuals and makeshift communities to locally take on the roles of medical and political institutions as they learn to interact with and navigate in expert domains. These individuals and groups use survival strategies that require extraordinary effort and self-transformation and, increasingly, undergo juridical initiation as they become formal subjects of rights and engage the ritual travails of the courts amid the growing privatization of health care (Biehl and Petryna 2011). In the process, the question of what is frugal and essential to health and well-being—what one can do without and what one must have to survive—is ever more tangled and contested.

The Return of the Juridical Subject

Despite the growing number of lawsuits for access to medicines in Brazil, and amid polarized debate about the phenomenon and its costs, there has been scant information concerning the content of lawsuits, the characteristics of patient litigants, and the legal strategies and rationales deployed by the various stakeholders. Research into right-to-health litigation has also been constrained by small samples, limited geographic coverage, and the few variables examined (Borges and Ugá 2010; Da Silva and Terrazas 2011; Messeder, Osorio-de-Castro, and Luiza 2005; Pepe et al. 2010; Vieira and Zucchi 2007). Most studies tend to corroborate the arguments of public health administrators that the judiciary is overstepping its role and that judicialization generates enormous administrative and fiscal burdens, distorts pharmaceutical policies, widens inequalities in health care access, and encourages irrational drug use within the public health care system.

To better understand the burgeoning number of right-to-health lawsuits in the state of Rio Grande do Sul, our research team first examined electronic registries of health-related lawsuits in the health secretariat (Biehl et al. 2012). We found that the number of new lawsuits grew more than 1,000 percent in just seven years, from 1,126 new cases in 2002 to 17,025 new cases in 2009. The majority of these judicial claims involved access to medicines, making up 70 percent of all health-related cases in 2008 and 2009.

As a second step, we created a database of medicinal lawsuits against the state of Rio Grande do Sul. Our data collection team worked in the Solicitor General's office, which is responsible for defending the state. From September 2008 to June 2009, we analyzed 1,080 lawsuits being reviewed by state prosecutors.[2]

Edgar's case was not among these lawsuits, but his travails are not exceptional. Among the plaintiffs who reported their employment status, more than half were retired, and about one-fifth were unemployed. Among those who reported income, over half earned less than the monthly national minimum wage (about US $300) and relied on the free legal services of public defenders.

Past research has suggested that right-to-treatment litigation is, for the most part, a practice of the financially better off (Chieffi and Barata 2009; Vieira and Zucchi 2007) and that low-income patients tend to sue for low-cost medicines, while higher-income patients tend to sue for very expensive medicines (Da Silva and Terrazas 2011,12). In contrast, our results suggest that patients who procure medicines through the courts are mostly poor individuals who are not working and who depend on the public system for both health care and legal representation.

Roughly two-thirds of the medicines requested were already on governmental drug formularies. About a quarter of lawsuits were exclusively for access to specialized high-cost medicines, though low-cost essential medicines were frequently requested alongside them. Off-formulary medicines requested by plaintiffs were also often low cost, and many had been available in the Brazilian market for a long time. This suggests that government pharmaceutical programs are failing to fulfill their role of expanding access and rationalizing use.

Moreover, judges at district and higher court levels almost universally grant access to all medicines requested, recognizing that their provision is consistent with Brazil's constitutional right to health. For example, in almost all cases, district judges granted plaintiffs an immediate injunction for access to medicines. In cases in which the initial ruling was in favor of the provision of medicines, the state's higher court usually upheld the decision.

While decentralization tried to establish clear responsibility at specific administrative levels—municipal, state, federal—our analysis found that plaintiffs tend to hold the regional state responsible for medicines, regardless of the designated responsible party, and that judges rarely disagree. State attorneys frequently argue that the state is not responsible for the provision of certain services. Judges, however, cite the principle of "unity" between levels of government to assert broad shared responsibility in guaranteeing the right to health. Lawsuits become the sites of a reluctant and undisciplined cooperation. In this way, the judicialization of the right to health momentarily instantiates the state as the singular governmental entity responsible for the provision of social rights.

Chronically Ill

Patients in our sample of 1,080 lawsuits were, for the most part, chronically ill. About half of the patients (48 percent) reported conditions such as cardiovascular disease, diabetes, lipid metabolism disorders, and pulmonary diseases. Some 16 percent of the patients reported neurologic and psychiatric conditions. Patient plaintiffs in our sample had various comorbidities and procured multiple drugs for their treatments. On average, they reported 1.5 diagnoses and requested 2.8 drugs. Among the twenty-five most requested drugs, twenty-three were medicines to treat chronic diseases, and only seven were not included in official drug formularies. However, we also found patients with a single disease who demanded one high-cost treatment.

Patients with chronic hepatitis C, for example, made up a significant number of cases. These patients typically demanded ribavirin and peginterferon alfa, both of which are on the federal government's specialized medicines formulary. The high frequency of requests for drugs to treat chronic hepatitis C in our sample stands in sharp juxtaposition to the rare request—a single case—for medicines for HIV/AIDS. Both pathologies have a similar prevalence in southern Brazil, and treatments for both are distributed by governmental programs at no cost.

What are some of the possible reasons for this sharp contrast?

It may reflect variations in the efficiency of governmental pharmaceutical distribution programs. While the strategic medicines program that distributes HIV/AIDS drugs is centrally managed and funded by the federal health ministry, with a single acquisition process for the entire country, the specialized medicines program is decentralized: it is managed by states, which are federally reimbursed. The latter program depends on administrative cooperation between federal and state government and is vulnerable to the vagaries of regional health policy and management.

The contrast may also result from the specific eligibility criteria and, in some cases, from the detailed treatment protocols through which specialized and special medicines must be accessed in the public health care system. When patients fall outside eligibility requirements and protocols, they may use lawsuits to access treatment. In addition, patients who were granted requests may use lawsuits to expedite treatment delivery or to guarantee provision of medicines when the government fails to provide them. As the case of a patient named Nelson Silva illustrates, the judiciary seems to offer citizens who are both diseased and politically injured the possibility of articulating a time-sensitive legal effort to make the state act biopolitically to guarantee the possibility of survival.

Head down, Nelson Silva walked into the Public Defender's office in August 2010 accompanied by his wife Sandra, who did most of the talking. "We cannot interrupt the treatment one more time," said Sandra. Her husband was a retired steel factory worker, and she still worked as a kindergarten teacher. Sandra begged the public defender to "treat us," for "we know that people who come here get the medicine they need."

Nelson had chronic hepatitis C, and he was greatly benefiting from a forty-eight-week treatment regimen of ribavirin and peginterferon alfa. His doctor said that he needed twenty-four extra weeks of treatment, but the state's medical expert denied the request for the medication, and "my doctor told me to come here," Nelson said. "It's just a matter of the judge releasing the treatment."

"Our first treatment," Sandra continued, "was in 2001 with regular interferon." Nelson added, "But after a while the state pharmacy did not have interferon, so I had to interrupt the treatment." In 2005, he fell ill and a doctor at Hospital Conceição prescribed ribavirin and peginterferon alfa. The health secretariat denied Nelson's treatment request, alleging that it would constitute "retreatment," which was not allowed by the medical protocol in place. "Then we had to file a lawsuit for him to get it," Sandra stated. In 2009, he was declared eligible for retreatment and now needed medicines for the twenty-four additional weeks.

"The doctor gave me the meds for two weeks," Nelson continued, "but I am afraid that the legal procedure will take too long and that by the time I get the meds, if I get them, I will have to stop treatment for it failed once again. I need it fast." Nelson was desperate to adhere to the treatment. For him and so many other patient plaintiffs facing a fatal condition, judicialization is a temporal lever. "We don't want to stop everything we started," lamented Sandra.

In line with the philosophy of "I will not let the citizen die," Souza gave the couple a road map of all they had to do and the documents they had to produce so that she could open the lawsuit the following day. Here, the court system—so often thought of as a place where claims go to die a quiet, bureaucratic slow-motion death—winds up being a surprising milieu of catalysis for the uncertainty and time-sensitivity of the body and its possibilities of repair and, ultimately, of survival. "Afterward," Souza told Nelson, "you open a lawsuit against the state for medical injury."

Open-Source Anarchy

According to legal scholar David Fidler, developments in health jurisprudence "have produced open-source anarchy and a more elastic relationship between power and ideas in global politics." In such an elastic relationship, "changes in material capabilities of state and non-state actors, and changes in the world of ideas, have more impact on each other than in the closed, state-centric system that prevailed during the Cold War" (Fidler 2008, 410). Fidler recognizes a "deeper importance for law in public health endeavors within and between countries" (2008, 394; see also Fidler 2007).

Anthropologists John Comaroff and Jean Comaroff have been attending to such a "judicialization of politics" in postapartheid South Africa and how it has affected social mobilization, particularly in the field of HIV/AIDS. Class struggles, they argue, "seem to have metamorphosed into class actions. Citizens, subjects, governments, and corporations litigate against one another, often at the intersection of tort law, human rights law, and the criminal law, in an ever mutating kaleidoscope of coalitions and cleavages" (Comaroff and Comaroff 2006, 26).

The judicialization of right-to-health litigation speaks to a productive "open-source anarchy" at both macro and micro levels in Brazil as well. Political scientist Luis Werneck Vianna (1999) would say that this is only one part of a broader pattern of the judicialization of politics in the country. For him, judicialization does not necessarily reflect judicial activism. Rather, it can be understood as a lever for multiple minority actors (from political parties to public defenders to civil society groups) to constitutionally challenge the political majority's efforts to determine the fundamental norms and objectives of government. In attending to these concrete and dynamic processes, the complex way in which the judiciary actively participates in everyday politicking in a large country with a young constitution comes to the foreground. The question is thus not who—the judiciary or the executive—is right in the debate over judicialization, but how to integrate their actions to best serve individuals and collectives while making democratic institutions more robust.

In this new chapter of the Brazilian history of citizenship and the right to health, then, the judiciary has become a powerful arbiter and purveyor of care and access to medical technology. Interviews we conducted with judges, attorneys, and health officials revealed divergent and conflicting views on the litigation pathway. Policy makers and administrators contend that the judiciary is overstepping its role and that judicialization skews budgets and increases inequalities in health care access. Some acknowledge, however, that legal pressure has improved the distribution of some medicines.

Many local judges working on right-to-health cases feel they are responding to state failures to provide needed medicines and that these waves of lawsuits are a milestone in the democratization of a culture of rights. For these judges, the poor Brazilians who are working through modes of legally arbitrated justice to access health care are not just fighting against legalized privileges and legitimated inequalities, as in James Holston's (2009) chronicle of "insurgent citizenship" practices in Brazil's urban spaces. Rather, they see widespread litigation as the expression of a distinct, equalizing legal system and of a novel rights-conscious society. Whether such a democratization of socioeconomic rights can be attained through individual claims and in courts, however, is contested. In fact, judges employ idiosyncratic rationales and create their own standards in adjudicating right-to-health cases. They tend to rule in terms of "risk of death" and "right to life" and base their rulings for the most part on constitutional interpretations and personal experiences—having specific tragic cases in mind.

The judiciary recognizes that the judicialization of health has the potential to attend to social inequality and to affirm citizens' rights. Like Souza at the Public Defender's office, Judge Eugenio Terra finds that lawsuits are largely filed by poor and desperate patients seeking treatments that should be available in the public system. He is in charge of all health-related cases in Porto Alegre.

"I am doing social justice, one by one," Terra told me in an interview in August 2011. "When I am issuing an injunction for cancer treatment provision, I am also indicting services that have not kept up with people's needs." It did not escape Terra that the high number of right-to-health lawsuits in southern Brazil might well speak of "a distinct political culture" fostered by numerous administrations of the Workers'

Party (Partido dos Trabalhadores) both in the capital and at the state level in the past two decades (the Workers' Party regained state power in 2011).

Patient-Citizen-Consumers

Even as judges recognize the constitutionality of individual lawsuits and grant requested medicines in the overwhelming majority of cases, the judiciary has repeatedly avoided directly mandating changes in policy or issuing decisions that would broadly affect the public health system. In April 2009, the Brazilian Supreme Court held a rare public hearing to examine the pressing challenges posed by right-to-health litigation (Supremo Tribunal Federal 2009). Public health officials, lawyers, physicians, activists, and academics testified before the court, providing varied viewpoints and recommendations on how to respond to the enormous judicial demand for medical goods. As an immediate outcome, there was a long overdue updating of governmental drug formularies. The Brazilian National Council of Justice also issued a set of recommendations for local judges, asking them to more systematically attend to scientific evidence and to strive for "more efficiency" when ruling on health-related cases (Conselho Nacional de Justiça n.d.).

If access to AIDS therapies was the litmus test of the right to health in the 1990s, access to genetic therapies now plays this role. Twelve-year-old Alexandre Lima de Moura suffers from mucopolysaccharidosis (MPS), an inherited metabolic disorder. Every week, the fourth grader travels with his mother, Cleonice, to Hospital de Clínicas in Porto Alegre, where he receives enzyme replacement therapy, a treatment that costs about US $200,000 per year. Because of his age, Alexandre was not allowed to enroll in a clinical trial taking place at the hospital. Without "the right to be researched," as the mother of another MPS patient put it, Alexandre became a patient litigant.

With the legal support of a well-organized patient association in São Paulo (partially funded by the drug manufacturer), the family won a court injunction forcing the federal government to begin providing the therapy. Like all parents of MPS children we spoke to, Cleonice suggested that not obtaining this treatment would be unconscionable and tantamount to killing her child. She knew that the federal attorneys would appeal and was ready for the struggle: "Besides entering the judiciary, we also entered the media." Cleonice has taken Alexandre's cause to all possible media outlets and is also using his condition to educate neighbors, local medical personnel, and officials about the meaning of, in her words, "citizenship" and a "normal life." "Ela é uma mãe boa" [She is a good mother], says Alexandre, who is thriving in school and seems to be responding positively to treatment.

One of the latest right-to-health landmark cases involves a request for a high-cost medicine for a genetic disease. This treatment was not recommended by the Ministry of Health's therapeutic guidelines and was not publically available. In March 2010, the court rejected the state's argument that it was not responsible for providing the medicine and decided in favor of the provision of the treatment. In his ruling, Justice Gilmar Mendes stated that once the disease was medically confirmed and treatment was indicated, "the Ministry of Health's guidelines can be questioned." Moreover, "the state has to provide resources, not only to support and fund the provision of universal care for its citizens, but also has to provide variable resources to attend to the needs of each individual citizen" (Supremo Tribunal Federal 2010).

The role of market forces in judicialization—a mix of clinical trials and marketing strategies that target physicians' prescriptions and fuel patient demand and of industry lobbying to have new treatments included in governmental drug formularies while facing limited regulatory oversight—must not be overlooked (Petryna 2009). Ample evidence shows that laboratories' monopoly on medico-scientific information and pharmaceutical marketing strongly inform physicians' prescriptive habits and patients' demands (Lakoff 2006). Additional qualitative studies are in order—they could help us chart how judicialization has become part of a pharmaceutical business plan in Brazil, with companies supporting patient associations and lawsuits for access to high-cost medicines specifically to open or enlarge markets (Diniz, Medeiros, and Schwartz 2012).

There is a heated debate in Brazilian courts on the positive duty the constitutional right to health imposes on the state and the extent to which the courts must enforce this right. But the country lacks a substantial public debate about the meaning of the right to health in light of medical advancements and financing, between what is possible and feasible and what is frugal and essential. As a "right to pharmaceuticals" is consolidated in Brazil, the various branches of government have yet to develop a systematic approach to tackling drug value and financing and the responsibilities of private health insurance plans to cover drug costs. Moreover, how can access to new medical technologies be reconciled with systems that foster the equitable inclusion of people in preventive as well as basic and sustained care initiatives? Is there a way to balance individuals' urgent demands for health care, often in the form of medicines, with the long-term, programmatic aspect of health care management and reform? Attention is also needed on the many other factors that play into the right to health, such as education, water quality, sanitation, vector control, air pollution, and violence prevention. These complementary concerns, which can be understood as social determinants of health, are critical to addressing the health needs of both the chronically ill and comorbid individuals in our database and the Brazilian population more generally.

Meanwhile, hard-to-pin-down *patient-citizen-consumers* draw from human rights language and jurisprudence and make governments work for them as they negotiate medical inclusion and the vagaries of the market and survival. The judicialization of health has, indeed, become a para-infrastructure in which various public and private health actors and sectors come into contact, face off, and enact limited "one by one" missions.

Conclusion

There is no pregiven biopolitical population in Brazil today to which Edgar, Nelson, Alexandre, and thousands of other atomized subjects of rights belong. Yet, in their private efforts to become such subjects, individuals have to rely on social relations and temporary collectivities that crop up at the intersection of patient-family demand, state institutions, therapeutic markets, and law.

Seen from the perspective of these medical subjects—undesirable, according to actual care delivery policies, budgets, and state public relation efforts—biopolitics is an insecure enterprise indeed, more a symptom of the limits of government than a marker of its presence and control. The ethnographic realities presented throughout this chapter also suggest that the subject of rights and the economic subject may actually be included or excluded according to shared or similar logics, practices, technologies, and knowledge and that the pursuit and enforcement of rights may be a key means by which one becomes part of a market segment.

If, for Foucault, "the question of the frugality of government is indeed the question of liberalism" (2008, 29), then in Brazil's late-liberal moment, one could argue, the biopolitical question is not necessarily about the "futility" of the rehabilitation of diseased and underserved poor subjects (Biehl 2005) but about the expansion of frugal government in the form of pharmaceutical access in lieu of infrastructural reform. Thus, in this contemporary republic of interests, we see the consolidation of an "inclusionary state activism without statism" (Arbix and Martin 2010) coupled with extraordinary market expansion and the vanishing of "civil society" as a viable transactional reality.

I have written elsewhere about ambiguous political subjects in light of the country's pioneering policy of universal HIV/AIDS treatment access as it was actualized in the context of poor urban communities: "Their political subjectivity is articulated through pastoral means, disciplinary practices of self-care, and monitored pharmaceutical treatment" (Biehl 2007b, 324–325). For Souza and her patient-citizens at the Public Defender's office, politics is not a sphere but a lack, a technology, and a process all at once. In Brazil today, medical commodities work in tandem with other ways of claiming citizenship, and desperate and creative interactions occasion novel public sites in which rights and health are privatized alongside the emergence of novel political subjectivities.

In the face of this situation, the public defender puts up a fight. Souza's pragmatic critique of the state brings attention to the symbiotic relationship between a hybrid government of social protection and market expansion and the ways that public institutions, in their frugality or futility, acquiesce to the social and biological death of those too ill or too poor to live in the new economy. Yet, as abandoned and injured as they are by various levels of actual government, some people still understand themselves to be the subjects of present rights and try to access care via the judiciary. People refuse to be stratified out of existence.

Souza's humanism and in-your-face politics produce a pathway to improving patients' situations. Against institutional realities that undermine health, control, and effectiveness, public defenders utilize medical and legal modes of veridiction and the framework of constitutional rights and human dignity to sustain their work and demand that the state act biopolitically.

Chronically ill and poor people find their way into the judiciary reluctantly, tinkering with available human and material resources. They are neither governable nor disruptive of the system. This minimum biopolitical belonging is part and parcel of the immanent field people invent to live in and by as they navigate the vagaries of market inclusion and survival in wounded cities. So, although right-to-health litigation is not by itself a solution for health problems or social injustice, it is an opening for citizens to engage creatively with public and private institutions and to reshape the social practice of care and demand that the state guard its citizens' human rights. It is not a substitute for health policies or health care systems, nor does it contribute to the privatization and fragmentation of those systems; on the contrary, right-to-health litigation allows for the reentry of human voices in these debates and institutions.

It is, paradoxically, by revealing the fragility of biopolitical interventions, showing how they are constantly entangled with and shaped by other (often economic) imperatives, that the stories of these patient litigants point to the temporal dimensions of medical technologies and to their own power to remake subjectivities and social worlds as they open up new spaces for claim making, contestation, and ethical problematization. It is at the intersection of the therapeutic imperative, the biotechnical embrace, and market reasoning that the intensity of survival becomes visible and the political battle over what is frugal and what is vital is played out.

Acknowledgments

I wish to express my deepest gratitude to Joseph J. Amon, Mariana P. Socal, and Adriana Petryna for their wonderful engagement with this research project. Thank you also to Ramah McKay, Peter Locke, Amy Moran-Thomas, Alexander Wamboldt, Igor Rubinov, Alex Gertner, Joshua Franklin, Jeferson Barbosa, Raphael Frankfurter, and Naomi Zucker for their insights and help. This chapter draws from João Biehl, "The Judicialization of Biopolitics," *American Ethnologist* 40(3):419–436. © 2013 Wiley/American Anthropological Association. Reproduced with permission of Blackwell Publishing Ltd. The Ford Foundation and Princeton's Health Grand Challenges Initiative and Woodrow Wilson School of Public and International Affairs generously supported research. Whenever requested, names of informants were changed to protect anonymity.

Notes

1. This chapter derives from a 2008–2012 multidisciplinary investigation of the judicialization of the right to health in southern Brazil (see Biehl n.d.).
2. See http://www.princeton.edu/grandchallenges/health/research-highlights/aids/Database_project.pdf.

References

Anand, Nikhil. 2012. Municipal disconnect: On abject water and its urban infrastructures. *Ethnography* 13 (4):487–509.

Appel, Hannah. 2012. Walls and white elephants: Oil extraction, responsibility, and infrastructural violence in Equatorial Guinea. *Ethnography* 13 (4):439–465.

Arbix, Glauco, and Scott M. Martin. 2010. Beyond developmentalism and market fundamentalism in Brazil: Inclusionary state activism without statism. Available at http://law.wisc.edu/gls/documents/paper_arbix.pdf (accessed April 21, 2013).

Azevedo, Solange. 2007. Remédios nos tribunais. *Revista Época*. December 12. Available at http://revistaepoca.globo.com/Revista/Epoca/0,EDG80696-8055-501,00-REMEDIOS+NOS+TRIBUNAIS.html (accessed April 21, 2013).

Biehl, João. 2005. *Vita: Life in a Zone of Social Abandonment*. Berkeley: University of California Press.

Biehl, João. 2007a. Pharmaceuticalization: AIDS treatment and global health politics. *Anthropological Quarterly* 80 (4):1083–1126.

Biehl, João. 2007b. *Will to Live: AIDS Therapies and the Politics of Survival*. Princeton: Princeton University Press.

Biehl, João. N.d. Right to health litigation. Available at http://joaobiehl.net/global-health-research/right-to-health-litigation/ (accessed April 21, 2013).

Biehl, João, Joseph J. Amon, Mariana P. Socal, and Adriana Petryna. 2012. Between the court and the clinic: Lawsuits for medicines and the right to health in Brazil. *Health and Human Rights* 14 (1):1–17.

Biehl, João, Byron Good, and Arthur Kleinman, eds. 2007. *Subjectivity: Ethnographic Investigations*. Berkeley: University of California Press.

Biehl, João, and Ramah McKay. 2012. Ethnography as political critique. *Anthropological Quarterly* 85 (4):1209–1227.

Biehl, João, and Adriana Petryna. 2011. Bodies of rights and therapeutic markets. *Social Research* 78 (2):359–386.

Biehl, João, Adriana Petryna, Alex Gertner, Joseph J. Amon, and Paulo D. Picon. 2009. Judicialisation of the right to health in Brazil. *Lancet* 373 (9682):2182–2184.

Borges, Danielle da Costa Leite, and Maria Alicia Dominguez Ugá. 2010. Conflitos e impasses da judicialização na obtenção de medicamentos: As decisões de 1a instância nas ações individuais contra o Estado do Rio de Janeiro, Brasil, em 2005. *Cadernos de Saúde Pública* 26 (1):59–69.

Chieffi, Ana Luiza, and Rita Barradas Barata. 2009. Judicialização da política pública de assistência farmacêutica e eqüidade. *Cadernos de Saúde Pública* 25 (8):1839–1849.

Collier, Stephen. 2011. *Post-Soviet Social: Neoliberalism, Social Modernity, Biopolitics*. Princeton: Princeton University Press.

Collucci, Cláudia. 2009. Triplicam as ações judiciais para obter medicamentos. *Folha de São Paulo*, January 9. Available at http://www1.folha.uol.com.br/fsp/saude/sd0901200901.htm (accessed April 21, 2013).

Comaroff, John, and Jean Comaroff. 2006. Law and disorder in the postcolony: An introduction. In *Law and Disorder in the Postcolony*, ed. Jean Comaroff and John Comaroff, 1–56. Chicago: University of Chicago Press.

Comaroff, John, and Jean Comaroff. 2011. *Theory from the South, or, How Euro-America Is Evolving toward Africa*. Boulder, CO: Paradigm.

Conrad, Peter. 2007. *The Medicalization of Society: On the Transformation of Human Conditions into Treatable Disorders*. Baltimore: Johns Hopkins University Press.

Conrad, Peter, Thomas Mackie, and Ateev Mehrotra. 2010. Estimating the costs of medicalization. *Social Science and Medicine* 70 (12):1943–1947.

Conselho Nacional de Justiça. N.d. Encontro anual da Enasp avalia trabalho e discute metas. Available at http://www.cnj.jus.br/index.php?option=com_content&view=article&id=10547:recomendacao-no-31-de-30-de-marco-de-2010&catid=60:recomendas-do-conselho&Itemid=515%29 (accessed April 21, 2013).

Constituição Federal do Brasil. 1988. Available at http://dtr2004.saude.gov.br/susdeaz/legislacao/arquivo/01_Constituicao.pdf (accessed April 21, 2012).

Das, Veena. 2007. *Life and Words: Violence and the Descent into the Ordinary*. Berkeley: University of California Press.

Da Silva, Virgilio Alfonso, and Fernanda Vargas Terrazas. 2011. Claiming the right to health in Brazilian courts: The exclusion of the already excluded? *Law and Social Inquiry* 36 (4):825–853.

Diniz, Debora, Marcelo Medeiros, and Ida Vanessa D. Schwartz. 2012. Consequences of the judicialization of health policies: The cost of medicines for mucopolysaccharidosis. *Cadernos de Saúde Pública* 28 (3):479–489.

Ecks, Stefan. 2008. Global pharmaceutical markets and corporate citizenship: The case of Novartis' anti-cancer drug Glivec. *Biosocieties* 3 (2):165–181.

Economist. 2011. Health care in Brazil: An injection of reality. *Economist*, July 30. Available at http://www.economist.com/node/21524879 (accessed April 21, 2013).

Edmonds, Alexander. 2010. *Pretty Modern: Beauty, Sex, and Plastic Surgery in Brazil*. Durham: Duke University Press.

Farmer, Paul. 2003. *Pathologies of Power: Health, Human Rights and the New War on the Poor*. Berkeley: University of California Press.

Fassin, Didier. 2007. *When Bodies Remember: Experiences and Politics of AIDS in South Africa*. Berkeley: University of California Press.

Fassin, Didier, and Mariella Pandolfi. 2010. *Contemporary States of Emergency: The Politics of Military and Humanitarian Interventions*. New York: Zone Books.

Ferguson, James. 1994. *The Anti-Politics Machine: "Development," Depoliticization, and Bureaucratic Power in Lesotho*. Minneapolis: University of Minnesota Press.

Ferraz, Octavio Luiz Motta. 2009. The right to health in the courts of Brazil: Worsening health inequities? *Health and Human Rights* 11 (2):33–45.

Fidler, David. 2007. Architecture amidst anarchy: Global health's quest for governance. *Global Health Governance* 1 (1):1–17.

Fidler, David. 2008. Global health jurisprudence: A time of reckoning. *Georgetown Law Journal* 96 (2):393–412.

Foucault, Michel. 2008. *The Birth of Biopolitics: Lectures at the Collège de France, 1978–1979*. Ed. Michel Senellart. Trans. Graham Burchell. New York: Palgrave Macmillan.

Gauri, Varun, and Daniel M. Brinks, eds. 2008. *Courting Social Justice: Judicial Enforcement of Social and Economic Rights in the Developing World*. Cambridge: Cambridge University Press.

Gertner, Alex. 2010. Science of uncertainty: Making cases for drug incorporation in Brazil. *Anthropological Quarterly* 83 (1):97–122.

Good, Mary-Jo DelVecchio, Sandra Teresa Hyde, Sarah Pinto, and Byron J. Good, eds. 2008. *Postcolonial Disorders*. Berkeley: University of California Press.

Greene, Jeremy A. 2010. When did medicines become essential? *Bulletin of the World Health Organization* 88:483.

Greene, Jeremy A. 2011. Making medicines essential: The emergent centrality of pharmaceuticals in global health. *Biosocieties* 6 (1):10–33.

Han, Clara. 2012. *Life in Debt: Times of Care and Violence in Neoliberal Chile*. Berkeley: University of California Press.

Holston, James. 2009. *Insurgent Citizenship: Disjunctions of Democracy and Modernity in Brazil.* Princeton: Princeton University Press.

Lakoff, Andrew. 2006. *Pharmaceutical Reason: Knowledge and Value in Global Psychiatry.* Cambridge: Cambridge University Press.

Lantz, Paula M., Richard L. Lichtenstein, and Harold A. Pollack. 2007. Health policy approaches to population health: The limits of medicalization. *Health Affairs* 26 (5):1253–1257.

Lock, Margaret. 2003. Medicalization and the naturalization of social control. In *Encyclopedia of Medical Anthropology: Health and Illness in the World's Cultures.*Vol. 1, ed. Carol R. Ember and Melvin Ember, 116–125. New York: Springer.

Marques, Silvia Badim, and Suell Gandolfi Dallari. 2007. Garantia do direito social à assistência farmacêutica no Estado de São Paulo. *Revista de Saúde Pública* 41 (1):101–107.

McKay, Ramah. 2012. Afterlives: Humanitarian histories and critical subjects in Mozambique. *Cultural Anthropology* 27 (2):286–309.

Messeder, Ana Márcia, Claudia Garcia Serpa Osorio-de-Castro, and Vera Lucia Luiza. 2005. Mandados judiciais como ferramenta para garantia do acesso a medicamentos no setor público: A experiência do Estado do Rio de Janeiro, Brasil. *Cadernos de Saúde Pública* 21 (2):525–534.

Nguyen, Vinh-Kim. 2010. *The Republic of Therapy: Triage and Sovereignty in West Africa's Time of AIDS.* Durham: Duke University Press.

Ong, Aihwa. 2006. *Neoliberalism as Exception: Mutations in Citizenship and Sovereignty.* Durham: Duke University Press.

Ong, Aihwa, and Stephen J. Collier. 2005. *Global Assemblages: Technology, Politics, and Ethics as Anthropological Problems.* Boston: Wiley-Blackwell.

Pepe, Vera Lúcia Edais, Miriam Ventura, João Maurício Brambati Sant'ana, Tatiana Aaragão Figueiredo, Vanessa dos Reis de Souza, Luciana Simas, and Claudia Garcia Serpa Osorio-de-Castro. 2010. Caracterização de demandas judiciais de fornecimento de medicamentos "essenciais" no Estado do Rio de Janeiro, Brasil. *Cadernos de Saúde Pública* 26 (3):461–471.

Petryna, Adriana. 2009. *When Experiments Travel: Clinical Trials and the Global Search for Human Subjects.* Princeton: Princeton University Press.

Petryna, Adriana, Andrew Lakoff, and Arthur Kleinman, eds. 2006. *Global Pharmaceuticals: Ethics, Markets, Practices.* Durham: Duke University Press.

Rabinow, Paul, and Nikolas Rose. 2006. Biopower today. *Biosocieties* 1 (2):195–217.

Scheffer, Mario, Andrea Lazzarini Salazar, and Karina Bozola Grou. 2005. *O remédio via justiça: Um estudo sobre o acesso a novos medicamentos e exames em HIV/Aids no Brasil por meio de ações judiciais.* Brasília, DF: Ministério da Saúde.

Scheper-Hughes, Nancy. 1992. *Death without Weeping: The Violence of Everyday Life in Brazil.* Berkeley: University of California Press.

Sindicato da Indústria de Produtos Farmacêuticos no Estado de São Paulo (SINDUSFARMA). 2012. Indicadores econômicos: Vendas em dólares (US$). Available at http://www.sindusfarmacomunica.org.br/indicadores-economicos (accessed April 21, 2013).

Spencer, Jonathan. 2007. *Anthropology, Politics and the State: Democracy and Violence in South Asia*. Cambridge: Cambridge University Press.

Supremo Tribunal Federal. 2000. Acórdão. RE 271286 AgR/RS. Agravo Regimental no Recurso Extraordinário. Relator: Celso de Mello. Diário da Justiça Eletrônico. November 24. Brasília, DF: Supremo Tribunal Federal.

Supremo Tribunal Federal. 2009. Principal. Available at http://www.stf.jus.br/portal/cms/verTexto.asp?servico=processoAudienciaPublicaSaude (accessed April 22, 2013).

Supremo Tribunal Federal. 2010. Acórdão STA/175 AgR/CE- Suspensão de Tutela Antecipada. Ministro Gilmar Mendes. April 30. Brasília, DF: Supremo Tribunal Federal.

Vianna, Luiz Werneck. 1999. *A judicialização da política e das relações sociais no Brasil*. Rio de Janeiro: Revan.

Vieira, Fabiola Sulpino. 2009. Ministry of Health's spending on drugs: Program trends from 2002 to 2007. *Revista de Saúde Pública* 43 (4):674–681.

Vieira, Fabiola Sulpino, and Paola Zucchi. 2007. Distorções causadas pelas ações judiciais à política de medicamentos no Brasil. *Revista de Saúde Pública* 41 (2):1–8.

Yamin, Alicia Ely, and Siri Gloppen, eds. 2011. *Litigating Health Rights: Can Courts Bring More Justice to Health?* Cambridge, MA: Harvard University Press.

Zero Hora. 2012. Saúde conquistada na justiça. *Zero Hora*, August 29.

Contributors

Pedro Ignacio Alonso is an architect, holds a MSc in architecture from the Pontificia Universidad Católica de Chile (2000), and completed his PhD at the Architectural Association in London (2008). Together with Hugo Palmarola, he curated the Chile Pavilion at the 14th International Architecture Exhibition (Biennale di Venezia, 2014) and authored the book *Panel* (2014). Alonso has taught at the Architectural Association since 2005, currently as a Visiting Tutor in the History and Critical Thinking M.A. Program. He also teaches theory of architecture and design at the Pontificia Universidad Católica de Chile, where he directs the Master's Program in Architecture and is the program director of the Architectural Association's Visiting School to Santiago.

Morgan G. Ames is a postdoctoral research fellow at the Intel Science and Technology Center for Social Computing in the Donal Bren School of Information and Computer Sciences at the University of California, Irvine. She completed her PhD in communication at Stanford University in 2013, where she received the Nathan Maccoby Outstanding Dissertation Award and was a National Science Foundation graduate fellow. Ames holds a PhD minor in anthropology from Stanford. She also holds a master's in information science and a bachelor's degree in computer science from UC Berkeley.

Javiera Barandiarán is assistant professor in the Global Studies program at the University of California, Santa Barbara. Barandiarán received her PhD in 2013 from the University of California, Berkeley, in environmental science, policy, and management. She holds a master's in public policy also from Berkeley. Her research has been awarded support from the Social Science Research Council and the National Science Foundation.

João Biehl is Susan Dod Brown Professor of Anthropology and faculty associate of the Woodrow Wilson School of Public and International Affairs at Princeton University. He is also the codirector of Princeton's Program in Global Health and Health Policy. Biehl is the author of the award-winning books *Vita: Life in a Zone of Social Abandonment* (revised edition, 2013) and *Will to Live: AIDS Therapies and the Politics of*

Survival (2007). He is the coeditor (with Byron Good and Arthur Kleinman) of *Subjectivity: Ethnographic Investigations* (2007) and (with Adriana Petryna) of *When People Come First: Critical Studies in Global Health* (2013).

Anita Say Chan is assistant research professor of communications and assistant professor of media studies in the Department of Media and Cinema Studies at the University of Illinois, Urbana-Champaign. Her book on the competing imaginaries of global connection and information technologies in contemporary Peru, *Networking Peripheries: Technological Futures and the Myth of Digital Universalism*, was published by the MIT Press in 2013. Her research has been awarded support from the Center for the Study of Law and Culture at Columbia University's School of Law and the National Science Foundation, and she has held postdoctoral fellowships at the CUNY Graduate Center's Committee on Globalization and Social Change and at Stanford University's Introduction to Humanities Program.

Amy Cox Hall holds a PhD in cultural anthropology from the University of Florida and a master of science in history of art from the University of Edinburgh. Her research focuses on photography, indigeneity, cultural heritage, and scientific expeditions in Peru and was supported by the National Science Foundation and Fulbright. She is currently working on a manuscript on the photography and science of the Yale Peruvian expeditions and teaches at the University of North Carolina, Charlotte.

Henrique Cukierman is professor at the Universidade Federal do Rio de Janeiro with affiliations in the Systems Engineering and Computer Science Department of the Alberto Luiz Coimbra Institute–Graduate School and Research in Engineering, the Graduate Program in the History of Sciences and Techniques and Epistemologies, and the Computer Engineering and Information Department of the Polytechnic School. He is the author of *Yes, We Have Pasteur: Manguinhos, Oswaldo Cruz and the History of Science in Brazil* (2007). He has held visiting positions at Stanford University (2001–2002), the Deutsches Museum (2009), and the Universität Konstanz (2009–2010) and is an Alexander von Humboldt Foundation fellow.

Ana Delgado is a postdoctoral fellow at the Centre for the Study of the Sciences and the Humanities at the University of Bergen, Norway. She is a social anthropologist by training and holds a PhD from the Autonomous University of Barcelona and the University of Bergen. Delgado is interetsed in how biological objects are stabilized, legitimized, and made public. She has done research on environmental conflicts, particularly on social movements and agro-biodiversity conservation, and also on emerging technosciences such as nanotechnology and synthetic biology.

Rafael Dias is a professor at the School of Applied Sciences and in the Science and Technology Policy Program of the State University of Campinas, Brazil. He is the director of the Innovation Policy Analysis Research Group and a research associate of the Group for Social Studies of Science and Technology (Brazil) and of the Technology Policy Research Center (US). He is a member of the Advisory Council of the Technological Incubator of Popular Cooperatives (ITCP) of the State University of Campinas. Rafael holds a master's and a PhD in science and technology policy (State University of Campinas, 2009).

Adriana Díaz del Castillo H. holds a master's degree in medical anthropology from the University of Amsterdam and an MD from the Universidad Nacional de Colombia. Her STS research has to do with human population genetics, forensics, and information systems and their relation to nation-building processes in Colombia. She currently works as an independent researcher and consultant in the study of urban infrastructures, equality, and well-being.

Mariano Fressoli is a sociologist and research assistant at the Consejo Nacional de Investigaciones Científicas y Técnicas in Argentina. He is currently working at the Institute of Studies on Science and Technology, Universidad Nacional de Quilmes, Argentina. He also has taught sociology of technology at the Universidade Federal da Integração Latino-Americana in Brazil and at other universities in Argentina. Fressoli holds an MA in cultural studies from Goldsmiths College, University of London, and a PhD in social sciences from the University of Buenos Aires. He is a member of the board of directors of *Revista Redes—Revista de estudios sociales de la ciencia*.

Jonathan Hagood is an assistant professor in the Department of History at Hope College, a position he began in 2008. He holds a PhD (2008) and an MA (2005) in Latin American history from the University of California, Davis. His undergraduate work was in both architecture and Latin American studies at the University of Texas, Austin (1998).

Christina Holmes is adjunct professor in the Department of Anthropology at St. Francis Xavier University, Antigonish, Nova Scotia, and coinvestigator in the Technoscience and Regulation Research Unit at Dalhousie University. Holmes holds a PhD (2008) from Dalhousie University in social anthropology and was a postdoctoral fellow at the École des Hautes Études en Sciences Sociales in Paris in 2010–2012. Her current areas of research are agricultural biotechnology, knowledge translation and science, health services quality, and local food production.

Matthieu Hubert is a researcher at the Consejo Nacional de Investigaciones Científicas y Técnicas and member of the Centro de Ciencia, Tecnología y Sociedad in Buenos Aires. He received his PhD in sociology from the University of Grenoble (France) in 2009, and was a postdoctoral fellow at the École des Hautes Études en Sciences Sociales in Paris in 2010–2011. His main research topics deal with the sociology of scientific work, social studies of nanotechnology, and ethnography of technoscientific platforms (book in press with Editions des Archives Contemporaines).

Noela Invernizzi is an Uruguayan anthropologist and holds a master's and PhD in science and technology policy (State University of Campinas, Brazil). She works at the education undergraduate program and the public policy graduate program of the Federal University of Paraná in Curitiba, Brazil. She had been a postdoctoral researcher at the Consortium for Science, Policy and Outcomes, Columbia University (US); a professor at the Development Studies Program at Zacatecas Autonomous University, Mexico; and a fellow at the Science, Technology and Innovation Program of the Woodrow Wilson International Center for Scholars. She coordinates the Latin American Network on Nanotechnology and Society (ReLANS).

Michael Lemon is a PhD student in history at Indiana University, Bloomington. He received the inaugural Pamela Laird communications research grant from the Society for the History of Technology (SHOT) in support of his dissertation research on the relationship between television technology, media, and politics during the presidency of Salvador Allende in Chile. He holds a BA in history from Oberlin College and an MA in Latin American studies from Indiana University, Bloomington.

Ivan da Costa Marques received his PhD in electrical engineering and computer science from the University of California, Berkeley in 1973. From 1977 to 1980 he was the coordinator for computer industry policy at CAPRE, the Brazilian government agency in charge of defining rules for computer manufacturing in Brazil. From 1981 to mid-1986 he was the CEO and main shareholder of a small private computer manufacturer in Rio de Janeiro (EBC-Embracomp), and from July 1990 to July 1992 he was president and CEO of the Brazilian state-owned computer manufacturer (COBRA). In 1990 he switched fields to the social sciences and held a two-year research fellowship at the Committee for Historical Studies of the New School for Social Research in New York. He is currently an associate professor at the Universidade Federal do Rio de Janeiro. Marques was twice elected vice-president of the Brazilian Association of Historians of Science (2008–2012) and is the first president of the recently formed Brazilian Association for Social Studies of Science and Technology (2011–2013).

Gisela Mateos obtained her PhD in the history of science at the Autonomous University of Barcelona, Spain in 2001. Since 2003 she has been working as a full-time researcher at the Centro de Investigaciones Interdisciplinarias en Ciencias y Humanidades, Universidad Nacional Autónoma de México. Her main research focuses on the history of twentieth-century physics. Currently she and Edna Suárez-Díaz are conducting a research project on cold war science in Mexico.

Eden Medina is associate professor of informatics and computing, adjunct associate professor of history, and director of the Rob Kling Center for Social Informatics at Indiana University, Bloomington. Medina received her PhD from MIT in the history and social study of science and technology and holds a master's in studies of law from

Yale Law School. She is the author of *Cybernetic Revolutionaries: Technology and Politics in Allende's Chile* (MIT Press, 2011), which received the 2012 Edelstein Prize for outstanding book in the history of technology and the 2012 Computer History Museum Prize for outstanding book in the history of computing.

María Fernanda Olarte Sierra holds a PhD in social sciences from the University of Amsterdam, a master's in medical anthropology from the University of Amsterdam, a BA in anthropology from the Universidad de los Andes, and held a postdoctoral fellowship in social anthropology at the University of Manchester. Her research interests revolve around the ensemble of science, technology, and society in ways that talk about nation building and citizenship processes in Colombia, through the lens of individual bodies and communities' experiences. She is an assistant professor of the Department of Design at Universidad de los Andes in Colombia.

Hugo Palmarola holds a master's in history and theory of industrial design from the Universidad Nacional Autónoma de México (2010) and studied Design at the Pontificia Universidad Católica de Chile, where he is a professor and researcher. Palmarola received the International Scholar Award from the Society for the History of Technology (SHOT). Together with Pedro Alonso, he curated the Chile Pavilion at the 14th International Architecture Exhibition (Biennale di Venezia, 2014) and authored the book *Panel* (2014). He is currently completing a PhD in Latin American studies at the Universidad Nacional Autónoma de México on technological imaginaries in the golden age of Mexican cinema.

Tania Pérez-Bustos Tania Pérez-Bustos has a PhD in education with a master's in development studies and did undergraduate studies in social anthropology and communication studies. She is currently working as an assistant professor at the Department of Anthropology in the Pontificia Universidad Javeriana. Her areas of research are related to the feminist politics of knowledge circulation, especially in popular settings (media, nonformal education, initiatives of scientific engagement with broad publics), and to the cultural feminization of certain practices such as matters of care in science and technology.

Julia Rodriguez teaches history at the University of New Hampshire. She studied at the New School and at Columbia University, where she completed a PhD in history specializing in Latin American history and the history of science and medicine. Rodriguez is the author of *Civilizing Argentina: Science, Medicine, and the Modern State* (University of North Carolina Press, 2006), and has published articles in the *American Historical Review*, *Isis*, *Science in Context*, and the *Hispanic American Historical Review*. She is also editor of the open-source teaching website HOSLAC: History of Science in Latin America and the Caribbean (www.hoslac.org). A National Science Foundation CAREER awardee, Rodriguez has received fellowships from the ACLS and the American

Association for the History of Medicine. Her current research focuses on the history of Americanist anthropology in Europe and the Americas.

Israel Rodríguez-Giralt is a lecturer in social psychology at the Open University of Catalonia, Barcelona, Spain. He was a Beatriu de Pinós Postdoctoral researcher at the Sociology Department, Goldsmiths College, University of London (2010–2012). He received his PhD in 2008 from the Universitat Autònoma de Barcelona. His research focuses on the study of science, technology, and democracy.

Edna Suárez-Díaz obtained her PhD on history and philosophy of science at the Universidad Nacional Autónoma de México in 1996, where she is currently professor in the School of Sciences. She has published abundantly on the history of biological sciences in the second half of the twentieth century, in particular on the history of molecular evolution and the origins of genomics. She has done research at the University of California, Irvine, Harvard University, and the Max Planck Institute for the History of Science in Berlin. Currently, she and Gisela Mateos are conducting a research project on cold war science in Mexico.

Hernán Thomas has a PhD in politics of science and technology from the Universidad Estadual of Campinas, Brazil. He is a principal researcher at the Consejo Nacional de Investigaciones Científicas y Técnicas and associate professor and director of the Institute of Studies on Science and Technology at the National University of Quilmes, Argentina. Thomas is also a professor at different universities at Argentina elsewhere in Latin America, including the Facultad Latinoamericana de Ciencias Sociales, Universidad Estadual of Campinas (Brazil), Universidad de La Habana (Cuba), and Universidad de la República (Uruguay).

Manuel Tironi is assistant professor in the Department of Sociology at the Pontificia Universidad Católica de Chile and coordinator of the research group CMAS (Controversias, Medioambiente y Sociedad). He was 2013–2014 visiting research fellow in the Department of Sociology, Goldsmiths College, University of London. He holds an MSc from Cornell University and a PhD from the Polytechnic University of Catalonia.

Dominique Vinck is professor of sociology and STS in the University of Lausanne, member of the Institute for Social Sciences (Lausanne), and member of the Tecnología y Sociedad group of the Universidad de los Andes (Bogotá). He is director of the *Revue d'Anthropologie des Connaissances*. He published *Everyday Engineering: An Ethnography of Design and Innovation* (MIT Press, 2003) and *The Sociology of Scientific Work* (2010).

Index

Page numbers in italic indicate illustrations; "t" after a page number indicates a table.

AABB Community Program (Banco do Brasil Foundation), 94
ABC Color (article), 207
Achahuanco, Neyder, 194, 197, 200, 201
Acosta, Armando, 167
Actor-network theory (ANT), 105
Adaptation, 123–125
Adelante cubanos (*Cubans Forward*) (film), 163
Adorno, Theodor W., 4
AEC. *See* Atomic Energy Commission
AFA. *See* Argentine Physics Association
Agre, Philip, 200
Agreement on Trade Related Aspects of Intellectual Property Rights (TRIPS), 340
Agricultural modernization, in Latin America, 121
Agriculture
 conflicting tension between standardization and diversification, 333, 345
 creole seeds in Brazil, 16, 331–347
Agroecology, 334, 345
AIDS, healthcare in Brazil, 350, 351, 356, 358, 364
Akrich, Madeline, 189
Alba, Francisco, 294
Alemán, Miguel, 292, 293
Alemán law, 292
Alimentação alternativa (Brandão), 105
Allende, Salvador, 167
Alliances, 48, 60
Almandos, Luis Reyna, 151–152
Almeida, M. P., 347
Alonso, Pedro Ignacio, 13, 112, 141, 159
Alvarez, Eduardo M., 139
Álvarez, Santiago, 161, 163
Amaya-Farfán, Jamie, 93
Ames, Morgan G., 14, 102, 103, 184, 207, 212, 221, 246
ANA. *See* National Agroecology Association
Anand, Nikhil, 353
Anderson, Warwick, 3, 4
Anecdotal evidence, 98, 106
Annan, Kofi, 187
Anomaly, 106
ANT. *See* Actor-network theory
Anthropology of politics, 352
Anthropometry, 208, 248, 258
"Anti-conquest," 31–32
Antrosio, Jason, 116
Appadurai, Arjun, 220
Appel, Hannah, 353
ARDEE group of companies, 54
Arellano Hernández, Antonio, 8, 11
Argentina
 atomic energy in, 15, 209, 267–283
 Chagas disease, 209
 fingerprint science in, 3, 13, 139–155
 foreign policy, 272–273

Argentina (cont.)
  growth of science, 144
  history, 145
  Huemul Island, 267–268
  immigration to, 145, 153–154
  industrial expansion, 271–272
  innovation in, 12, 49–52
  mandatory fingerprinting of immigrants, 153–154
  nanoscience policy in, 14, 228, 232t, 235, 236, 237
  nanosectoral fund (FS-NANO), 228
  nanotechnology history, 230–231t
  national fingerprint identification register, 152
  National Institute of Industrial Technology (INTI), 51
  national STS policy, 9
  Nazi Germany and, 275–276, 281
  neoliberalism in, 307
  nuclear projects in, 293, 298
  Peron's "New Argentina," 277, 278, 282
  Proyecto Huemul, 267–270, 275, 276–283
  public investment in R&D, 61
  Pulqui II, 280
  social innovation movement in, 49
  suppression of *La Prensa*, 273–274
  Technologies for Social Inclusion Network (TSIN), 48, 58, 59, 60
Argentinean Nanotechnology Foundation (FAN), 228, 237
Argentine Physics Association (AFA), 274
Arns Neuman, Zilda, 93, 94, 95
Ashcroft, Bill, 18
Atomic energy. *See* Nuclear energy
Atomic Energy Commission (AEC, US), 294
Atoms for Peace initiative, 292, 294
"Atoms in Action" (exhibition), 294
Aymara (language), 193, 201
Aymara (people), 185, 199
Azevedo, Fernando de, 8
Azze, Edmund, 168

Babo, Lamartine, 29, 44
Bachelor, Steven, 290
Balseiro, José A., 268
Bank of Brazil Foundation, 50, 94
Barandiarán, Javiera, 15, 280, 293, 305, 346
Barberena Vega, Miguel Angel, 300
Barbosa, Ruy, 90
Basalla, George, 17
Baticum, 35, 37
Batteau, Allen, 276
Beazley, Francisco, 151
Beccaria, Cesare, 143
Becerra, Oscar, 186–187, 188, 200
Beltrán, Virgilio, 294
Bender, Walter, 190
Berger, Bennett, 210
Bertillonage, 151, 153
Bestor, Paul, 255
Biehl, João, 16, 349
Bijker, Wiebe, 87
Bi-National Sustainability Laboratory (LBNS), 232t
Bingham, Hiram
  about, 245, 258
  choice of camera, 250
  Eastman and, 259
  Inca Land, 258
  Kodak and, 248, 250
  Lost City of the Incas, 247
  on Markham, 257
  Yale Peruvian Expeditions, 14–15, 245–261, *249, 253, 256*
Biodiversity, 332–333
Biomedicine, 357
Bionatur project, 342, 346, 347
Biopolitics, 352, 353
Birn, Anne-Emmanuelle, 290
Block (film), 172
Bode, Germán, 164
Bohío, El (cartoon), 163
Bolivia, 307
Boone, Christopher G., 124

Index

Bottom of the Pyramid (BoP) movement (India), 48, 52–53, 56–57, 58
Bracker, Milton, 273
Brandão, Dr. Clara, 92–93, 94, 95, 105
Brandão, Dr. Rubens, 93
Brazil
    Canadian Light and Power Company and, 124
    cell phones in, 17
    creole seeds, 16, 331–347
    Cruz letter, 28
    Decree 5153/2004, 342
    "farmers' privilege," 342
    folk healing, 33–34, 44
    genetically modified organisms (GMO), 333
    history of STS in, 8
    innovation in, 49–52
    "invention" of, 28, 29, 32, 34
    Jeca Tatu (fictional figure), 88–91, 99–100, 102, 105
    languages, 34–36
    malaria in, 38, 40
    Manguinhos scientific expeditions, 12, 27–45, *30*, *31*, *33*, *35*
    multimistura, 13, 91–99, 100–104, 105
    nanoscience policy in, 14, 227, 232t, 235, 236, 237, 240
    nanotechnology history, 230t, 233
    national STS policy, 9
    neoliberalism in, 307
    New Act on Seeds (2003), 333
    nuclear projects in, 293, 298
    One Laptop per Child (OLPC) project, 221
    One Million Cisterns Program, 50, 59, 61
    ontological politics and local knowledge, 85–107
    Plant Variety Protection Law, 345, 346
    public investment in R&D, 61
    "Regional Validation," 339, 346
    right-to-health legislation, 16, 349–368
    Science without Borders Program, 17
    sense of unity of country, 32
    sertanejo people, 31, 44
    sertão, 44
    social innovation movement in, 49, 57
    Social Technology Network (STN), 48, 49–52, 58, 59, 60
    La Via Campesina, 333–334, 336, 337
Brody, Tomás, 294
Brown, J. Andrew, 127
Bulboulian, Silvia, 294, 295
Bush, Vannevar, 292

Cabral, Pedro Álvares, 29, 44
Callon, Michel, 38, 103
Cameras. *See* Photography
Capitalism, science and, 38–39
Cárdenas, Lázaro, 299
Cardoso, Fernando Henrique, 19
CARE, 54–55
Caring, and forensic genetics, 69, 73–75
Carrillo, Nabor, 288, 291, 292, 293, 295, 299
Castellano, Nicéforo, 152
Castro, Fidel, 162, 165, 170, 171
Castro, Rafael, 168, 170
CBD. *See* Convention on Biological Diversity
Center for Research on Patagonian Ecosystems (CIEP), 322, 325
Central America, nanotechnology history, 230–231t
Central Research Institute for the Experimental Planning of Housing (CNIIEP), 162, 166, 167
CFE. *See* Comisión Federal de Electricidad
Chagas, Carlos, 30, *30*, 38, 40
Chambi, Martin, 257
Chan, Anita Say, 14, 102, 103, 181, 185, 191, 199, 201, 209, 246
Charisma, 208, 213–216, 221
Charismatic objects, 208, 218–221
Chevalier, Michel, 7
Chicago Boys, 305, 306, 308, 311
Children's Pastoral Service of the Brazilian National Bishops' Conference (CNBB), 93, 94, 95, 105, 106

Chile
  Chicago Boys and, 305, 306, 308, 311
  cybernetic computer system and socialist revolution, 126
  energy policy, 15–16, 305–325
  energy politics, timeline, 309t
  Environmental Impact Assessment (EIA), 309t, 316–323, 320t, 325
  HidroAysén, 309t, 310, 317, 319–322, 325
  labor in textile mill, 123
  nanoscience policy, 228, 235, 236
  nanotechnology history, 231t, 233
  national STS policy, 9
  neoliberalism in, 305
  Nuclear Energy Plan, 313
  nuclear energy program, 293, 305–325, 309t
  "social benefit," 312
  Soviet gift of large-panel housing, 167
  Start Up Chile, 17
Chilectra, 309t, 310, 311, 316
Ciclón (*Cyclone*) (film), 161, 163
Ciencia, política y cientificismo (Varsavsky), 9
Cientificismo, use of term, 9
CIEP. *See* Center for Research on Patagonian Ecosystems
Circulation, 298
Cisterns, 50, 59, 61
Claxton, Robert H., 119
CLEI. *See* Latin American Conference on Informatics
CNBB. *See* Children's Pastoral Service of the Brazilian National Bishops' Conference
CNEA. *See* National Atomic Energy Commission
CNEN. *See* National Commission on Nuclear Energy
CNI. *See* National Investigation Commission
CNIIEP. *See* Central Research Institute for the Experimental Planning of Housing
Coatsworth, John, 121
Coexistence, 335
Colbún, 317
Cold War, 287, 298
Collier, Stephen J., 307, 324, 353

Colloredo-Mansfeld, Rudi, 116
Colombia
  care in, 69
  engineering schools in, 119–120
  forensic genetics in, 12, 67–81
  Law 975, 71, 80–81
  nanoscience policy in, 228, 237
  nanotechnology history, 230–231t
  One Laptop per Child (OLPC) project, 221
"Colonial frontier," term, 12
Colonialism, 5–6, 18
Comaroff, John and Jean, 362
Comisión Federal de Electricidad (CFE, Mexico), 299
Comisión National de Energía Nuclear (CNEN, Mexico), 292
Commercial seeds, 334
Compañía de Luz y Fuerza Motriz de Mexico, 299
Computers. *See also* One Laptop per Child projects; XO laptops
  digital education projects, 183
  FLOSS engineers, 184, 185, 190
  ICT for Education (ICT4E) programs, 183
  as "imported magic," 1–2
  low-cost ultraportable machines, 199
CONACYT. *See* National Council for Science and Technology
Concrete panel systems
  adaptation to Cuba, 165, *166*, 175
  construction of José Martí district, *160*, 168, 171
  Cuban factories for, 165, 173
  for Cuban housing, 3, 14, 141, 159–177, *160*
  Cuban workers, 167, 168, 171
  given to Chile by Soviet Union, 167
  given to Muslim Soviet republics, 165–166
  given to Yugoslavia by Soviet Union, 161
  modified large panel, 174
  Noriega, Marta, 174
  rationalized large panel, 173
  transporting panels, 165
  types of panels, 164
Conrad, Peter, 357

Constitutive co-production, 307
Constructionism, 210, 216, 218, 221
Contact zones, 12, 32–33, 36, 44
Convention on Biological Diversity (CBD), 332, 347
Co-production, 307, 324
Coronil, Fernando, 4
Costa, Julio, 146
Costa Rica, 123, 221, 228, 230t, 237
Cotton, cultivation of, 40
Council for Mutual Economic Assistance, 170
Couto Soares, María Clara, 49
Cox Hall, Amy, 14–15, 245, 259
Craig, Mr., 255, 260
"Creole," use of term, 331, 345
Creole seeds
  accounting, 339–340
  in Brazil, 331–347
  commercializing, 340, 342
  community-based catalog, 340
  registering, 338–340, *341*
  selection of, 335–337, 346
  transient standardization, 334–335, 343–344
"Criminal anthropology," 143
Criminal identification cards, 149, *150*
Criminality
  biological origins of, 143
  growth and modernity and, 141, 143
Criminal Man (Ferri), 143
Criminology. *See also* Fingerprinting
  "classical school" of, 143
Cristia, Julián, 183
CR-USA Foundation, 237
Cruz, David, 201
Cruz, Oswaldo, 28, 39, 44
CSI effect, 71, 74, 77
Cuba
  bohíos, 163
  concrete panel factory, 3, 14, 141, 159–177, *160*
  housing films, 161–164
  Hurricane Flora (1983), 160–161, 162
  Ministry of Construction (MICONS), 161
  nanoscience technology policy, 229
  nanotechnology history, 230t
  national STS policy, 9
  Special Period, 173–174, *174*
  Urban Reform Law, 163
Cuba Internacional (magazine), 172
Cuban Institute of Cinematographic Art and Industry (ICAIC), 163, 171
Cuban Revolution, 159
  housing equality and, 162–163
Cubans Forward (*Adelante cubanos*) (film), 163
Cueto, Marcos, 10, 114, 276, 281, 290
Cukierman, Henrique, 12, 27, 44, 60, 141, 257
Cyclone (*Ciclón*) (film), 161, 163

D'Acosta, Hugo, 164, 165
Dactiloscopía, 146
Dactiloscopía comparada (Vucetich), 149
"Dactyloscopic Register of Immigrants" (Argentina), 152–153
Dactyloscopy. *See* Fingerprinting
Dagnino, Renato, 61
Davidson, William, 53
Davyt, Amilcar, 61
de Greiff, Alexis, 197, 291
de la Peña, Carolyn, 127
Delgado, Ana, 16, 331, 345–346
Dependency theories, 9, 18, 19, 120, 289–290
de Pierola, Nicolas, 258
de Rosas, Juan Manuel, 282
Derry, Gregory, 278, 279
de Vecchi Appendini, Bruno, 295, 300
DeWalt, Billie R., 124
D'Hooghe, Alexander, 161
Dias, Rafael, 12, 47
Díaz del Castillo, Adriana, 12, 67
Diffusion narratives, 2, 17, 87
Digital education projects, 183. *See also* Computers; One Laptop per Child projects
Diprose, Rosalyn, 324
"Dirty wars," 290, 291
Diversification, tension between standardization and, 333, 345
DNA, forensic genetics, 70

Domínguez Martínez, Raúl, 299
Dominican Republic, 228–229, 231t
Duarte, Rolando, 167, 174
Duque, Rick, xvi
Dürck, Hermann, 28, 44
Dussel, Enrique, 5

Eastman, George, 248, 250, 259
Ecenarro, Eduardo, 163
ECLA. *See* United Nations Economic Commission on Latin America
Economic growth, in Rostow's theory, 17
Ecuador, 231t
Edgerton, David, 128, 172
Edwards, E., 260
EIA. *See* Environmental impact assessment
Eibenschutz, Juan, 295
Einstein, Albert, 274
Eisenhower, Dwight, 292
Electric power
  atomic research for, 274
  in Mexico, 299
Empire Writes Back, The (Ashcroft, Griffiths, and Tiffin), 18
Endesa, 309t, 310, 311, 313, 314, 316, 325
Enel, 325
Energy policy, in Chile, 15–16, 305–325
Ensmenger, Nathan, 115
Environmental impact assessment (EIA), 309t, 316–323, 320t, 325
Erdis, Ellwood, 259
Escuelab Puno (association), 193–195
ESOCITE. *See* Latin American Conference on the Social Studies of Science and Technology
"Essential" medications, 357, 358
"Euro-American metaphysics," 97

FAN. *See* Argentinean Nanotechnology Foundation
Feibleman, James, 278
Fein, Seth, 290
"Feministing," use of term, 80, 81

Feminist perspective
  forensic genetics in Colombia, 12, 67–81
  science and technology, 4
Fenoglio, Valeria, 61
Ferri, Enrico, 143
Fetishism, 208
Fidler, David, 362
Film, on Cuban housing and prefabrication, 161–164, 170–172
Fingerprinting, 139, 140, 145–146, 149, 150, 153, 154
  in Argentina, 3, 13, 139–155
  Bertillonage, 151, 153
  Galton system, 146
  Henry system, 153
  history of, 146
  Vucetich system, 146, *147*, 148, 153, 154
Fishing, exploitation of, 40
FLOSS engineers, Peru, 184, 185, 190
Folk healing, in Brazil, 33–34, 44
Ford, David, *249*, 258, 259
Forensic genetics
  in Colombia, 12, 67–81
  feminized work, 69
  predominance of women in, 69–71, 80
Forensics, in Colombian news, 71
Formadores program, 212–213
Foucault, Michel, 307, 352, 366
Frames of reference, 85, 103
Fressoli, Mariano, 12, 47, 61, 78
Freud, Sigmund, 100, 101
Frontiers, 32
FUMEC. *See* Mexico-United States Science Foundation

Galton, Sir Francis, 143, 146, 148, 149
Gálvez Cruz, Luis, 300
Gans, Richard, 275
García Cuenca, José Antonio, 163
García Márquez, Gabriel, 17, 111, 127
García Robles, Alfonso, 298
Garreaud, Fernando, 258
Garrido, Santiago, 61

Gaviola, Enrique, 274–275, 279
Gender studies, and forensic genetics, 12, 67–81
General Atomics Co., 295
"General Instructions for the 'Province of Buenos Aires' System of Filiation" (Vucetich), 148
Genetically modified organism (GMO), 333
Genetics. *See* Forensic genetics
GIAN. *See* Grassroots Innovation Augmentation Network
"Give 1, Get 1" program, 222
Glick, Thomas, 277
Globalization, and nanotechnology research, 225
Globo, O (newspaper), 17
GMO. *See* Genetically modified organism
Gobat, Michel, 8
"God's-eye" trick, 87, 91
Goodwin, Paul B., Jr., 121
Gortari, Eli, 8
Graef, Carlos, 291
Grandin, Greg, 287
Grassroots Innovation Augmentation Network (GIAN), 54
Grassroots Innovation (GI) movement (India), 48, 52–53, 56–57, 58
Greene, Jeremy, 358
Griffiths, Gareth, 18
Gross, Hans, 143, 149
Grosvenor, Gilbert, 260
Groundnut digger/separator, Indian program, 54
Guaraní (language), 222
Guatemala, 221, 231t
Guevara, Ernesto (Che), 164, 176
Gupta, Anil K., 52, 58

Hagood, Jonathan, 15, 102–103, 141, 209, 267, 293
Halme, Minna, 56
Haraway, Donna, 68, 81, 260
Harding, Sandra, 4, 29

Harwood, J., 290
Hayek, Friedrich, 325
Healthcare. *See also* Illness
 right-to-health litigation in Brazil, 16, 349–368
Hecht, Gabrielle, 288, 314
Heller, Edmund, 259
Henry, Edward, 153
Hepatitis C, case history, 361–362
Herrera, Amílcar, 8, 9, 18, 19, 47
Hess, David J., 115, 277
HidroAysén, 309t, 310, 317, 319–322, 325
High-tech, 60, 61
High Voltage Engineering Corp., 299
Hispanic American Historical Review (journal), xv, 114, 116–118, *117*, *118*
Hispanic Society of the Americas, 260
Historiographical inscriptions, 92, 105
History and Technology (journal), 114, 116–118, *117*, *118*, 119, 127
HIV/AIDS, and healthcare in Brazil, 350, 351, 356, 358, 364
Hobart, M., 100
Holanda, Sérgio Buarque de, 36
Holston, James, 363
Honey Bee Network (HBN) (India), 53, 54, 55–56, 57, 61
Housing
 bohíos, 163
 concrete panel systems, 3, 14, 141, 159–177, *160*
 khrushchovkas, 162, 167
 kommunalkas, 162, 167
 Soviet Union, 161–163
Housing (La vivienda) (film), 162, 163
Housing blocks. *See* Concrete panel systems
Houssay, Bernardo, 280, 283
Hrdlička, Aleš, 258
Hubert, Matthieu, 14, 225
Hughes, Thomas, 87
Hurricane Flora (1983), 160–161, 162
Hurtado de Mendoza, Diego, 293

IAEA. *See* International Atomic Energy Agency
IANEC. *See* Inter-American Nuclear Energy Commission
ICAIC. *See* Cuban Institute of Cinematographic Art and Industry
Icnofalangométrica, 146
ICT4D project. *See* Information and Communication Technologies for Development project
ICT for Education (ICT4E) programs, 183
IDB. *See* Inter-American Development Bank
Ideological work, 210
Illness. *See also* Healthcare
 healthiness of tropics, 41–42
 language to describe, 34–36, 37
 malaria in Brazil, 38, 40
 tropical dysphagia, 41
 vexame, 34, 35, 37, 38, 41
Immigrants, mandatory fingerprinting of, 153–154
"Immigrant's book" (Argentina), 152
Immigration, 42
Imperial Eyes (Pratt), 31–32
"Imported magic," use of phrase, 1–2, 225, 238, 281, 332
Import substitution industrialization (ISI), 120
Inca Land (Bingham), 258
Incas, 256, 261. *See also* Machu Picchu; Yale Peruvian Expeditions
Inclusive innovation, 48, 60
 in India, 52–57
 shortcomings of, 58
"Indecent," use of term in Chile, 315, 325
India
 Bottom of the Pyramid (BoP) movement, 48, 52–53, 56–57, 58
 Grassroots Innovation (GI) movement, 48, 52–53, 56–57, 58
 groundnut digger/separator, 54
 Honey Bee Network (HBN), 53, 54, 55–56, 57, 61
 inclusive innovation and social entrepreneurship in, 52–58, 61

"Indigenous innovation," 124
INEN. *See* National Institute of Nuclear Energy
Information and Communication Technologies for Development (ICT4D) project (MIT), 181
Innocenti, Bernie, 185, 190–193, 197, 200, 201
Innovation, 12, 47–59
 in Argentina and Brazil, 49–52
 Cuban concrete panel factory, 3, 14, 141, 159–177, *160*
 defined, 197
 fingerprinting, 3, 13, 139–155
 high-tech, 60, 61
 "indigenous innovation," 124
 knowledge inequality and, 57
 local movements in, 60
 microfunding, 55
 nanoscience policy, 14, 225–240
 and OLPC, 14
 patents from, 55
 scouting and documenting, 53, 61
 traditional knowledge and, 57, 58
Inscription, 338, 339, 346
In situ biodiversity, 332–333
Instituto Aerotécnico (Argentina), 280
Instituto Nacional de Energía Nuclear (INEN, Mexico), 296
Instructionism, 210
INTA. *See* National Institute of Agricultural Technologies
Intel, Classmate PC netbooks, 183
Inter-American Development Bank (IDB), 55, 181–182, 183, 188, 198
Inter-American Nuclear Energy Commission (IANEC), 293, 294
Interferences, 334
"Internal brain drain," 9, 19
International Atomic Energy Agency (IAEA), 293, 294, 295, 296
International Treaty on Plant Genetic Resources for Food and Agriculture (IRFAA-FAO), 345

International Union for the Protection of New Varieties of Plants (UPOV), 340
International Union of Architects (UIA), 162
"In the Wonderland of Peru" (National Geographic), 256, 260
INTI. *See* National Institute of Industrial Technology
Invernizzi, Noela, 14, 225
Iraolagoitía, Pedro, 268
IRFAA-FAO. *See* International Treaty on Plant Genetic Resources for Food and Agriculture
Irony of Fate, The (Ryazanov), 172
ISI. *See* Import substitution industrialization
Isomorphism, 229
Ives, W. C., 259

Jaiswal, Anand Kumar, 56
Jasanoff, Sheila, 307, 324
Jáuregui, Carlos A., 5
Jeca Tatu (fictional figure), 88–91, 99–100, 102, 105
Journal of Latin American and Caribbean Anthropology, xv
Journal of Latin American Studies, 114, 116–118, *117, 118*
Judicialization of health, 16, 350

Kaplinsky, Raphael, 56
Karnani, Aneel, 56
Khan, Yusuf, 54
Khrushchev, Nikita, 161, 164
Kipling, Rudyard, 258
Klor de Alva, Jorge, 5
Knowledge economy, nanoscience policy and, 226, 233
Knowledge entities, 102, 281
Knowledge inequality, innovation and, 57
Kodak
    Brigham and, 248, 250, 286
    camera equipment for Yale expeditions, 250, 258
    camera technology, 246, 256
    panorama camera, 250, *256*, 260

Kreimer, Pablo, 8, 11, 61, 209
Kruse, Corina, 71
Kuhn, Thomas S., 106

Labor and technology, histories of, 121–123
Laguna Verde plant, 296
Lakoff, A., 307, 324
Lalouf, Alberto, 61
Landless Rural Workers' Movement (MST), 347
Language
    Manguinhos expeditions and, 34–36
    sympathy for, 36, 37
LANOTEC facility (Costa Rica), 228, 230t
Laptops. *See* XO laptops
Large-panel technology, 159–160. *See also* Concrete panel systems
Latin America. *See also* Science and technology studies in Latin America
    Cold War science in, 289, 291, 298
    defined, 7–8
    "dirty wars," 290, 291
    history of technology and Latin American studies, 111, 127
    languages in, 7–8
    Latin American scholarship, 112
    modernization in, 121
    nanoscience and nanotechnology research, 225–240
    nanotechnology, history in, 230–231t, 233
    neoliberalism in, 306–307
    ontological political perspectives, 87–88
    postcolonial studies, 4
    railroads in, 121
    as region, 6–7
    technology scholarship in, 13, 111–128
    Tratado para la Prohibición de Armas Nucleares en America Latina y el Caribe, 297
    use of term, 7–8
Latin American Conference on Informatics (CLEI), 6

Latin American Conference on the Social Studies of Science and Technology (ESOCITE), xv, 11, 19
Latin American Research Review (journal), 114, 116–118, *117*, *118*
Latin American School of Agroecology of Via Campesina, 334, 345
Latour, Bruno, 37, 75, 166, 325
Law, John, 97, 343, 345
LBNS. *See* Bi-National Sustainability Laboratory
"Learning learning" (Papert), 210
Leloir, Luis, 280, 287
Lemon, Michael, 13, 34, 111, 209
Lemos, Edgar, 349, 358, 359, 366
Libro Negro de la Segunda Tiranía (CNI), 282
Lima, Rocha, 28, 44
Lima de Moura, Alexandre, 364, 366
Lindeman, Sara, 56
Linna, Paula, 56
Lipartito, Kenneth, 282
Local knowledge, ontological politics and, 85–107
Local seeds. *See* Creole seeds
Lombroso, Cesare, 143, 149
London, Ted, 53
Lopez, Michelle, 49
López Mateos, Adolfo, 295, 299
López Sánchez, José, 8
Lost City of the Incas (Bingham), 247
Loyo González, Juan, 292
Lucena, Juan C., 119
Lugo, Fernando, 207
Lutz, Adolpho, 40, 42

Machado, Astrogildo, 40, 42
Machu Picchu, 14–15, 245–251, 255–257, *256*
"Machu Picchu: A Lost City Uncovered: Photographs from the Hiram Bingham Expeditions 1931–1935," 260
"Machu Picchu 102 years: A Glimpse of the Expedition that Astonished the World," 260

Macias, Angel, 168
Macondo (fictional town), 111, 127
Magaña Plaza, Antonio, 300
Magical realism, 2, 111
Majority Program, 55
Malaria, in Brazil, 38, 40
Maldonado, Oscar Javier, 197
Malnutrition, 87, 92–93, 97, 100, 104
Mamani Pacho, Eleazar, 201
Manguinhos Institute, 27
Manguinhos scientific expeditions, 12, 30–45, *30*, *31*, *33*, *35*
  contact zones, 12, 32–33, 36, 44
  describing illness, 34–36, 37
  expansion of markets and, 38–40
  language and, 34–36
  money use by tribes, 39–40
  "perfecting" the races, 41–43
  women in, 45
Mariscotti, Mario, 269–270, 271, 283
Markham, Sir Clements, 257
Marquardt, Steve, 123
Marques, Ivan da Costa, 13, 39, 47, 85, 354
Martí, José, 168
Martinez, Tomas Eloy, 275
Martínez Vergne, Teresita, 122
Martins de Oliveira, Edileusa, 99
Massachusetts Institute of Technology (MIT), Information and Communication Technologies for Development (ICT4D) project, 181
Mateos, Gisela, 15, 112, 141, 159, 282, 287
Mattoon, Robert H., Jr., 121
Mazzaropi, Amácio, 105
McAllister, Carlota, 325
McIntosh, Donald, 208
McOndo Movement, 127
Media, and Cuban housing and prefabrication, 163, 170–172
Medicalization, 357
Medina, Eden, 13, 34, 47, 111, 124, 126, 209
Memoirs of the American Anthropological Association, 258

Mendes, Gilmar, 365
Mercosur, 240
Mexican Revolution, 287
Mexico
  academic exchanges with US institutions, 291, 292
  Alemán Law, 292
  binational nuclear project, 296
  Cold War science in, 289, 290, 298
  Laguna Verde plant, 296
  leftist ideology in universities, 298
  modernization and nuclearity, 291–295
  nanoscience policy in, 14, 227–228, 232t, 235, 237
  nanotechnology history, 230t, 233
  nonalignment of, 290
  nuclear-free position, 290, 297
  nuclear program during the Cold War, 15, 141, 282, 287–300
  nuclear reactors in, 295, 296
  oil industry in, 299
  postrevolutionary period, 287, 299
  Programa Nacional de Reactores de Potencia, 296
  radioisotopes, 294–295, 299
  tilling and planting tool, 124
  Tlatelolco Treaty, 298
  Treaty on the Non-Proliferation of Nuclear Weapons, 297
  Van de Graaff accelerator purchase, 293, 299
  video production in, 126
Mexico-United States Science Foundation (FUMEC), 237
Meyer, Lorenzo, 299
MICONS. See Cuba: Ministry of Construction
Microfunding, 55
Mignolo, Walter, 5, 7
Milanés, Pablo, 171
Millennium Institute (Chile), 228, 233
Miller, Clark A., 291
Minimum cultivation, 107
Minor García, Andrea, 293, 299
Miranda, Isabelle, 49

Mirowski, Philip, 306
"Modernity's Others," 4
Mol, Annemarie, 68, 345
Money, use of, 39–40
Monroy, Ruben, 201
Monteiro Lobato, José Bento Renato, 88–91, 102, 105
Moore, A., 106
Morales Amado, Arnulfo, 300
Moraña, Mabel, 5
Moreira, Tiago, 346
Moreno, Augusto, 294, 295
Moshinsky, Marcos, 295
"Mourning and Melancholy" (Freud), 100
Movimento dos Trabalhadores Sem Terra (MST), 347
Moya, José C., 5
Mucilon (Nestlé), 94
Multimistura, 13, 91–99, 100–104, 105
Multinational companies, 56
Muniz, R. P. A., 293
Muñoz, Ernesto Weigel, 150–151

Nanoscience and nanotechnology
  history of in Latin America, 230–231t, 233
  knowledge economy and, 226, 233
  nanotechnology defined, 240
  policy, 14, 225–240
  publications, 235–237
  research, 234–238
Nanosectoral fund (FS-NANO, Argentina), 228
National Agroecology Association (ANA, Brazil), 336, 337, 338
National Atomic Energy Commission (CNEA, Argentina), 268, 269, 270, 271
National Commission on Nuclear Energy (CNEN, Mexico), 292–293, 294, 295, 296, 297
National Council for Science and Technology (CONACYT, Mexico), 236
National Energy Commission (Chile), 306, 311, 313

National Geographic, 247, 248, 255, 256, *256*, 259, 260
National Innovation Foundation (NIF, India), 54, 55
National Institute of Agricultural Technologies (INTA, Argentina), 51
National Institute of Industrial Technology (INTI, Argentina), 51
National Institute of Legal Medicine and Forensic Sciences (NILM, Colombia), 70
National Institute of Nuclear Energy (INEN, Mexico), 296, 297
National Investigation Commission (CNI, Argentina), 269, 270, 280, 282
National Mapping of Risky Climatic Zones (Brazil), 338
National Reactors Program (Mexico), 296
National Registry (of creole seeds, Brazil), 339
National University (UNAM, Mexico), 292, 293, 294, 295, 297, 299
Navegar (software), 216, 217
Nazi Germany, Argentina's relationship with, 275–276, 281
Necochea López, Raúl, 290
Negroponte, Nicholas, 187, 188, 200
Neiva, Arthur, 31, *31*, 32, 43, 44
Neiva-Penna report, 30–31, 32, 33, 36, 37, 38, 41, 42
Neoliberal economic model, 10, 112, 127
Neoliberalism
  about, 306–307
  as political technology, in Chile, 305–325
New Act on Seeds (Brazil, 2003), 333
NGOs. *See* Nongovernmental organizations
Nicaragua, 221
Nieto, Mauricio, 291
NIF. *See* National Innovation Foundation
NILM. *See* National Institute of Legal Medicine and Forensic Sciences
Nongovernmental organizations (NGOs), 207, 218

"North-South Dialogue on Nanotechnology: Challenges and Opportunities" (2005 conference), 233
No tenemos derecho a esperar (*We Have No Right to Wait*) (film), 170–172, 176
Novoa (prefabricated building systems), 163
Nuclear energy
  in Argentina, 15, 141, 209, 267–283
  in Chile, 305–325, 309t
  during the Cold War, 15, 141, 282, 287–300
  Mexican exhibitions, 294
  Mexican research facility (Salazar Center), 295, 300
  in Mexico, 295, 296
  positive uses of, 294
Nuclear energy policy
  in Chile, 15–16, 305–325
  international character of, 292–293
  in Mexico, 15, 141, 282, 287–300
Nuclearity, 287, 288
Nuclear weapons, nonproliferation of, 297, 298
Nunes, Guillermo J., 149

Oak Ridge National Laboratory, 294
OAS. *See* Organization of American States
O'Bryen, Rory, 127
Oil industry, in Mexico, 299
Olarte Sierra, María Fernanda, 12, 67
One Hundred Years of Solitude (García Márquez), 114, 127
One Laptop per Child (OLPC) projects
  about, 14, 102, 103, 201
  broken laptops, 212, 221
  conversion, 193–196
  formadores program, 212–213
  history, 200
  IDB report, 55, 181–182, 183, 188, 197
  mission statement, 200
  Paraguay, 14, 102, 184, 207–222
  pedagogical experts for, 192
  Peru, 14–15, 102, 181–201, *182*, 209
  rescripting design, 190–193

scale of, 197, 201
scripting reform, 186–190, 200
teacher training, 210–213, 221
Uruguay, 187, 221
One Million Cisterns Program (Brazil), 50, 59, 61
Ontological politics, 13, 85–107
O'Reilly, Vitervo, 161, 164
Organization of American States (OAS), nanoscience and, 233
Ortiz, Eduardo, 279
Ortiz Tirado, José María, 293

Pacheco, Felix, 153
Palacios, Adelaida, 294–295
Palmarola, Hugo, 13, 112, 141, 159
Panelstory (film), 172
Papert, Seymour, 210
Paraguay
 One Laptop per Child project, 14, 102, 184, 207–222
 XO laptops, 14, 184, 207–222
Paraguay Educa, 207, 209, 210, 212, 213, 214, 216, 217–218, 219, 221
Para-infrastructure, 353, 365–366
Pastoral da Criança. *See* Children's Pastoral Service of the Brazilian National Bishops' Conference
Paternity, and forensic genetics, 70, 74
Patient-citizen-consumers, 365
Patriotic and People-Oriented Science and Technology movement (India), 53
PAV. *See* Vacant Areas Program
PEMEX, 299
Penna, Belisário, 31, *31*, 32, 34, *35*, 43, 44
People's Science Movement (India), 52–53
Pereza, Carlota, 240
Pérez Bustos, Tania, 12, 67
Perón, Juan, 15, 267, 270, 271–282
Peru
 annual deployment of XO laptops, 183, 199
 educational spending per student, 182–183, 198

Escuelab Puno (association), 193–195
FLOSS engineers, 184, 185
Machu Picchu, 14–15, 245–251, 255–257, *256*
nanotechnology history, 231t
One Laptop per Child (OLPC) project, 14–15, 102, 181–201, 209, 221
public school teachers, 186, 200
Teacher Career Law, 186, 187
Yale Peruvian Expeditions, 14–15, 245–261, *249*, *253*, *256*
Petrobras, 50
Pharmaceuticalization, 357–358
Pharmaceuticals, and right-to-health litigation in Brazil, 349–368
Photography. *See also* Kodak
 expeditionary science and, 246–247, 255, 257, 258
 expedition circulars on, 251–252, 259
 hand-tinted lantern slides, 252, 259
 panorama camera, 250, *256*, 260
 as scientific tool, 245–246
 sociotechnical relationships in, 256, 260
 and Yale Peruvian Expeditions, 14–15, 245–261
Picabea, Facundo, 61
Pinch, Trevor, 87
Pinhão-barbados nut (*Jatropha curcas*), 45
Pinto, Aníbal, 19
Plan Ceibal (Uruguay), 221
Planck, Max, 274
Plan Huascaran, 201
Plant Variety Protection Law, 345, 346
Plehwe, Dieter, 306–307
Polanco, Xavier, 19
Political technology, 280
 examples of other, 324
 neoliberalism and, 305–308
Poole, Deborah, 251
Portugal, colonial management by, 4
Postcolonial perspective, 3–5, 18
Poverty, technological fixes for, 52
Prahalad, C. K., 52, 53, 56, 58

Pratt, Mary Louise, 12, 32, 40
Prebisch, Raúl, 8
Prefab People, The (film), 172
Prefabrication, of concrete panel systems, 3, 14, 141, 159–177, *160*
Prensa, La (newspaper), suppression of, 273–274
Programa Nacional de Reactores de Potencia (Mexico), 296
Pro-poor innovation, 52, 53, 54, 57–58, 61
Proyecto Huemul, 267–268, 275, 276–283
Puig de la Bellacasa, Maria, 68, 73, 77, 79
Puno (Peru), 193, 209
"Purifications," of Chile nuclear power plan, 310–316, 317, 325

Quechua (language), 193, 201
Quechua (people), 185, 199, 248, *249*
Quipu (journal), 10

Radioisotopes, 294–295, 299
Radiwosky, Simon, 151
Railroads, in Latin America, 121
Rajayoga (Castro), 170
Ramos, Eduardo, 171
Reductionist development model, 87
RedVNano (Venezuelan Nanotechnology Network), 228, 237
"Regional Validation," 339, 346
Registro Nacional de Cultivares (RNC, Brazil), 338, 346
Registro Nacional de Sementes e Mudas (RENASEM, Brazil), 346
Reparations, and forensic genetics, 70, 75, 78
Ribeiro de Andrade, Ana M., 293
Richter, Ronald, 15, 267–268, 276, 277, 279, 280, 283
Right-to-health legislation, in Brazil, 16, 349–368
RNC. *See* Registro Nacional de Cultivares
Roa, Raúl, 171
Roche, Marcel, 19
Rodriguez, Julia, 3, 13, 139

Rodríguez, Silvio, 171
Rodríguez-Giralt, Israel, 16, 331
Rose, Hilary, 73
Rosenthal, Anton, 122–123
Rosner, Daniela K., 212
Rossi, José, 151
Rostow, W. W., 17
Ruiz Cortines, Adolfo, 293, 295

Sábato, Jorge, 8
Saldaña, Juan José, 6, 8
Sánchez, Jorge, 161
Sandino (prefabricated building systems), 163
Sandoval-Vallarta, Manuel, 288, 291, 292, 293, 299
Sarewitz, Daniel, 308
Sarmiento, Domingo Faustino, 128
Schmink, Marianne, 122
Schmitt, Claudia J., 347
Schumacher, E. F., 52
Science
  capitalism and, 38–39
  in the Cold War, 288–289, 298
  defined, 278
  history of, 8, 18
  social history of in Latin America today, 289–290
  transnational histories of, 289
Science and technology studies (STS)
  history of, 10, 18, 87
  scholarship in Latin America, 10–12, 19, 111–128
  social dimensions of, 8–9
Science and technology studies (STS) in Latin America. *See also under names of individual countries*
  about, xv, 1, 6–11
  atomic energy, 15, 141, 209, 267–283
  concrete panel systems in Cuba, 3, 14, 141, 159–177, *160*
  creole seeds in Brazil, 16, 331–347
  critical stance of, 47
  dependency theory, 9, 18, 19

diffusion narratives, 2, 17
feminism, 4, 67–68
fingerprinting, 3, 13, 139–155
forensic genetics, 12, 67–81
future areas of study, 11
history of, 8–11, 87
as "imported magic," 2
and Indian initiatives, 56–57
innovation and social inclusion, 12, 47–61
isolationist tendency, 29
Manguinhos scientific expeditions, 12, 27–45, *30*, *31*, *33*, *35*
nanoscience policy, 14, 225–240, 232t
neoliberalism, 10, 305–325
nuclear energy during the Cold War, 15, 141, 282, 287–300
nuclear energy policy in Chile, 15–16, 305–325
One Laptop per Child, 14, 102, 181–201, *182*, 207–222
ontological politics, 13, 85–107
in politics, 15
postcolonial perspective, 3–5, 18
right-to-health legislation in Brazil, 16, 349–368
scholarship in US vs. Latin America, 10–11, 19
STS as coming from elsewhere, 2, 17
technology scholarship review, 13, 111–128
Yale Peruvian Expeditions, 14–15, 245–261, *249*, *253*
Science without Borders Program, 17
Scranton, Philip, 115, 127
Scratch and Turtle Art (software), 216, 218
Secreto atómico de Huemul, El (Mariscotti), 270
Seeds
  commercial seeds, 334
  creole seeds in Brazil, 331–347
  "farmers' privilege," 342
  genetically modified organisms (GMO), 333
  Plant Variety Protection Law, 345, 346
  registering, 338–340, 346
  selection of, 336–337, 346

"Seeds: Heritage of the People for the Good of Humanity" (campaign), 337, 346
Self-invention, 36
Semi-Arid Association, 50
Serment Cabrero, Vinicio, 300
Serres, Michel, 43, 346
Sertanejo people, 31, 44
Sertão, 44
Sesto, Carmen, 124, 126
SHOT. *See* Society for the History of Technology
Silva, Nelson, 361–362, 366
Singleton, Vicky, 68
"Single window," 317, 325
Slave labor, 122
Smith, Adrian, 61
Smith, Laurel C., 124, 126
Social entrepreneurship, in India, 52–56
Social inclusion, 12, 47
"Socialismo y el hombre en Cuba, El" (Guevara), 176
Social Technology Network (STN, Brazil), 48, 49–52, 58, 59, 60
Social topology, 345, 346
Sociedad Latinoamericana de Estudios Sociales de la Ciencia y la Tecnología. *See* Latin American Conference on the Social Studies of Science and Technology
Sociedad Latinoamericana de Historia de las Ciencias y la Tecnología, 10
Society for Research and Initiatives for Sustainable Technological Institutions (SRISTI), 54, 55
Society for the History of Technology (SHOT), 127
Sociotechnical alliances, 48, 60
Soete, Luc, 240
Solidarity economy, 49, 61
Sora (Indian peoples), 100–101, 107
Sorensen, Diana, 7
Soto Laveaga, Gabriela, 290
Souza, Paula Pinto de, 354–355, 366, 367

Soviet Union
   concrete panels used in, 161, 167
   Cuban concrete panel factory, 3, 14, 141, 159–177, *160*
   housing in, 161–163
   Khrushchev's new building technologies, 161–162, 167
   large-panel systems for Chile, 167
   large-panel systems in Muslim Soviet republics, 165–166
   large-panel technology for Yugoslavia, 161
Spain, colonial management by, 4
Spencer, Jonathan, 352
Spenser, Daniela, 290
SRISTI. *See* Society for Research and Initiatives for Sustainable Technological Institutions
Standardization, tension between diversification and, 333, 345
Start Up Chile, 17
Staudenmaier, John, 121–122, 127
Stavrinaki, Maria, 208
Stengers, Isabelle, 184
Stilgoe, J., 106
STN. *See* Social Technology Network
STS. *See* Science and technology studies
Study for the Social Studies of Science (4S), xv, xvi
Suárez-Díaz, Edna, 15, 112, 141, 159, 282, 287
Suchman, Lucy, 189, 200
Sugar Labs, 190
Sugar workers, 122
Sunkel, Osvaldo, 8, 19
Swift Group, 209
Synchro-cyclotron, 277
Syngenta, 334

Tank, Kurt, 267, 280
Tarde, Gabriel, 346
Teacher training, 210–213, 221
Technological adaptation, 123–125
Technological artifacts, 120

Technologies for Social Inclusion Network (TSIN, Argentina), 48, 58, 59, 60
Technology. *See also* Science and technology studies; Science and technology studies in Latin America
   defined, 115, 128, 278
   García Márquez and, 111
   history of, 111, 112, 127, 128
   labor and technology, histories of, 121–123
   politicizing, 278–279
Technology and Culture (journal), 111, 114, 116–118, *117*, *118*, 119, 121–122, 123, 127, 128
Technology scholarship, in Latin America, 13, 111–128
Technology transfer, 16, 312
"Technototem," 277
Tejera, Ariel, 294, 295
Terra, Eugenio, 363–364
Textile mill, labor in, 123
Thirring, Hans, 269, 270
Thomas, Hernán, 12, 47, 61
"Thought collective," use of term, 307
Tiffin, Helen, 18
Tironi, Manuel, 15, 280, 293, 305, 346
Tlatelolco Treaty, 298
Topology, 332, 345
Torin, Hilda, 93
Torres, Bibiano S., 150
Traditional knowledge, innovation and, 57, 58
Transient standardization, 334–335, 343–344
"Transitions," in Chile nuclear power plan, 309t, 317–323, 325
Tratado para la Prohibición de Armas Nucleares en America Latina y el Caribe, 297
Treaty on the Non-Proliferation of Nuclear Weapons, Mexico, 297
Treviño Arizpe, Roberto, 300

Index 395

TRIGA MARK III reactor, 295
TRIPS. *See* Agreement on Trade Related Aspects of Intellectual Property Rights
TSIN. *See* Technologies for Social Inclusion Network
Tsing, Anna, 208, 221
Tux Paint (software), 215–216

UIA. *See* International Union of Architects
UNAM. *See* National University
UNASUR. *See* Union of South American Nations
UNDP. *See* United Nations Development Program
UNIDO. *See* United Nations Industrial Development Organization
Unilever Shakti initiative, 54
Union of South American Nations (UNASUR), 233
United Nations Development Program (UNDP), 55
United Nations Economic Commission on Latin America (ECLA), 8, 18
United Nations Industrial Development Organization (UNIDO), 233
United Nations Nuclear Energy Commission, 292
University of Córdoba, Social Habitat group, 51
UPOV. *See* International Union for the Protection of New Varieties of Plants
Uruguay
  nanoscience and technology, 228, 231t, 237
  One Laptop per Child (OLPC) project, 187, 221
  streetcars in Montevideo, social importance of, 123
Urupês (Monteiro Lobato), 88

Vacant Areas Program (PAV, Argentina), 232t
Valderrama, Andrés, 119, 126
Vals, Maira, 170
Van de Graaff accelerator, in Mexico, 293, 299
Vara, Ana María, 293
Vargas, Getúlio, 276
Varsavsky, Oscar, 8, 9–10, 19
Vazquez, Josefina Zoraida, 299
Vázquez Reyna, Mario, 295
Vélez Ocón, Carlos, 293, 300
Velho, Lea, 11, 233
Venezuela, nanoscience and nanotechnology, 228, 231t, 233, 235, 237
Venezuelan Nanotechnology Network (RedVNano), 228, 237
Ventura, Roberto, 32–33
Verran, Helen, 88
Vessuri, Hebe, 10–11, 209
Vexame, 34, 35, 37, 38, 41
Via Campesina, 333–334, 336, 337
Vianna, Luis Werneck, 362–363
Victims of armed conflict, and forensic genetics, 70, 75, 78
Villanueva, Eduardo, 184
Vinck, Dominique, 14, 225
Vitebsky, Pier, 100–101
Vivienda, La (*Housing*) (film), 162, 163
Vucetich, Juan, 139–140, 141, *142*, 144, 145, 146, 148–149, 151, 152, 153–154
Vucetich classification system, 146, *147*, 148, 153, 154
"Vucetichism," 154–155

Wajcman, J., 68
"Washington consensus," use of term, 112, 127
Weber, Max, 208
We Have No Right to Wait (*No tenemos derecho a esperar*) (film), 170–172, 176
Wells, Allen, 121
Will to Live: AIDS Therapies and the Politics of Survival (Biehl), 356
Winn, Peter, 123

Women, and forensic genetics in Colombia, 12, 67–81
Wood, Robert, 274
World Bank, 233
Wormseed, 34, 44

XO laptops. *See also* One Laptop per Child projects
  about, 181, 187, 201
  broken hardware, 212, 221
  as charismatic objects, 208–209, 218–221
  most-used activities, 216, 217
  Paraguay, 14, 184, 207–222
  Peru, 183, 199, 221
  as toy, 211–212
  trackpad, 222
  understanding of function of, 211
  uses of, 211–212

Yale Peruvian Expeditions, 14–15, 245–261, *249*, *253*, *256*
  anthropometry, 208, 248, 258
  choice of camera, 250
  circulars on photography, 251–252, 259
  collected ethnographic objects, 257, 258
  coloring slides, 252
  developing negatives, 254, 255, 259, 260
  expeditionary science and photography, 246–247, 258
  expedition circulars, 251, 259
  exposure of photographs, 254, 259
  hand-tinted lantern slides, 252, 259
  instructions on taking photos, 252, 259
  marking photographs, 252
  National Geographic and, 247, 248, 255, 256, *256*, 259, 260
  number of images, 247, 258
  printing photographs, 255, 260
  sponsorship, 248, 250, 256, 258, 259
  taking photographs, 251–255, 259
  test images, 254, 259
  types of images, 247–248
Yo, Juan Domingo Perón (Perón), 275

Yugoslavia, Skopje earthquake (1983) and housing, 161

Zabala, 209
Zamalloa, Walter, 201
Zolov, Eric, 290
Zoneamento Agrícola de Risco Climático de MAPA, 338, 346